Plant Genetic Conservation

The in situ *approach*

Edited by

N. Maxted, B.V. Ford-Lloyd and J.G. Hawkes

The University of Birmingham, UK

CHAPMAN & HALL
London · New York · Tokyo · Melbourne · Madras

Published by Chapman & Hall, 2–6 Boundary Row, London SE1 8HN

Chapman & Hall, 2–6 Boundary Row, London SE1 8HN, UK

Chapman & Hall GmbH, Pappelallee3, 69469 Weinheim, Germany

Chapman & Hall USA, 115 Fifth Avenue, New York, NY 10003, USA

Chapman & Hall Japan, ITP-Japan, Kyowa Building, 3F, 2-2-2 Hirakawacho, Chiyoda-ku, Tokyo 102, Japan

Chapman & Hall Australia, 102 Dodds Street, South Melbourne, Victoria 3205, Australia

Chapman & Hall India, R. Seshadri, 32 Second Main Road, CIT East, Madras 600 035, India

First edition 1997

© 1997 Chapman & Hall

Typeset in 10/12pt Palatino by Columns Design Ltd, Reading, England
Printed in Great Britain by TJ International Ltd, Padstow

ISBN 0 412 63400 7 (Hb) 0 412 63730 8 (Pb)

A catalogue record for this book is available from the British Library

Library of Congress Catalog Card Number: 96-86665

∞ Printed on permanent acid-free text paper, manufactured in accordance with ANSI/NISO Z39.48-1992 and ANSI/NISO Z39.48-1984 (Permanence of Paper).

Contents

Contents

Contributors

Anikster, Y.
George S. Wise Faculty of Life Sciences
Institute for Cereal Crop Improvement
Ramat Aviv
Tel Aviv 69978
Israel

Bellon, M.R.
Genetic Resources Center
International Rice Research Institute
PO Box 933
1099 Manila
Philippines

Boshier, D.
Oxford Forestry Institute
University of Oxford
South Parks Road
Oxford OX1 3RB
UK

Brush, S.B.
Genetic Resources Conservation Program
University of California
Davis, CA 95616
USA

Damania, A.B.
Genetic Resources Conservation Program
University of California
Davis, CA 95616
USA

Dinoor, A.
Department of Plant Pathology
The Hebrew University of Jerusalem
Rehovot 76100
Israel

Dulloo, M.E.
National Parks and Conservation Service
Ministry of Agriculture and Natural Resources
Réduit
Mauritius

Ertug Firat, A.
Aegean Agricultural Research Institute
PO Box 9
Menemen
Izmir 35661
Turkey

Eshed, N.[†]
The Hebrew University of Jerusalem
Rehovot 76100
Israel

Feldman, M.
The Weizmann Institute of Science
Jerusalem
Israel

Ford-Lloyd, B.V.
School of Biological Sciences
The University of Birmingham
Edgbaston
Birmingham B15 2TT
UK

Gillman, M.
Department of Biology
The Open University
Walton Hall
Milton Keynes MK7 6AA
UK

Guarino, L.
IPGRI Sub-Saharan Regional Office
c/o ICRAF
Nairobi
Kenya

Hawkes, J.G.
School of Continuing Studies
The University of Birmingham
Edgbaston
Birmingham B15 2TT
UK

Hawtin, J.C.
IPGRI
Via delle Sette Chiese 142
00145 Rome
Italy

Hodgkin, T.
IPGRI
Via delle Sette Chiese 142
00145 Rome
Italy

Horovitz, A.
Department of Botany
The Hebrew University of Jerusalem
Givat Ram
Jerusalem 91904
Israel

Jackson, M.J.
Genetic Resources Center
International Rice Research Institute
PO Box 933
1099 Manila
Philippines

Kanowski, P.
Department of Forestry
Australian National University
Canberra
ACT 0200
Australia

Keesing, V.
Department of Entomology and Animal Ecology
Lincoln University
PO Box 84
Lincoln
New Zealand

Lawrence, M.J.
School of Biological Sciences
The University of Birmingham
Edgbaston
Birmingham B15 2TT
UK

Lunyova, N.
The Herbarium
N.I. Vavilov Research Institute of Plant Industry
44 Bolshaya Morskaya Street
St Petersburg 190000
Russia

Marshall, D.F.
School of Biological Sciences
The University of Birmingham
Edgbaston
Birmingham B15 2TT
UK

Maxted, N.
School of Biological Sciences
The University of Birmingham
Edgbaston
Birmingham B15 2TT
UK

Newbury, H.J.
School of Biological Sciences
The University of Birmingham
Edgbaston
Birmingham B15 2TT
UK

Ortega, R.
Centro Regional de Recursos Genéticos de Tuberosas y Raíces
National University of Cusco
Apartado 295
Cusco
Peru

Pham, J.-L.
Genetic Resources Center
International Rice Research Institute
PO Box 933
1099 Manila
Philippines

Prance, G.T.
Royal Botanic Gardens, Kew
Richmond
Surrey TW9 3AE
UK

Qualset, C.O.
Genetic Resources Conservation Program
University of California
Davis
CA 95616
USA

Tan, A.
Aegean Agricultural Research Institute
PO Box 9
Menemen
Izmir 35661
Turkey

Ulyanova, T.
The Herbarium
N.I. Vavilov Research Institute of Plant Industry
44 Bolshaya Morskaya Street
St Petersburg 190000
Russia

Williams, J.T.
International Network for Bamboo and Rattan
DRC
New Delhi
India

Worede, M.
Seeds for Survival
PO Box 5760
Addis Ababa
Ethiopia

Wratten, S.D.
Department of Entomology and Animal Ecology
Lincoln University
PO Box 84
Lincoln
New Zealand

Zanatta, A.C.A.
Genetic Resources Conservation Program
University of California
Davis
CA 95616
USA

Zohary, D.
Department of Genetics
The Hebrew University of Jerusalem
Givat Ram
Jerusalem 91904
Israel

Preface

The recent development of ideas on biodiversity conservation was already being considered almost three-quarters of a century ago for crop plants and the wild species related to them, by the Russian geneticist N.I. Vavilov. He was undoubtedly the first scientist to understand the importance for humankind of conserving for utilization the genetic diversity of our ancient crop plants and their wild relatives from their centres of diversity. His collections showed various traits of adaptation to environmental extremes and biotypes of crop diseases and pests which were unknown to most plant breeders in the first quarter of the twentieth century.

Later, in the 1940s–1960s scientists began to realize that the pool of genetic diversity known to Vavilov and his colleagues was beginning to disappear. Through the replacement of the old, primitive and highly diverse land races by uniform modern varieties created by plant breeders, the crop gene pool was being eroded. The genetic diversity of wild species was equally being threatened by human activities: over-exploitation, habitat destruction or fragmentation, competition resulting from the introduction of alien species or varieties, changes and intensification of land use, environmental pollution and possible climate change.

FAO (Food and Agriculture Organization of the United Nations) began in 1961 to look for ways of counteracting this trend (Whyte and Julen, 1963), but little action was taken until the two FAO conferences on genetic resources held in Rome in 1967 and 1972 (Frankel and Bennett, 1970; Frankel and Hawkes, 1975), combined with the establishment of the International Board for Plant Genetic Resources in 1974. This was the period when the so-called 'green revolution' varieties of wheat, rice and other crops were being adopted by farmers all over the world. With the resultant changes in farming methods, land races of old cultivars and wild species were rapidly and increasingly disappearing.

It was clear that swift action needed to be taken to preserve the genetic diversity of both crop and wild species, and at the two conferences mentioned above the ground work for the ways and means of doing so was established. The answer to this problem was to collect samples based on an understanding of population biology and to store them in what came to be known as gene banks, using methods devised by seed physiologists. This was the beginning of what we have now come to call *ex situ* conservation, and was absolutely necessary at that time to conserve the major part of crop genetic diversity which was disappearing before our very eyes. Crop gene banks were thus established on regional and national scales whilst others in the CGIAR system, such as IRRI, CIMMYT, etc. were focusing often on a single or a few crop species.

It soon became evident that although the *ex situ* methods of drying and cooling seeds of most of the major crops (wheat, rice, maize and other cereals) were technically satisfactory, other crops, and particularly tree crops, did not possess the 'orthodox' seeds of the crops mentioned above, but what came to be called 'recalcitrant' ones, which died quickly when treated with the storage methods suitable for the orthodox seeds. There were also practical seed conservation problems associated with vegetatively propagated crops such as cassava, potatoes, bananas and many others, mostly tropical ones. Samples of these could of course be transplanted to what came to be called 'field gene banks' but the area required to conserve even a reasonable amount of their genetic diversity needed to be so large as to be often impractical.

Another problem was concerned with the conservation of wild and weedy crop relatives. Seeds of those species that were not classified as recalcitrant could be collected and stored *ex situ* in the usual way, and this method has in fact been widely adopted. However, with finite resources available for conservation, this meant that priorities were established: certain species were selected for active conservation while others were not and the latter species continued to suffer genetic erosion or extinction. Yet another factor that had been left out of the equation was the lack of ongoing evolutionary change of the crops and wild relatives, in response to the crop pests and diseases, which of course were not stored away in gene banks but were constantly evolving new pathotypes to which the material in the gene banks had not been exposed. This is in fact a potentially serious problem, which has become more and more evident in the last decade, for although regeneration of seeds stored in gene banks should be undertaken from time to time, it is at such long intervals and for only one year or less each time that there will obviously be no opportunity for resistant crop biotypes to emerge in such short periods.

In situ conservation (literally in its 'original' place, from the Latin) of the crop diversity that is still found in traditional farming conditions is one answer to this problem, though for many parts of the world such farming methods along with the diversity of local land races have long since ceased to exist. Rather more hopefully, however, wild and weedy crop relatives or other wild species are found throughout the world and by active conservation of these areas the species found in them can be effectively conserved *in situ*.

The use of *in situ* techniques for conservation was emphasized by the United Nations Conference on the Environment and Development (UNCED) held in Rio in 1992, which spawned the Biodiversity Convention and the implementation plan, Agenda 21. The articles of both of the latter recognize the economic, political and social consequences of the steady loss of plant genetic diversity along with the potential benefits to humankind that can result from conserving and exploiting the world's plant genetic resources.

The whole concept of *in situ* conservation has become a political 'hot potato' in the last decade. The concept has become fashionable and has often been advocated by enthusiasts who may not comprehend fully the scientific bases of genetic conservation. Although both *ex situ* and *in situ* techniques have their advantages and disadvantages, it is wrong to see them as alternatives (Ford-Lloyd and Maxted, 1993). They should be seen as being complementary techniques, as is stated in Article 9 of the Convention on Biological Diversity (UNCED, 1992). Frequently one method of conservation will act as a back-up to another. Each has an important role to play, and one or other may receive more emphasis according to the crop or wild species concerned and the actual situation amongst farming communities, peasant or high-tech agriculture and forestry, together with the pests and diseases that threaten each crop.

It is undoubtedly true that although the two strategies are complementary, much more research has been focused on the scientific bases of *ex situ* genetic conservation. In this book we have attempted to redress the balance and analyse the problems concerned with the *in situ* conservation of crops (on-farm conservation) and wild species (genetic reserve conservation). The challenge is to provide simple but appropriate *in situ* conservation protocols and methodologies that will enable conservationists to conserve and utilize the world's flora.

We have tried to look at these problems from theoretical and practical viewpoints, drawing the ideas together from the rather scanty published sources and with the help of colleagues from many different parts of the world. Therefore the book is divided into four parts: Part One introduces and provides a background to conservation, and more specifically genetic conservation; Part Two focuses on the theory and practice of farmer-based and genetic reserve *in situ* conservation, including chapters discussing the planning, establishment and management of conservation regimes for crops, crop relatives or wild species; Part Three provides practical case studies written by those actively engaged in the *in situ* conservation of genetic diversity; and finally, Part Four discusses models that might be helpful to those involved in *in situ* conservation and looks toward the future development of genetic conservation.

It has been our wish in producing this book to open a debate focusing on the scientific principles underlying *in situ* conservation. If this is accomplished we shall be well satisfied.

N. Maxted, B.V. Ford-Lloyd and J.G. Hawkes

The University of Birmingham

List of acronyms and abbreviations used in the text

AGIS United States Agricultural Genome Information Server
BGCI Botanic Gardens Conservation International
CABI Commonwealth Agricultural Bureaux International (also abbreviated as CAB International)
CBD Convention on Biological Diversity
CGIAR Consultative Group on International Agricultural Research
CIAT Centro Internacional de Agricultura Tropical
CIMMYT Centro Internacional de Mejoramiento de Maiz y Trigo
CIP Centro Internacional de la Papa
CITES Convention on International Trade in Endangered Species of Wild Fauna and Flora
CSIRO Commonwealth Scientific and Industrial Research Organization, Australia
DBMS Database Management System
ECP/GR European Cooperative Programme for Crop Genetic Resources Networks
ERIN Environmental Resources Information Network
ETI Expert-Centre for Taxonomic Identification
EUCARPIA European Association for Research on Plant Breeding
FAO Food and Agriculture Organization of the United Nations
GATT General Agreement on Tariffs and Trade
GEF Global Environment Facility
GIS Geographical Information Systems
GPS Geographical Positioning Systems
GRAIN Genetic Resources Action International
GRID Global Resources Information Database (UNEP)
GRIN United States Genetic Resources Information Network
IARC International Agricultural Research Centre
IBPGR International Board for Plant Genetic Resources (now IPGRI)

ICARDA International Centre for Agricultural Research in the Dry Areas

ICRAF International Centre for Research in Agroforestry

ICRISAT International Crops Research Institute for the Semi-Arid Tropics

IITA International Institute for Tropical Agriculture

IK Indigenous Knowledge

ILCA International Livestock Centre for Africa

IPGRI International Plant Genetic Resources Institute (formerly IBPGR)

IRRI International Rice Research Institute

IUCN International Union for the Conservation of Nature and Natural Resources

NGO Non-governmental organization

PGRP United States National Plant Genome Research Program

UKPGRG United Kingdom Plant Genetic Resources Group

UNCED United Nations Conference on the Evironment and Development

UNEP United Nations Environment Programme

UPOV Union for the protection of new varieties of plant

WCMC World Conservation Monitoring Centre

WRI World Resources Institute

WWF World Wide Fund for Nature

WWW World Wide Web

Part One

Introduction

1

The conservation of botanical diversity

G.T. Prance

1.1 INTRODUCTION

Plants are the basis for life on Earth and without their capacity to capture the sun's energy through the process of photosynthesis there would be no life on our planet. Therefore the conservation of plants is vital to the continued existence of life. There are between 250 and 300 thousand species of flowering plants, an estimated 200 thousand species of algae (Groombridge, 1992), as well as many species of ferns, conifers, mosses and liverworts, giving a total of over half a million plant species in the world (Table 1.1). Each species has different habitat requirements, performs different ecological functions in different ecosystems and has different uses or potential uses for humankind. It is not then enough to preserve a few selected plant species. For the survival of a quality of life, and the maintenance of the physical processes such as climate patterns, atmosphere and soil, the majority of plant species are needed. It is, therefore, of utmost importance that conservation techniques for plants aim at preserving as many species as possible.

The only way to achieve this is to place a strong emphasis on *in situ* conservation and so I am pleased to introduce this volume that concentrates on that approach. Although *ex situ* conservation certainly plays a most important role in the conservation of plants, especially through botanical gardens and seed and tissue banks, it is insufficient to maintain the many physical and biological processes that must continue for the survival of life. As director of a botanical garden I certainly place much importance on the role that botanic gardens and seed banks play in plant

Plant Genetic Conservation.
Edited by N. Maxted, B.V. Ford-Lloyd and J.G. Hawkes.
Published in 1997 by Chapman & Hall. ISBN 0 412 63400 7 (Hb) and 0 412 63730 8 (Pb).

Table 1.1. Estimated numbers of plant species in the world (data from Prance, 1977; Hawksworth, 1991; Groombridge, 1992)

Taxon	Described species	Estimated number of species
Algae	40 000	200–350 000
Lichens	13 500	20 000
Bryophytes: mosses	8 000	9 000
liverworts	6 000	7 000
Ferns and fern allies	12 000	12 500
Gymnosperms	650	650
Angiosperms	250 000	300 000
TOTALS	320 150	549 150

conservation, but, as a tropical botanist working in the Amazon rainforest, I know the vital necessity for *in situ* conservation. I am taking for granted the importance of *ex situ* techniques and commenting in detail only on the topic of this volume, the *in situ* approach to plant conservation.

1.2 WHY IS THERE A NEED FOR CONSERVATION OF PLANT SPECIES?

Firstly, plant conservation is urgently needed because species are going extinct and many others are threatened and endangered. Tropical rainforest covers only just over 7% of the land surface of the planet, yet it harbours about 50% of the species. Since 50% of the world's tropical rainforest has already been destroyed or seriously perturbed, many species have already been lost. Some of the areas with most endemism of tropical forest species have been the most destroyed – for example, Atlantic Coastal Brazil (Mori *et al.*, 1981), Pacific Coastal Ecuador (Gentry, 1986) and Madagascar. Certainly many plant species have been lost in these areas (Myers, 1988a, 1990; Campbell and Hammond, 1989). Raven (1988) estimates that 200 plant species were being lost per year in the tropics and subtropics, based on the loss of half the species in the area likely to be deforested by 2015. Myers (1988b) calculated that 7% of the plant species of the world could be lost in the next decade, based on the destruction of natural habitat in hot spots such as the three mentioned above. Although these are probably overestimates, nevertheless there is no doubt that much plant extinction is occurring and it is increasing as small remnants of natural ecosystems continue to be destroyed. The World Conservation Monitoring Centre 1992 lists contain 596 extinct plant taxa, many of which are from oceanic islands (Table 1.2) such as

Hawaii, Mauritius, Rodrigues, St Helena (Table 1.3) and Cuba. However, many continental countries also feature, especially South Africa and Australia (Groombridge, 1992). Table 1.4 lists the total number of threatened plant species for 14 countries. It is notable that the number is often higher in countries where the flora has been well studied such as the United States, Australia and South Africa and in most cases at least 10% of the plant species are threatened. These few statistics should be enough to demonstrate that the plants are under threat and that there is a great need for their conservation. (For further details see Lucas and Synge, 1978; Davis *et al.*, 1986; Campbell and Hammond, 1989; Groombridge, 1992.)

Secondly, we need to conserve plant species because of human dependence on them for many different uses and indeed for our survival. Plants provide our basic food crops, building materials and medicines as well as oils, lubricants, rubber and other latexes, resins, waxes, perfumes, dyes and fibres. So far only about 10% of plants have ever been evaluated for their medicinal or agricultural potential and so there are certainly many new drugs and new crops yet to be discovered. We need to leave our future options open by having as many living plant species available as possible.

Table 1.2 Documented extinction of plant species on islands (source: Groombridge, 1992)

Island	Extinct	Endangered	Threatened or Vulnerable	Rare	TOTAL Endemic spp.
Hawaii	108	153	126	9	–
St Helena	7	19	–	–	46
Bermuda	3	4	7	–	15
Rodrigues	8	27 (some extinct)	16	–	45
Norfolk Island	1	11	23	–	36

Table 1.3 Extinct species from St Helena (source: Maunder, 1995)

Species	Family
Heliotropium pannifolium Burchell ex Hemsl.	Boraginaceae
Wahlenbergia burchellii A. DC.	Campanulaceae
Wahlenbergia roxburghii A. DC.	Campanulaceae
Commidendrum rotundifolium (Roxb.) DC.	Asteraceae
Acalypha rubra Roxb.	Euphorbiaceae
Mellissia begonifolia (Roxb.) Hook.f.	Solanaceae
Bulbostylis neglecta (Hemsl.) C.B. Clarke	Cyperaceae

Table 1.4 Number of threatened plant species for selected countries (source: IUCN Red Data Books)

Country	Number
Australia	1016
China	350
Colombia	327
Costa Rica	419
Cuba	860
Greece	526
India	1336
Jordan	752
Malaysia	522
Mexico	883
South Africa	1016
Spain	936
Turkey	1944
United States	2262

Thirdly, a great diversity of plants is needed to keep the various natural ecosystems functioning stably. No organism exists alone but all depend on a multitude of interactions that relate them together such as pollination and dispersal dependencies between plants and animals, bacteria which fix nitrogen in the nodules of leguminous plants, plants which are protected by ants, etc. Natural forests have many other less obvious but most essential practical uses such as the protection of watersheds to provide potable water for many cities, to stabilize soil preventing erosion and often to reduce flooding because of their capacity to absorb and release slowly much of the rainfall.

Fourthly, the plants that we already use as crops are still dependent upon the broad genetic base that exists in their wild relatives. Crops are often improved or better protected by using genes from wild species. Modern genetic engineering techniques even allow the transfer of genes from completely unrelated organisms. In 1970 the maize crop of the United States was severely threatened by corn blight. This was quickly corrected before the next year's crop through the use of blight resistance genes from Mexican wild, less inbred varieties of maize. Only in 1978 was a new perennial species of maize (*Zea diploperennis*) discovered (Iltis *et al.*, 1979).

Lastly, it is important to observe that natural ecosystems and the diversity of plant species which they contain bring much pleasure and inspiration to many people. Life would be very dull without the aesthetic experience of walking in a forest or a savanna, or without seeing a rare orchid or smelling the perfume of a jasmine or honeysuckle.

1.3 THE ADVANTAGES OF THE *IN SITU* APPROACH

Although many plant species are being rescued by *ex situ* methods and reintroductions, the single most important way to conserve a plant species is through the protection of the habitat in which it lives. This conserves the associated animals upon which it may depend for pollination and dispersal of its diaspores and also the animals, particularly insects, that might depend upon the plant species (Chapter 14). An individual species is never its own isolated ecosystem; it interacts with others in many ways and usually supports several others. Conserving the mycorrhizal fungi upon which many plants depend is seldom considered but they are automatically preserved by the habitat approach. Perhaps the greatest importance of the conservation of an ecosystem rather than an individual species *ex situ* is that it allows the process of evolution to continue. No species is static but is continually interacting with the physical environment and is competing with the other species in the ecosystem. For a species to have a viable future it must be able to compete and it will only maintain its competitive ability if the evolutionary process is allowed to continue. This aspect is becoming particularly important in light of the worldwide climate changes that are taking place as the result of global warming. As climate patterns change, plants will need to adapt as well as migrate to new areas (Lynch and Landé, 1993). Unfortunately the fragmentation of habitat is already making this process harder, and much human assistance will have to be given to the dispersal of less vagile species.

Another advantage of *in situ* conservation, which is most relevant to the evolutionary development of a species, is that it is much easier to conserve a viable population of a species in its natural habitat rather than in an *ex situ* situation. This is particularly true of tree species. It requires a lot of space to conserve a thousand individual trees of a species in a botanical garden or other *ex situ* area. Estimates of a minimum viable population have varied from 500 (Franklin, 1993) to 5000 (Nunney and Campbell, 1993). A good discussion of minimum viable population size was given by Boyce (1992; see also Chapter 6). Seed banks, however, do endeavour to bank enough seeds of a species to contain a future population (Prendergast *et al.*, 1991). Part of the initial survey to select areas for *in situ* conservation should include data on the population size of the species to be conserved so that it contains an adequate sample of its genetic diversity. This often includes the need for two or more reserves in different places (see Chapter 7 for population ecology and Chapter 6 for population genetics).

In many tropical countries where the flora is still poorly studied there are only small lists or even no lists of endangered species. For example, Brazil, with a much larger flora (ca. 50 000 species) and many endangered habitats, lists only 318 threatened plants whereas the United States, with well under half the species of Brazil, lists 2262 (Groombridge, 1992). In species-rich tropical countries such as Brazil and Colombia the flora is not known well enough to produce adequate lists of endangered species. It is therefore impossible to adopt an *ex situ* strategy for conservation apart from a few obviously threatened species. The *in situ* approach is essential when a flora has not been adequately inventoried. Also, in somewhere like the Amazon rainforest, many species are still unknown and they can only be protected by *in situ* methods. Approximately one plant collection in every hundred made by the author in Amazonia over the last 30 years has turned out to be a new species. Also about one in every hundredth specimen sent to Kew for identification is a new species (Prance and Campbell, 1988). With such a poor knowledge of the flora, the only viable approach to conservation is *in situ*.

The majority of orthodox seeds, which can be stored in seed banks, are either from temperate regions, where dormancy is enforced by cold freezing winters, or from arid regions, where dormancy is enforced by long dry spells. In rainforests, which contain about half of all the plant species in the world, the majority of seeds are recalcitrant. *In situ* conservation is particularly important in areas where there is a high incidence of recalcitrance.

The Convention on Biological Diversity (Article 9) certainly favours *in situ* conservation and regards *ex situ* as second best (italics are author's emphasis):

> Each contracting party shall, as far as possible and as appropriate, and *predominantly for the purpose of complementing* in situ *measures*:
>
> (a) Adopt measures for the *ex situ* conservation of components of biological diversity, preferably in the country of origin of such components;
> (b) Establish and maintain facilities for the *ex situ* conservation of and research on plants, animals and micro-organisms, preferably in the country of origin of genetic resources;
> (c) Adopt measures for the recovery and rehabilitation of threatened species and for their reintroduction into natural habitats under appropriate conditions.

The emphasis here is on the need to develop in-country facilities even when *ex situ* methods are adopted.

1.4 *IN SITU* CONSERVATION DOES NOT ONLY INCLUDE PRISTINE AREAS

In situ conservation is often thought of purely as conservation in biological reserves of relatively undisturbed natural ecosystems. It is, however, important to stress that many disturbed areas harbour a large amount of biodiversity. Disturbance can also maintain and increase diversity (Connell, 1978; Hubbell, 1979; Denslow, 1985; Pickett and White, 1985; Clark, 1991). Areas that we often think of as natural are often artificially maintained. For example, the conservation of many of the orchids and other rare species of chalk grasslands in England depends on grazing by sheep in what is essentially a man-made environment. In the tropics the diversity in some areas of mature secondary forest is often as high as that in primary forest nearby, as was shown by Balée (1994) in a study comparing old fallow of the Ka'apor Indians with that of the primary forest nearby where the total species diversity was no different.

In situ conservation does not necessarily exclude the use of a reserve in some way for economic purposes. The conservation of plant species in many areas is far more likely to succeed if the area is yielding some form of income to the local people, whether it be from recreation, ecotourism, sustainable logging or the controlled hunting of game. An important example of this, which has halted deforestation in some areas, is the extractive reserves that have been set up in some neotropical countries, especially in Brazil. In these reserves the local people are allowed to extract non-timber forest products such as Brazil nuts, rubber latex, resins and fruit, but not to fell the forest (Prance, 1989; Schwartzman, 1989; Allegretti, 1990). The pros and cons of extraction reserves have been discussed in detail elsewhere (Prance, 1994) and a detailed evaluation of them was made by Salafsk *et al.* (1993). In spite of the difficulties, there is no doubt that the sustainable use of non-timber forest products is likely to be a major help for the conservation of rainforest species.

The buffer zones of biosphere reserves are another good example of extending conservation beyond a fully protected area. Many of the wild relatives of crop species occur in disturbed habitats rather than in natural areas and these are of particular importance for conservation because of the genetic erosion of many crop species through the use of genetically uniform hybrids. Seed banks exist for most major annual crops and they are successfully conserving a broad spectrum of genetic material, but this does not work well for many woody perennial crops with recalcitrant seeds and so *in situ* conservation is essential. Often these species can be conserved in disturbed habitats.

Vital to the future potential of many crops are the land races of them that are grown by many traditional farmers. Since the tendency has been to replace such traditional systems with modern agriculture, serious genetic erosion is occurring. The best people to preserve these traditional varieties *in situ* are the farmers and native inhabitants themselves. The *in situ* conservation of native systems of agriculture could be vital for the future of various crops such as maize and potatoes.

Where *in situ* and *ex situ* techniques for conservation come together most closely is in reintroductions and recreation of habitat for rare and endangered species. As more natural habitat is lost and species are preserved only *ex situ*, it will be necessary to restore suitable habitats. In future many endangered species will not be in undisturbed natural habitats, but in new man-made habitats that have been created to move species back into *in situ* situations. (For examples in tropical rainforest, see Gomez-Pompa, 1991; Leith and Lohmann, 1993; for an oceanic island, see Maunder *et al.*, 1995.)

1.5 RESERVE SIZE

There are now sufficient studies to show that reserve size is critical to adequate conservation of both plants and animals. The rate of species loss and genetic deterioration for most species is inversely proportional to reserve size (Soulé, 1986, 1987). In addition to the problems of maintaining viable interactive populations in small reserves there is also the fact that they are much more prone to destruction. For example, a small reserve could be entirely destroyed in a fire or by a storm, whereas in a larger area the destruction is unlikely to be total. The edge effect (Lovejoy *et al.*, 1986) can alter all of a small reserve, but only a small proportion of a larger one. A small reserve is also much more prone to invasion by exotic species. Many new data about the importance of reserve size are coming from the 'Dynamics of Forest Fragments' project north of Manaus, Brazil (Lovejoy *et al.*, 1984, 1986; Bierregaard *et al.*, 1992). It is also helping to refute the hypothesis of Simberloff and Abele (1976) who stated that the theory of island biogeography is neutral versus a series of smaller reserves. The danger of this statement is that developers and conservation planners will leave only small areas of forest as reserves (Diamond, 1976, 1984). Rankin de Mérona *et al.* (1990) showed the severe edge effect on trees in small plots of the Manaus project. Even after two years of isolation many dead and broken trees occurred around the windward margin of the reserve. For plants the particular disadvantage of small reserves is the loss of agents of pollination and dispersal. Powell and Powell (1987), also in the Manaus project, showed the loss of the important neotropical pollinators, englossine bees, in the smaller fragments. Terborgh (1986, 1992) observed that the loss of large seed predators from an area would affect the distribution of tree species in a rainforest and that certain plant species are keystone resources for animals. There is little doubt that, for optimum conservation of rainforest plants, reserve areas large enough to maintain biological interactions as well as adequate population size are essential.

1.6 ISSUES FOR PLANNING RESERVES TO PROTECT PLANT SPE-CIES

1. The inventory of plant species is far from complete and much further work is needed in this area to enable both *in situ* and *ex situ* conservation (Prance, 1977; Prance and Campbell, 1988). It is not easy to select the best areas for conservation when the inventory of species is so incomplete. This does not just involve the description of new undescribed species but even more important is local inventory of the species content and population dynamics of poorly known ecosystems.
2. Areas of high species diversity are of high conservation priority, whether they be in rainforest or arid regions (Prance and Brown, 1987).
3. Areas of high species endemism are of high priority. Measurements which quantify both diversity and endemism are of great importance for the selection of reserves (Vane Wright *et al.*, 1991; Williams *et al.*, 1992).
4. Areas that have been important for the origin of crop species do not always coincide with categories 2 and 3 above and are of particular importance, especially the Vavilov centres (Vavilov, 1951).
5. It will sometimes be necessary to place selective conservation priorities on different taxa. Therefore techniques that consider the phylogenetic relationships among taxa will become increasingly important to achieve the conservation of as much genetic and evolutionary diversity as possible (May, 1990; Vane Wright *et al.*, 1991; Williams *et al.*, 1992).

1.7 CONCLUSION

I am firmly of the belief that it is preferable to save both species and ecosystems *in situ*. However, it is obvious that this will not be possible for many species where habitat destruction has been total or for island species where the island has been invaded by exotics or goats. Therefore there is a most important role for *ex situ* conservation of plants which often have much stronger habitat preferences than animals. At present 13 of the species held at the Royal Botanic Gardens, Kew, are known to be extinct in the wild (Figure 1.1) and there are at least 20 extinct taxa in cultivation in British botanical gardens (Table 1.5). We have a greater responsibility for keeping these alive, together with the over 1000 threatened species which we hold. However, whenever it is possible we are also involved in species reintroductions to islands like Rodrigues, St

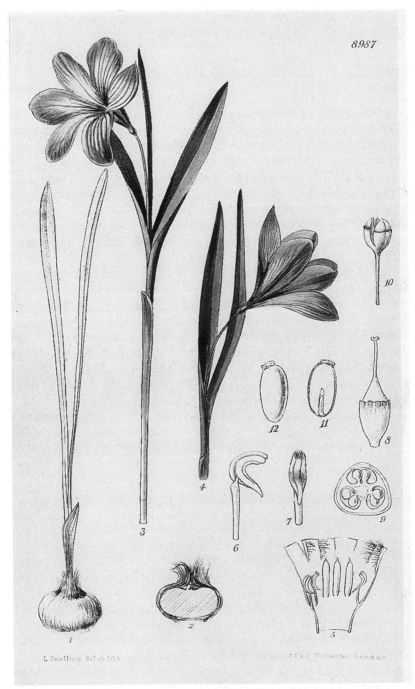

8987

Figure 1.1 *Tecophilaea cyanocrocus*, the Chilean blue crocus, a plant that is now extinct in the wild but survives in cultivation.

Table 1.5 Taxa 'Extinct in the Wild' cultivated in British Botanic Gardens (status as recorded by IUCN/WCMC; source: Michael Maunder)

Taxon	No. of collections	Country of origin	Status confirmed	International conservation programmes	In-country conservation activities
Anthurium leuconeurum	1	Mexico	?	?	?
Arctostaphylos uva-ursi ssp *loebreweri*	3	California, USA	propagated in USA	no	yes
*Bromus verticillatus**	4	UK	yes	yes	?
*Calandrinia feltonii**	3	Falkland Islands	yes	RBG Kew and Falklands Gov.	yes
Ceratozamia hildae	14	Central America	?	no	no
*Commidendrum rotundifolium**	1	St Helena	reintroduction into wild	no	yes
*Cosmos atrosanguineus**	5	Mexico	yes	RBG Kew with UNAM, Mexico	yes
Dombeya acutangula	1	Rodrigues	yes	yes	yes
D. mauritiana	?	Mauritius	yes	yes	yes
*Erica verticillata**	10	Republic of South Africa	propagated and reintroduced	UK material repatriated	yes
*Encephalartos woodii**	7	Republic of South Africa	confirmed extinct, secure in cultivation	no	yes
*Franklinia alatamaha**	22	USA	confirmed extinct, secure in cultivation	no	yes (CPC)
Graptopetalum bellus	2	Mexico	?	?	?
Helichrysum selaginoides	1	Tasmania	?	?	?
*Lysimachia minoricensis**	10	Minorca	reintroduced by Brest BG and others	Brest BG and Medio Ambiente, Spain	yes
Opuntia lindheimeri	1	Mexico?	?	?	?
Paphiopedilum delelanatii	3	?	?	?	?
P. druryi	5	India	rediscovered in wild	?	yes
*Sophora toromiro**	3	Easter Island	confirmed extinct	Toromiro Management Group	yes
*Tecophilaea cyanocrocus**	2	Chile	confirmed extinct by Chilean authorities	RBG Kew and CONAF	yes
*Trochetiopsis erthroxylon**	3	St Helena	reintroduced	University of Oxford	yes
*Tulipa sprengeri**	>10	Turkey	?	RBG Kew	no

* Verified by M. Maunder.

Helena and Easter Island. One of the problems of plant conservation to date has been the relatively poor integration of *in situ* and *ex situ* techniques. Botanical gardens, such as the Royal Botanic Gardens, Kew, which are often studying both aspects are uniquely able to improve this situation and ensure that the correct balance between methods is obtained (Maunder, 1994a). It would be most unfortunate if future conservation efforts were to concentrate on either in-site or off-site conservation to the exclusion or reduction of the efforts in the other. Working closely together the two methods have the potential to conserve for the future many of the plant species that are still alive today (Falk and Holsinger, 1991). However, at present human-induced extinction of plants proceeds at a far greater rate than the natural one, and it is urgent to slow down this trend by whatever are the best means, whether it is by banking the seeds or growing the species in botanical gardens and arboreta, or whether it is by creating biological, extractive or recreated reserves.

In situ plant conservation will only be successful when it combines traditional biogeographical methods of species distribution, patterns of endemism and diversity with the more modern approaches of population dynamics and genetic structure and the interactions within and between populations, species and ecosystems.

2

Complementary conservation strategies

N. Maxted, B.V. Ford-Lloyd and J.G. Hawkes

2.1 INTRODUCTION

The challenge facing the world's biological and conservation scientists is threefold: to classify the existing biological diversity; to halt the rate of ecosystem, habitat, species and genetic loss; and to feed the ever increasing human population. It is generally agreed that a catastrophic loss of plant genetic diversity is occurring at this moment: species, gene combinations and alleles are being lost for ever and this process of genetic erosion is likely to become even more grave in the future. The conservation of plant diversity is of critical importance, because of the direct benefits to humans that can arise from its exploitation in new agricultural and horticultural crops, the development of medicinal drugs and the pivotal role played by plants in the functioning of all natural ecosystems. The economic, political and social consequences that would result from a steady loss of plant diversity combined with rapid population growth is likely to be devastating if unchecked. The importance of these issues to humankind is underlined in Article 1 of the objectives of the Convention on Biological Diversity (UNCED, 1992):

> The objectives of this convention ... are the conservation of biological diversity, the sustainable use of its components and the fair and equitable sharing of the benefits arising out of the utilisation of genetic resources ...

Increased awareness of the current threats to biodiversity and the scientific problems associated with conservation and sustained utiliza-

Plant Genetic Conservation.
Edited by N. Maxted, B.V. Ford-Lloyd and J.G. Hawkes.
Published in 1997 by Chapman & Hall. ISBN 0 412 63400 7 (Hb) and 0 412 63730 8 (Pb).

tion have highlighted the need for more efficient and effective protocols and methodologies. This requirement is particularly important in those regions of the world with the highest levels of botanical diversity: such regions are almost invariably where the flora is poorly known, under threat and where the conservationists are too few and often not adequately trained. There is an urgent requirement to clarify and enhance the methodologies that currently enable scientists to classify, conserve, manage and utilize their native flora.

2.2 BACKGROUND TO PLANT GENETIC CONSERVATION

Although there is a currently increased awareness and concern over the accelerating loss of plant genetic diversity, attempts to conserve and utilize plant diversity are as old as humankind itself. Even as hunter–gatherers in pre-agricultural times, humans explored their environment for food. The use and conservation of medicinal plants was established very early and was usually associated with older 'wiser' persons (wise woman, witch, and later herbalist). Historical records show that ancient Greeks, Persians, Chinese, Assyrians, Egyptians and others used specific plants for their medicinal, benign or malign properties.

These cultures were essentially local plant gatherers; they used the plants growing in their immediate region. Plant exploration, which developed later, was more sophisticated and purposeful, and involved the location of plants in one area and their transfer *ex situ* to a second area where they may be used. The earliest record of targeted collecting is that undertaken by Sargon in 2500 BC, who crossed the Taurus mountains to the Anatolian Plateau to collect figs, vines and olives (Juma, 1989). Another early example of exploration is that of Queen Hatsheput of Egypt in 1495 BC, who directed an expedition to Punt (= Somalia) looking for trees that could yield frankincense (*Boswellia carteri*) from their resin (Coats, 1969). The expedition is recorded in wall carvings at her palace in Thebes. The living plants were transplanted into pots, brought down the Nile on barges and planted in the palace garden. This was possibly the first 'target oriented' mission and was to be followed by many other examples, such as Alexander the Great in the fourth century BC bringing the pomegranate from Armenia and peach and apple from Central Asia back to Greece, and the Arabs taking coffee from Ethiopia to Arabia in AD 900.

Following the waves of European explorers to the New World in the late sixteenth and early seventeenth centuries came the idea of searching for plants and natural products that could be of use to humans. For example, there is the report of Thomas Heriot in 1588, 'A brief and true report of the new found land of Virginia'; he studied the plants of Roanoak (now in North Carolina, USA) and brought back samples to

Europe for study. Each of the colonial powers in Europe had their own explorers who discovered and transported newly discovered plant species for exploitation.

Parallel to the birth of long-distance plant exploration came the establishment of the first physic or botanical gardens in the sixteenth and seventeenth centuries to provide the plants required by medicine. By 1545 there were already gardens in Padua, Florence and Pisa. Leiden was established in 1593, Paris in 1635 and Edinburgh in 1690. The Dutch colonialists established a garden in Cape Town in 1694; the French in Mauritius in 1733 and during the eighteenth century the British established gardens in St Vincent and Jamaica, Calcutta and Penang. By 1841, when Sir William Hooker became the first official Director of the National (later Royal) Botanic Gardens based at Kew, the real economic benefits of plant exploration as a means of improving crops both at home and in the colonies were established. This was the era when rubber (*Hevea brazilliensis*) was transferred from Brazil via Kew to Malaysia, bread fruit (*Artocarpus laciniata*) from the East to the West Indies, oil palm from West Africa to South East Asia, cocoa from Central America to West Africa, etc. However, all these introductions were based on a few plants, with a narrow genetic base, and there was thus a danger that the lack of diversity might be a threat in the future.

It was N.I. Vavilov who first realized the importance of collecting the diversity of crops and their wild relatives, rather than sampling single or a few representatives of each species. He directed or stimulated numerous expeditions throughout the world during the 1920s and 1930s for wheat, rye, potato, barley and many other crops such as leguminous forages and grain crops. His studies led him to formulate his hypothesis of the 'centres of origin' of crop plants. His ideas stimulated a generation of subsequent genetic conservationists, such as Harry V. Harlan (barley), Jack R. Harlan (wheat), Erna Bennett (cereals), Hermann Kuckuck (wheats), Jack Hawkes (potatoes) and others, to survey and systematically conserve the wealth of plant genetic resources. The early scientific developments of *ex situ* conservation techniques took place in the 1960s and 1970s (Frankel and Bennet, 1970; Frankel, 1973; Frankel and Hawkes, 1975), which in turn led to the establishment of the International Board for Plant Genetic Resources (now IPGRI) in 1974. These initiatives were related to the urgent need to save crop genetic resources from genetic decimation, especially in the Vavilov centres of diversity. Scientific techniques for sampling and storage were devised and standards of evaluation were made readily available. The prime focus of these activities was *ex situ* conservation of genetic diversity. It is only in the last decade that interest has begun to focus more on *in situ* genetic conservation techniques.

This change of direction is acknowledged in the Convention on

Biological Diversity, Article 9 of which stresses the complementary nature of *in situ* and *ex situ* techniques (UNCED, 1992). It is undoubtedly true, however, that much more scientific research has been focused on the scientific basis of *ex situ* genetic conservation, and, as pointed out by Hawkes (1991a), *in situ* techniques are still very much in their infancy.

2.3 A NEED FOR CLARIFICATION

The broad objective of genetic conservation is to ensure that gene pools are secure, efficiently held and readily available for sustainable utilization, but how is this to be achieved? The conservationist must clearly define and understand the processes involved, and then attempt to develop practical techniques to achieve this objective. Conservationists, when undertaking a particular conservation exercise, use their knowledge of genetics, ecology, geography, taxonomy and many other disciplines to understand and manage the genetic diversity they wish to conserve. It is these procedures that must be made explicit and incorporated into any model if it is to be of practical assistance to conservation scientists working in the field.

2.4 DEFINITION

As Spellerberg and Hardes (1992) define it:

Biological conservation aims to maintain the diversity of living organisms, their habitats and the interrelationships between organisms and their environment.

These authors also stress that conservation is not just about individual plant and animal species, but includes all aspects of biodiversity from ecosystems through communities, species and populations to genetic diversity within species. In recent years an attempt has been made to differentiate between conservation at the ecosystem and at the genetic levels and these may be referred to as ecological and genetic conservation, respectively. While much interest has been focused on developing models for ecosystem and habitat conservation (Shafer, 1990; Spellerberg *et al.*, 1991; Fiedler and Jain, 1992; Groombridge, 1992; Forey *et al.*, 1994), and for defining various aspects of genetic conservation (Allard, 1970; Bennett, 1970; Marshall and Brown, 1975; Brown, 1978; Yonezawa, 1985; Chapman, 1989; Guarino *et al.*, 1995), less progress has been made in the development of an overall system for genetic conservation that includes both *in situ* and *ex situ* strategies.

2.5 A METHODOLOGY FOR PLANT GENETIC CONSERVATION

The raw materials of genetic conservation are genes within gene pools, the total diversity of genetic material of the particular taxon being conserved. The product of the gene pool is either preserved or utilized genetic diversity. The processes linking the raw matter and the utilized gene pool is conservation. This may be summarized in the simple model proposed in Figure 2.1.

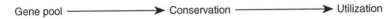

Gene pool ⟶ Conservation ⟶ Utilization

Figure 2.1 Simple model of plant genetic conservation.

Conservation is the process that actively retains the diversity of the gene pool with a view to actual or potential utilization; utilization is the human exploitation of that genetic diversity.

The basis of genetic conservation is genetic diversity, the sum of the allelic variation found in nature: it is this genetic diversity that is conserved and utilized. There is clearly an essential and intimate link between conservation and utilization: humans conserve because they wish to utilize. Conservation has an economic cost and it is difficult to persuade society to meet the cost unless it is seen as being of some value. It is relatively easy to argue the economic benefit that might accrue from the conservation and subsequent utilization and exploitation of land races or wild relatives of crops in breeding programmes, but it is more difficult to ascribe economic value to truly 'wild' species. However, it is argued that virtually all plants are likely to be of some value, whether in terms of immediate crop breeding potential or for pharmaceutical, recreation, eco-tourism and educational use or for less overt forms of utilization, such as making people feel 'good' to think that nature is 'safe'. Like all biodiversity, genetic diversity is part of any nation's heritage alongside its art and culture. Therefore, it is important to make an explicit link between conservation and utilization in any conservation strategy.

The simple model can be expanded to produce the more explicit conservation model proposed in Figure 2.2.

2.5.1 Selection of target taxa

Conservation activities will always be limited by the financial, temporal and technical resources available. It would be impossible to conserve actively all species, so it is important to make the most efficient selection of species to conserve. This choice should be objective, based on logical, scientific and economic principles related to the perceived value of the species. The attribution of comparative value to taxa is discussed in Chapter 3. Although it is important that 'value' is attributed objectively

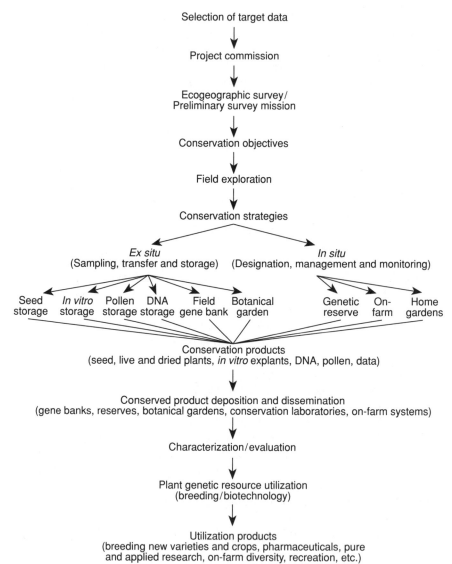

Selection of target data

↓

Project commission

↓

Ecogeographic survey/
Preliminary survey mission

↓

Conservation objectives

↓

Field exploration

↓

Conservation strategies

Ex situ   *In situ*
(Sampling, transfer and storage)   (Designation, management and monitoring)

Seed   *In vitro*   Pollen   DNA   Field   Botanical   Genetic   On-   Home
storage   storage   storage   storage   gene bank   garden   reserve   farm   gardens

Conservation products
(seed, live and dried plants, *in vitro* explants, DNA, pollen, data)

↓

Conserved product deposition and dissemination
(gene banks, reserves, botanical gardens, conservation laboratories, on-farm systems)

↓

Characterization/evaluation

↓

Plant genetic resource utilization
(breeding/biotechnology)

↓

Utilization products
(breeding new varieties and crops, pharmaceuticals, pure
and applied research, on-farm diversity, recreation, etc.)

Figure 2.2 Proposed model of genetic conservation.

and scientifically, the actual value given will not be universal; it will
depend on the priorities of the agency, the country or even the individual.

The sort of factors that provide a species with 'value' are current conservation status, potential economic use, threat of genetic erosion, genetic distinction, ecogeographic distinction, national or conservation
agency priorities, biologically important species, culturally important

species, relative cost of conservation, conservation sustainability, and ethical and aesthetic considerations. Each of these factors is discussed in detail in Chapter 3. Rarely will one of them on its own lead to a taxon being given conservation priority; more commonly, all or a range of the factors will be scored for a particular taxon, which will then be given a certain level of national, regional or world conservation priority. Factors related to potential economic use will commonly be given higher comparative value, especially in poor economies where income generation is of the highest priority. If the overall score passes a threshold level or is higher than for competing taxa, the taxon will be conserved; then either it will be collected and stored *ex situ*, or a genetic reserve will be established or on-farm conservation proposed.

2.5.2 Project commission

In practice, once taxa are selected for conservation, the actual conservation activities are necessarily preceded by some form of commission. This may take the form of a formal statement, which establishes the objectives of the conservation and justifies its selection, specifying the target taxa and target areas, how the material is to be utilized, where the conserved material is to be safety duplicated, etc., and perhaps indicating which conservation technique is to be employed. It is worth taking the time to formulate a clear, concise commission statement as it will help to focus subsequent conservation activities. The commission may vary in breadth from individual conservationists establishing genetic reserves for given species within their own country, to an international conservation agency (BGCS, IPGRI, IUCN, WWF, etc.) commissioning a systematic collection programme for a range of target taxa from throughout their geographical range, e.g. onion (*Allium*) species of Central Asia, groundnut (*Arachis*) species worldwide, or sweet potato (*Ipomoea batatas*) from Irian Jaya. So the commissioner can vary from an individual to an international agency, from one species to hundreds of species and from one region of a country to the whole world. In each case, to warrant active conservation a particular group of taxa from a defined geographical area must be considered currently insufficiently conserved (either *in situ* or *ex situ*), of sufficient actual or potential use, and/or endangered.

2.5.3 Ecogeographic survey and preliminary survey mission

Once the target taxon or group of taxa has been selected and delimited, the conservationist begins to amass and synthesize fundamental biological data to help formulate an appropriate conservation strategy. The synthesis and analysis of these data enable the conservationist to make vital decisions concerning, for example, which taxa to be included in the target group, where to find these taxa, which combination of *ex situ* and *in situ* conservation to use, what sampling strategy to adopt, and where to store the germplasm or site the reserve. If the basic biological data for a particular species – for example, the perennial relative of the garden pea, *Vavilovia formosa* – indicate that the species has been previously found on limestone scree slopes above 2000 m in south-west Asia, then further material of this species is likely to be found currently under similar constraints and is less likely to be found in different habitat types or far distant regions.

The process of collating and analysing geographical, ecological and taxonomic data is referred to as ecogeography and was defined by Maxted *et al.* (1995) as:

> an ecological, geographical and taxonomic information gathering and synthesis process for a particular taxon. The results are predictive and can be used to assist in the formulation of collection and conservation priorities.

Ecogeographic studies involve the use of large and complex data sets obtained from the literature and from the compilation of passport data associated with herbarium specimens and germplasm accessions. The data compiled are essentially of three kinds: ecological, geographical and taxonomic. These data are synthesized to produce three basic products: the database, which contains the raw data for each taxon; the conspectus, which summarizes the data for each taxon; and the report, which discusses the contents of the database and conspectus, as well as proposing future collection and conservation strategies. Ecogeographic techniques enhance the efficiency of crop relative and wild species conservation, because they enable the conservationist to identify clearly the geographical regions and ecological niches that the taxon inhabits, and so not only identify areas with high numbers of target taxa, but also areas that contain high taxonomic or genotypic diversity of taxa, uniqueness of habitat, economic or plant breeding importance, etc. (Chapter 4).

If the ecogeographic data for the target taxon are unavailable or limited, the conservationist will not have sufficient background biological knowledge to formulate an effective conservation strategy. In this case it would be necessary to undertake an initial survey mission to gather the fresh ecogeographic data required on which to base the conservation strategy. The survey mission may be in the form of 'coarse grid sampling', which involves travelling throughout a likely target region and sampling sites at relatively wide intervals over the whole region. The precise size of the interval between sites depends on the level of environmental diversity across the region, but Hawkes (1980) suggests sampling every 1–50 km. The population samples and data collected during this mission can then be used to formulate further conservation priorities and develop an appropriate strategy, thus providing the same result as the ecogeographic survey for better biologically understood groups.

2.5.4 Conservation objectives

The products of the ecogeographic survey or survey mission provide a basis for the conservationist to formulate future conservation priorities and strategies for the target taxon. Within the target area, zones of particular interest may be identified, e.g. areas with high concentrations of diverse taxa, low or very high rainfall, high frequency of saline soils, or extremes of altitude or exposure, etc. In general these areas are those likely to contain plants with distinct genes or genotypes. If a taxon is found throughout a particular region then the researcher can use the ecogeographic data to positively select a series of diverse habitats to designate as reserves. If a taxon has been found at one location, but not at another with similar ecogeographic conditions, then the ecogeographer may suggest that these similar locations should be searched. Within the target taxon specific variants can be identified which warrant conservation priority, e.g. species that have previously unrecognized utilization potential, populations that are particularly in danger of genetic erosion, those that had not previously been conserved, etc.

The conservationist must set out a clear, concise statement of the proposed conservation strategy for the target taxon and, if appropriate, priorities. These may have been established in the project commission, but if not the conservationist should undertake the task. The statement should answer certain questions. For example: which populations require conservation? Can local farmers play a part in conservation activities? Do population levels require close monitoring? Should a national or international collecting team be directed to collect the priority target taxa? What conservation strategy is or strategies are appropriate? What combination of conservation techniques is appropriate? Or is a more detailed study required before any of these questions can be answered?

2.5.5 Field exploration

Once the conservation objectives have been clarified, whichever conservation strategy is to be applied, the ecogeographic information is used to locate and identify the general locality of the plant populations that are to be conserved. The ecogeographic data will rarely be sufficiently comprehensive to locate actual populations precisely. Therefore, the preparatory element of conservation activities will be followed by field exploration, during which actual populations are located. Ideally populations of the target taxon that contain the maximum amount of genetic diversity in the minimum number of populations will be identified, but how is this goal to be achieved? Commonly there will be too much diversity in both crops and wild species to conserve all their alleles, even if these were known then or at some future time. Thus the conservationist must attempt to conserve the range of diversity that best reflects the total genetic diversity of the species.

Marshall and Brown (1975) suggest that the objective is to conserve the plants that will contain 95% of all the alleles at a random locus occurring in the target population with a frequency greater than 0.05. In practice this will be impossible for the botanist to assess in the field (Chapter 6). How many plants must be sampled, which plant and what pattern of sampling is appropriate? To answer these specific questions the conservationist should know the amount of genetic variation within and between populations, local population structure, breeding system, taxonomy and ecogeographic requirements of the target taxon, as well as many other biological details. Some of this information will be supplied following the ecogeographic survey, but some will remain unavailable. Therefore, the practice of field exploration will be modified depending on what biological information is available on the target taxon and target area. For example, there is little or no interpopulation variation in the grass *Phalaris tuberosa* in south-east Australia; therefore, the appropriate strategy would be to conserve a large number of individuals from a few populations. However, if we know that a target species is highly differentiated, but each population contains few genotypes, as with subterranean clover *Trifolium subterraneum*, then we would conserve a few individuals from a large number of populations. Detailed discussion of how this may be achieved is provided by Hawkes (1980), Ford-Lloyd and Jackson (1986), Frankel *et al.* (1995) and Guarino *et al.* (1995).

The field botanist should positively select populations for conservation if they are found on the periphery of the target taxon's distribution or those which contain morphological or ecological variants. Atypical populations, or those growing under atypical conditions, may possess genes or alleles that are unknown or extremely rare in the target taxon's centre of diversity, and this material possibly contains genetic variation of special use to the breeders (e.g. disease or pest resistance; adaptation to soil or climate which is unknown in the crop itself).

2.5.6 Conservation strategies

There are two basic conservation strategies, each composed of various techniques, that the conservationist can adopt to conserve genetic diversity once it has been located. The two strategies are *ex situ* and *in situ*. Article 2 of the Convention on Biological Diversity (UNCED, 1992) provides the following definitions of these categories:

Ex situ **conservation** means the conservation of components of biological diversity outside their natural habitats.

In situ **conservation** means the conservation of ecosystems and natural habitats and the maintenance and recovery of viable populations of species in their natural surroundings and, in the case of
domesticates or cultivated species, in the surroundings where they
have developed their distinctive properties.

There is an obvious fundamental difference between these two strategies:
ex situ conservation involves the sampling, transfer and storage of target
taxa from the target area, whereas *in situ* conservation involves the designation, management and monitoring of target taxa where they are
encountered. Owing to this fundamental difference there is little overlap
between the two strategies, with at least one exception: the provision of
traditional varieties to farmers to cultivate within traditional systems.
Accurately this form of conservation is *ex situ* if the varieties are not from
the local region, even though the technique applied will essentially be
on-farm conservation.

The two basic conservation strategies may be further subdivided into
several specific techniques:

- *Ex situ*:
  - seed storage;
  - *in vitro* storage;
  - DNA storage;
  - pollen storage;
  - field gene bank;
  - botanical garden.
- *In situ*:
  - genetic reserve;
  - on-farm;
  - home gardens.

(a) Ex situ *techniques*

Genetic variation is maintained away from its original location. Samples
of a species, subspecies or variety are taken and conserved either as living collections of plants in field gene banks, botanical gardens or arboreta, or as samples of seed, tubers, tissue explants, pollen or DNA
maintained under special artificial conditions. The techniques are generally appropriate for the conservation of crops, crop relatives and wild
species.

Seed storage conservation
Ex situ seed collection and storage is the most convenient and widely used method of genetic conservation. Seeds are the natural dispersal and storage organs for the majority of species. This technique involves seed samples being collected from crop or wild populations and then transferred to a gene bank for storage, usually at subzero temperatures, after previously being dried to a suitable moisture content. This procedure has been adopted for the bulk of orthodox-seeded species (those species that have seed which can be dried and stored at low temperature without losing viability). The advantages of this technique are that it is efficient and reproducible, and feasible for secure storage in the short, medium and long term. The disadvantages are associated with problems in storing recalcitrant seeded species. These cannot be dried and cooled in the way used for orthodox seeds; often they rarely produce seed or are normally propagated by cloning. This technique has also been criticized because of the 'freezing of evolution'. Germplasm held in a gene bank is no longer continuously adapting to changes in the environment, such as new races of pests or disease, or major climatic changes.

In vitro *conservation*
In vitro conservation involves the maintenance of explants in a sterile, pathogen-free environment and is widely used for vegetatively propagated and recalcitrant-seeded species. *In vitro* conservation offers an alternative to field gene banks. It involves the establishment of tissue cultures of accessions on nutrient agar and their storage under controlled conditions of either slow or suspended growth. The main advantage is that it offers a solution to the long-term conservation problems of recalcitrant, sterile or clonally propagated species. The main disadvantage is the risk of somaclonal variation, the need still to develop individual maintenance protocols for the majority of species and the relatively high-level technology and cost required. The best answer for cheap, long-term *in vitro* conservation in the future may be cryopreservation (Hoyt, 1988); that is, the storage of frozen tissue cultures at very low temperatures, for example in liquid nitrogen at $-196°C$. If this technique can be perfected to reduce the damage caused by freezing and thawing, it may be possible to preserve materials indefinitely.

Pollen conservation
The storage of pollen grains is possible in appropriate conditions, allowing their subsequent use for crossing with living plant material. It may also be possible in the future to regenerate haploid plants routinely from pollen cultures, but no generalized schedules have been developed yet. It has the advantage that it is a relatively low-cost option, but the disadvantage is that only paternal material would be conserved and regenerated.

DNA storage conservation
The storage of DNA in appropriate conditions can be achieved easily and inexpensively, given the appropriate level of technology, but the regeneration of entire plants from DNA cannot be envisaged at present, though single or small numbers of genes could subsequently be utilized. The advantage of this technique is that it is efficient and simple, but the disadvantage lies in problems with subsequent gene isolation, cloning and transfer.

Field gene bank conservation
The conservation of germplasm in field gene banks involves the collecting of material from one location and the transfer and planting of the material in a second site. It has traditionally provided the answer for recalcitrant species (whose seeds cannot be dried and frozen without loss of viability) or sterile seeded species, or for those species where it is preferable to store clonal material. Field gene banks are commonly used for such species as cocoa, rubber, coconut, mango, coffee, banana, cassava, sweet potato and yam. The advantages of field gene banks are that the material is easily accessible for utilization, and that evaluation can be undertaken while the material is being conserved. The disadvantages are that the material is restricted in terms of genetic diversity, is susceptible to pests, disease and vandalism, and involves large areas of land. The latter point practically limits the genetic range of material that can be held, so the full range of ecogeographic conditions under which the species normally grows and the total genetic diversity cannot be reflected in a field gene bank. However, in certain cases there are no viable alternative techniques.

Botanical garden conservation
Historically, botanical gardens were often associated with physic gardens or displays of single specimens of botanical curiosities and, as such, did not attempt to reflect the genetic diversity of the species. However, in recent years with increased public awareness of environmental and conservation issues, there has been a movement toward the establishing of conservation units within botanical gardens (Maunder, 1994a,b). In this context, botanical gardens hold living plant collections of species that were collected in a particular location and moved to the garden to be conserved. The advantage of this method of conservation is that botanical gardens do not have the same constraints as other conservation agencies; for example, plant breeding institutes will by definition focus their activities on crop or crop-related species. Botanical gardens have the freedom to focus on wild species that may otherwise not be given sufficient priority for conservation. There are two disadvantage to this technique of conservation. The first is that the number of species that can be genetically conserved in a botanical garden will always be limited because of

the available space for growing plants. The majority of botanical gardens are located in urban areas in temperate countries. At their present site, expansion would be prohibitively expensive. The majority of botanical diversity is located in tropical climates, yet the majority of botanical gardens are located in temperate countries, so there is a need to keep the species included in expensive glasshouses, which will also necessarily limit the space available. The second disadvantage is related to the first: only a very few individuals can be grown of each species, thus neglecting the need to conserve the range of genetic diversity in the wild. However, if the target species is very near extinction and only one or two specimens remain extant, this objection of course does not hold.

Many botanical gardens also contain gene banks and tissue culture facilities, or their staff may be involved in conservation activities away from the gardens, but these activities are here considered as separate, distinct techniques from keeping collections of plants. Perhaps the most useful conservation activity that traditional botanical gardens can perform is in increasing public education, and making the public aware of the importance of botanical conservation.

There were always some serious limitations associated with *ex situ* conservation, as discussed above, such as the difficulty of conserving clonally propagated crops (e.g. bananas, tropical tuber crops, etc.) and certain tropical trees and shrubs, whose seeds cannot be stored by means of drying and cooling (the 'recalcitrant' and 'intermediate' seeded species) and the fact that species whose seeds are kept under long-term storage are not subjected to evolutionary pressures of environmental change and the development of new and possibly more aggressive strains of pests and diseases. In contrast, species maintained *in situ*, that is to say in farmers' fields or in natural wild species reserves, do not require drying and cooling and are subjected to selection pressures of new races or mutant forms of pests and diseases. Therefore in recent years attention has switched toward the conservation of genetic resources *in situ*.

(b) In situ *techniques*

These techniques involve maintenance of genetic variation at the location where it is encountered, either in the wild or in traditional farming systems. Much genetic conservation research (Allard, 1970; Bennett, 1970; Frankel and Hawkes, 1975; Brown, 1978; Yonezawa, 1985; Chapman, 1989) has concentrated on *ex situ* genetic conservation, while relatively little progress has been made in developing scientific principles appropriate for *in situ* genetic conservation.

It has also been the case that the majority of existing nature reserves or natural parks were established to conserve animals or protect aesthetically beautiful landscapes, but even today few have as their primary goal plant conservation, let alone the genetic conservation of plant species (Hoyt, 1988). As acknowledged above, *in situ* genetic conservation, as a conservation strategy, is still very much in its infancy and there remain many unknowns.

   In situ techniques involve the conservation of germplasm in its natural 'wild' or 'farmed' habitat. Confusingly, however, some authors do not distinguish genetic conservation in a genetic reserve from on-farm conservation, misleadingly referring to both solely as *in situ* conservation. As the two prime techniques for *in situ* conservation are quite distinct both in their targets (on-farm for crops, genetic reserve for wild species) and their mode of conservation, it is important to stress that genetic reserve and on-farm conservation are each distinct techniques of the *in situ* conservation strategy.

Genetic reserve conservation

Conservation of wild species in a genetic reserve involves the location, designation, management and monitoring of genetic diversity in a particular, natural location. This technique is the most appropriate for the bulk of wild species, whether closely or distantly related to crop plants, because it can, when the management regime is minimal, be inexpensive; it is applicable for orthodox and non-orthodox seeded species; it permits multiple taxon conservation in a single reserve and allows continued evolution of the species. The disadvantages are that the conserved material is not immediately available for agricultural exploitation and, if the management regime is minimal, little characterization or evaluation data may be available. In the latter case the reserve manager may even be unaware of the complete specific composition of the reserve.

On-farm conservation

Farmer-based conservation involves the maintenance of traditional crop varieties or cropping systems by farmers within traditional agricultural systems. On traditional farms, what are generally known as 'land races' are sown and harvested; each season the farmers keep a proportion of harvested seed for resowing. Thus the land race is highly adapted to the local environment and is likely to contain locally adapted alleles that may prove useful for specific breeding programmes. This is perhaps the most recent technique for genetic conservation recognized by scientists, but has obviously been practised by traditional farmers for millennia.

   On the basis of the actual material conserved, this technique can be subdivided into seed crops (seed and grain crops, vegetables, forages and fodder species), vegetatively propagated crops (potato, sweet potato, yams, cassava, taro, *Xanthosoma* and a range of other minor crops) and the wild and semi-cultivated species (the weedy or ruderal species that are unable to survive under natural habitat conditions and need open areas amongst crops, around dwellings and by walls, hedges, path sides and roadsides for their survival). The on-farm conservation of these three groups of species in traditional farming systems is discussed in detail in Chapter 22.

The overall advantage of this technique is that it ensures the maintenance of ancient land races and those wild species dependent on traditional agriculture. However, the land races may yield less than modern cultivars and so the farmer may require subsidizing and possibly monitoring to ensure continued cultivation. It should be noted that contemporary economic forces will tend to act against the continued cultivation of ancient land races and they are undoubtedly suffering rapid genetic erosion, if not facing extinction. A back-up system of *ex situ* conservation is therefore essential, as discussed below. There is clearly a need for much more detailed study of the dynamics of on-farm conservation; see Chapters 10, 17, 18, 19 and 21 for further discussion.

Home garden conservation

This technique is closely related to on-farm conservation and involves smaller-scale but more species-diverse genetic conservation in home, kitchen, backyard or door-yard gardens. The focus of this form of *in situ* conservation is medicinal, flavouring and vegetable species (e.g. tomatoes, peppers, coumarin, mint, thyme, parsley, etc.). Orchard gardens, which are often expanded versions of kitchen gardens, can be valuable reserves of genetic diversity of fruit and timber trees, shrubs, pseudo-shrubs such as banana and pawpaw, climbers and root and tuber crops as well as the herbs mentioned above (Hawkes, 1983; Chapter 10).

The relative advantages and disadvantages of the various techniques are summarized in Table 2.1. *Ex situ* conservation involves the conservation of a single target taxon in any one accession; collectors actively avoid taking mixed taxon collections. Germplasm collecting is a relatively expensive process and for that reason the bulk of *ex situ* conservation activities has been focused on crops or close crop relatives. Although *in situ* conservation may be no less expensive, if the populations are actively managed and monitored, it does allow the possibility of conserving more than a single target taxon in any one genetic reserve or group of traditional farms.

For example, the Ammiad reserve (Chapter 15) was established to conserve the wild relatives of cereal grasses but, because these target taxa are conserved within a community, non-target taxa such as genetically important legume species will also be conserved at the same site. Therefore, it could be argued that the overall expenditure per unit of taxon conservation may be lower in a genetic reserve, and lower conservation costs will mean that wild species not of immediate utilization potential have a greater chance of being conserved.

Table 2.1 Relative advantages and disadvantages of various conservation techniques

Strategy	Techniques	Advantages	Disadvantages
Ex situ	Seed storage	Efficient and reproducible Feasible for medium and long-term secure storage Wide diversity of each target taxon conserved Easy access for characterization and evaluation Easy access for utilization Little maintenance once material is conserved	Problems storing seeds of 'recalcitrant' species Freezes evolutionary development, especially that which is related to pest and disease resistance Genetic diversity may be lost with each regeneration cycle (but individual cycles can be extended to periods of 20–50 years or more) Restricted to a single target taxon per accession (no conservation of associated species found in the same location)
	In vitro storage	Relatively easy long-term conservation for large numbers of 'recalcitrant', sterile or clonal species Easy access for evaluation and utilization	Risk of somaclonal variation Need to develop individual maintenance protocols for most species Relatively high-level technology and maintenance costs
	DNA storage	Relatively easy, low cost of conservation	Regeneration of entire plants from DNA cannot be envisaged at present Problems with subsequent gene isolation, cloning and transfer
	Pollen storage	Relatively easy, low cost of conservation	Need to develop individual regeneration protocols to produce haploid plants; further research needed to produce diploid plants Only paternal material conserved but mixtures from many individuals could be envisaged

Table 2.1 Continued

Strategy	Techniques	Advantages	Disadvantages
	Field gene bank	Suitable for storing material of 'recalcitrant' species Easy access for characterization and evaluation Material can be evaluated while being conserved Easy access for utilization	Material is susceptible to pests, disease and vandalism Involves large areas of land, but even then genetic diversity is likely to be restricted High maintenance cost once material is conserved
	Botanical garden	Freedom to focus on wild plants Easy public access for conservation education Freedom to focus on non-economic plants	Space limits the number (generally only one or two individuals) and genetic diversity of the species conserved Involves large areas of land, so genetic diversity is likely to be restricted High maintenance costs in glasshouse once conserved
In situ	Genetic reserve	Dynamic conservation in relation to environmental changes, pests and diseases Provides easy access for evolutionary and genetic studies Appropriate method for 'recalcitrant' species Allows easy conservation of a diverse range of wild relatives Possibility of multiple target taxa reserves	Materials not easily available for utilization Vulnerable to natural and man-directed disasters, e.g. fire, vandalism, urban development, air pollution, etc. Appropriate management regimes poorly understood Requires high level of active supervision and monitoring Limited genetic diversity can be conserved in any one reserve

On-farm	Dynamic conservation in relation to environmental changes, pests and diseases Ensures the conservation of traditional land races of field crops Ensures the conservation of weedy crop relatives and ancestral forms	Vulnerable to changes in farming practices Appropriate management regimes poorly understood Requires maintenance of traditional farming systems and possible payment of premiums to farmers Restricted to field crops Only limited diversity can be maintained on each farm, so multiple farms in diverse regions are required to ensure the conservation of genetic diversity Easily confused with farmer-based breeding and selection activities
Home, orchard, etc. gardens	Dynamic conservation in relation to environmental changes, pests and diseases Ensures the conservation of traditional land races of minor crops, fruit and vegetables, medicinal plants, flavourings, culinary herbs, fruit trees and bushes, etc. Ensures the conservation of weedy relatives and ancestral forms	Vulnerable to changes in management practices Appropriate management regimes poorly understood Requires maintenance of traditional cultural systems, and possible subsidization of the farmer

2.5.7 Complementary conservation

There has been a recent trend among genetic conservationists toward *in situ* conservation techniques, both because of the urgent need to protect ecosystems threatened with imminent change and also for other more nebulous political reasons. In recent years there has been extensive political and ethical discussion concerning the export of germplasm from developing countries, which contain the bulk of biodiversity, to international or regional gene banks, which are primarily located in the northern developed countries (Juma, 1989; Fowler and Mooney, 1990; Cooper *et al.*, 1992). This discussion has become particularly heated since the application of biotechnological techniques and the advent of widespread bioprospecting in the tropics, which has resulted in increased economic value being placed on wild species. The transfer of germplasm out of the country of origin is generally associated with a transfer of political and economic control over the material, which has undoubtedly resulted in external exploitation of biodiversity without economic benefit to the country of provenance. The problems of sovereignty and patenting of biological diversity is currently a matter of extensive international debate.

Unfortunately, this debate has led to a competitive attitude between the proponents of the two basic conservation strategies. It may surprise some to realize, as pointed out by Cohen *et al.* (1991), that the United States spent $37.5 million on biodiversity conservation in 1987, but of this only just over 1% was spent on *ex situ* projects. So the expenditure of funding is clearly weighted, at least in the case of the United States, toward *in situ* conservation projects.

Leaving the largely political arguments to one side, both conservation strategies have advantages and disadvantages, as is shown in Table 2.1. Scientifically the two strategies should not be seen as alternatives or in opposition to one another (Ford-Lloyd and Maxted, 1993), but rather as being complementary, as is stated in Article 9 of the Convention on Biological Diversity (UNCED, 1992). One conservation strategy or technique will act as a back-up to another, the degree of emphasis placed on each depending on the conservation resources available, the aims and the utilization potential for that target taxon. This has led conservationists pragmatically to adopt a more 'holistic' approach to conservation (Withers, 1993). When formulating an overall conservation strategy, the conservationist should think in terms of applying a combination of different techniques available, including both *in situ* and *ex situ* techniques, where the different methodologies complement each other.

Each of the specific conservation techniques discussed above has as its objective the maintenance of plant genetic diversity. Thus the different techniques may be thought to slot together like pieces in a jig-saw puzzle to complete the overall conservation picture (Withers, 1993). The adoption of the holistic approach requires the conservationist to look at the characteristics and needs of the particular gene pool being conserved and then to assess which of the strategies or combination of techniques offers the most appropriate option to maintain genetic diversity within that taxon. To formulate the conservation strategy the conservationist may also need to address not only genetic questions but also the practical and political ones. For example, what are the species storage characteristics? What do we know about the species breeding system? Do we want to store the germplasm in the short, medium or long term? How important is the crop? Where is the germplasm located and how accessible is it or does it need to be? Are there legal issues relating to access? How good is the infrastructure of the gene bank? What back-up is necessary or desirable?

Given answers to questions such as these, the appropriate combination of techniques to conserve the gene pool can be applied in a pragmatic and balanced manner (Figure 2.3). A different balance would be

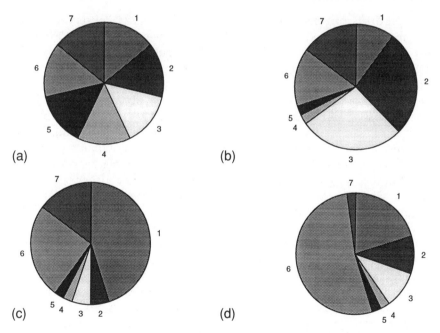

Figure 2.3 Hypothetical representation of the proportions of the gene pool conserved using seven different conservation techniques for different crops: (a) equal segments of the gene pool conserved using different techniques; (b) combination of conservation strategies applicable for root or tuber crop; (c) combination of conservation strategies for orthodox seed producer; (d) combination of conservation strategies for orthodox seed-producing forest species. 1 = seed storage; 2 = field gene bank; 3 = *in vitro* storage; 4 = pollen storage; 5 = DNA storage; 6 = genetic reserve; 7 = on-farm conservation.

required for different crops, wild species or regions. Figure 2.3a shows the situation where the entire gene pool is conserved equally by the different techniques, but in practice the combination of techniques will vary according to the target taxon. For example, if the target taxon was a root or tuber crop (e.g. potato), field gene banks and *in vitro* conservation would predominate (Figure 2.3b), with seed and pollen being held as a long-term option and wild relatives held in genetic reserves. On the other hand, for an orthodox seed producer (e.g. a cereal), seed storage would predominate (Figure 2.3c), with other techniques playing lesser roles. For an orthodox seed-producing forest species, *in situ* conservation would play as important a role as seed conservation (Figure 2.3d). The precise combination of techniques is formulated afresh for each species or group of species, demonstrating the flexibility of the holistic approach.

2.5.8 Conservation products

The products of conservation activities are primarily conserved germplasm, live and dried plants, cultures, and conservation data. Orthodox seed conserved *ex situ* is commonly held in gene banks at subzero temperatures and low moisture content to prolong its life. Live plants are conserved in genetic reserves, field gene banks, botanical gardens and research laboratories. Germplasm that is stored in a suspended form such as tissue, pollen or DNA is stored as cultures in specialist laboratory facilities. Dried voucher specimens are held in herbaria and tied to specific samples of germplasm, and are as much as possible representative of the conserved populations. If there are any queries concerning the identification of the species, the identification of the voucher specimens can be easily checked. Voucher seed samples, particularly of cereals and legumes, are used to check the seeds derived from a regeneration cycle. Conserved material is ideally associated with a range of passport data, which details the taxonomic, geographical and ecological provenance of the material. All passport data associated with conserved material should be entered into a database and made available for the management of the material, the formulation of future conservation priorities and strategies and utilization. The various conservation products, where they are stored and where they should be duplicated are summarized in Table 2.2. Evaluation data will also be added to the database from time to time when obtained.

2.5.9 Conserved product dissemination

The conservation products are either maintained in their original environment or deposited in a range of *ex situ* storage facilities. Whether

Table 2.2 Conservation products, their storage and duplication sites

Conservation product	Storage site	Duplication site
Germplasm (seed, vegetative organs, etc.)	Gene bank	National, regional and international gene banks, duplication with other conservation techniques
Live plants	Field gene bank, botanical garden, genetic reserve, on-farm	Duplication with other conservation techniques, e.g. gene bank storage of seed
Dried plants	Herbarium	National, regional and international herbaria
Explants or plantlets	Tissue culture	Duplication with other conservation techniques, e.g. gene bank storage of seed
DNA and pollen	Various cultures	Duplication with other conservation techniques, e.g. gene bank storage of seed
Conservation data	Conservation database	Duplication with other national, regional and international conservation agencies

the germplasm, voucher specimens, passport data, etc. are conserved *in situ* or *ex situ*, to ensure its safety the material should be duplicated in more than one location. The distribution of duplicate sets avoids accidental loss of the material due to fire, economic or political difficulties, warfare or other unforeseen circumstances. Duplication of the passport data is relatively easy from the conservation database and copies should be held by the commissioning agency, relevant host country institutes and other interested parties. The voucher specimens, once mounted, identified and labelled with passport data, should be distributed between the appropriate host country's herbaria and major international herbaria with an interest in the host country. The *ex situ* germplasm duplicates should be divided between the relevant host country institutes and internationally accredited centres. There is by definition a problem with the duplication of *in situ* conserved germplasm: if it is moved from its original location, it becomes *ex situ*. This is a purely semantic problem; it should not impede the duplication of germplasm conserved in a genetic reserve or on-farm, or in gene banks or other *ex situ* locations. Also, when planning *in situ* conservation either in a genetic reserve or on-farm, it would be wise to designate multiple sites for establishing the reserves or subsidizing the traditional farming or forestry system so that not all the conservation 'eggs' are in a single basket.

Conservation activities are deemed completed when not only the genetic diversity is adequately conserved, but also the expedition or conservation project reports and publications are finished. This is vital so that the wider community is made aware of the conservation activities undertaken, of any significant discoveries and as an aid to the promotion of the utilization of the conserved material.

2.5.10 Plant genetic resource utilization

As discussed above, there should be an intimate linkage between conservation and utilization. The products of conservation, whether they be 'living' or 'suspended', should be made available for utilization by humankind. Conservation can be seen as the safe keeping of preserved material, so that the material is available for utilization at a future date. In certain cases the material will be used directly, say in the selection of forage accessions, where little breeding is undertaken, or the reintroduction of primitive land race material following its local extinction during a period of civil unrest. More commonly the first stage of utilization will involve the recording of genetically controlled characteristics (characterization) and the material may be grown out under diverse environmental conditions to evaluate and screen for drought or salt tolerance, or the deliberate infection of the material with diseases or pests to screen for particular biotic resistance (evaluation). The biotechnologist will be screening for single genes which, once located, may be transferred into a host organism. The biochemist (bioprospector) will be screening for particular chemical products that may be of use to the pharmaceutical industry. The products of utilization are therefore numerous, including new varieties, new crops, pharmaceuticals, etc., as well as more nebulous products such as a 'good' environment for recreational activities.

ACKNOWLEDGEMENTS

We thank Luigi Guarino and Lyndsey Withers for their contributions to this work, and Debbie Dale for preparation of the illustrations.

Part Two

Theory and Practice of *In Situ* Conservation

3

Selection of target taxa

N. Maxted and J.G. Hawkes

3.1 INTRODUCTION

The activities of conservationists will always be limited by the financial, temporal and technical resources available to them. Therefore they must prioritize and select the taxa they are to conserve. The implication of this, in terms of *in situ* genetic conservation, is that particular species and representative populations will be selected for protection in a genetic reserve or in traditional farming systems while others will not receive active protection.

The choice of target taxa for active conservation should be objective, based on logical, scientific and economic principles related to the perceived value of the species. Several authors have recently attempted to address this question (McNeely, 1988; Pearce and Turner, 1990; Goldsmith, 1991a; Groombridge, 1992; Given, 1994; Pearce and Morgan, 1994). The latter authors conclude that 'biodiversity will be more prone to depletion when direct use values are not realized', i.e. if a species has little or no perceived value to humans then it is less likely to be given high conservation priority. It follows from this that each plant species is ascribed a comparative 'value'. The 'value' given will have a marked effect on commitment of conservation resources and therefore it is important that 'value' is ascribed objectively and scientifically.

3.2 FACTORS THAT INFLUENCE CONSERVATION VALUE

The sort of factors that provide a species with 'value' are current conservation status, potential economic use, threat of genetic erosion, genetic distinction, ecogeographic distinction, conservation agency priorities,

Plant Genetic Conservation.
Edited by N. Maxted, B.V. Ford-Lloyd and J.G. Hawkes.
Published in 1997 by Chapman & Hall. ISBN 0 412 63400 7 (Hb) and 0 412 63730 8 (Pb).

biologically important species, culturally important species, relative cost of conservation, conservation sustainability, and ethical and aesthetic considerations. It should be noted that rarely will one of the above factors alone lead to a taxon being given sufficient priority to justify establishment of a genetic reserve or continued cultivation in traditional farming systems. More commonly, each factor will be scored for a particular taxon and then given a certain level of national, regional or world conservation priority. If the overall score passes a threshold level, a reserve will be established or on-farm conservation enacted.

3.2.1 Current conservation status

Before selecting target taxa the conservationist should review current conservation activities relating to those taxa. If sufficient genetic diversity is already safely conserved from the full range of ecological habitats and geographical locations using both *in situ* and *ex situ* techniques, further active conservation may not be warranted. Details of the material currently being conserved can be obtained from the catalogues and databases of botanical gardens, gene banks and *in situ* conservation areas (Perry and Bettencourt, 1995).

However, care must be taken when interpreting information on current gene bank, botanical garden or genetic reserve holdings. The material held may be incorrectly identified, though it should be possible to check the identification by consulting voucher material or identifying living material. The actual quantities of germplasm available could also be misleading. Gene bank and botanical garden managers are encouraged to duplicate their collections in other institutions, as are managers of genetic reserves (Chapter 22). Therefore, unless detailed passport data are consulted, duplicated accessions can give a false impression of the genetic diversity actively conserved. The conservationist should also be aware that although accessions may be held in a collection, the material may for various reasons be unavailable to potential users and so create a false impression of a taxon's conservation status. Similarly, just because a species occurs within the confines of a reserve, it may not be the case that it is adequately protected, and the numbers of individuals of the species may be declining due to an inappropriate management plan. Further discussion of this point is provided in Chapter 4.

3.2.2 Potential economic use

It seems likely that when target taxa are selected for conservation, plants of economic importance to humankind that provide food, fuel, medicines, construction, technology (hunting, craft, adornment, transport), industrial products, recreation, etc., will each be given a comparatively high value and thus priority for conservation. Species whose value to humans is less immediately apparent will be given a lower level of priority than those of immediate use. In this context, it is worth emphasizing the total percentage of species that are of some human use; the figure may surprise those who are used to focusing their attention on the 20 major crops. Prance *et al.* (1995) showed that four groups of Amazonian Indians use up to 79% of the tree species in their home ranges, while Milliken *et al.* (1992) in a similar study found 81% of tree or vine species were utilized and this number rose to 86% if other categories of plants were added from the literature sources. It seems likely, however, that this high utilization rate of botanical diversity is at its extreme in underdeveloped societies and that as societies become more technologically advanced the percentage of species exploited will decrease.

Although there are many human uses of plant genetic resources, the highest priority will commonly be given to their value to agricultural exploitation. Ingram and Williams (1984) list six uses for the plant genetic resources of the wild relatives of crop plants: increased yield; increased disease resistance; improvement of growth patterns and growth rates; wider adaptability to environmental conditions for extending the range of crops; adaptation to changing agricultural practices; and improved nutritional quality. Therefore any species that can be used in this way will have enhanced value.

The gene pool concept proposed by Harlan and de Wet (1971) is often used to assist selection of target taxa. The conservationist may set priorities and targets for conservation depending on the genetic distance between the crop and the potential species to be conserved (Figure 3.1). The target taxa may belong to either the primary gene pool (GP-1) in which GP-1A are the cultivated forms and GP-1B are the wild forms; or the secondary gene pool (GP-2), cenospecies with which gene transfer is possible but difficult; or the tertiary gene pool (GP-3), where gene transfer is very difficult or impossible, but if possible requires sophisticated techniques, such as embryo rescue, etc. If the highest priority is given to species in the primary gene pool (GP-1), then the secondary gene pool (GP-2) and finally the tertiary gene pool (GP-3) are given a lower level of priority.

Harlan and de Wet's concept of three distinct gene pools has become more blurred with the advance of biotechnology and is obviously centred on individual crops. When considering more general plant genetic conservation, we are ideally attempting to conserve the total plant genetic variation which exists as plant genes, not just that variation as it relates to an individual crop. Therefore a single species may be in GP-1 for one crop, GP-2 for another and GP-3 for a third, so in terms of total conservation of genetic diversity a more appropriate representation may be that of the 'gene sea', where each species is at the centre of its own gene pool,

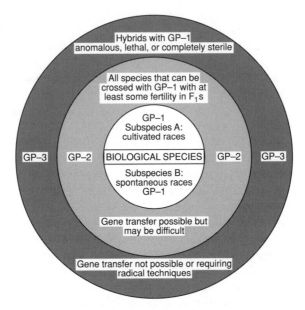

Figure 3.1 Schematic diagram of primary gene pool (GP-1), secondary gene pool (GP-2) and tertiary gene pool (GP-3). (Reproduced from Harlan and de Wet, 1971, with kind permission of the International Association for Plant Taxonomy.)

but all individual gene pools are interrelated in one sea of genetic diversity (Figure 3.2). The 'gene sea' concept may be more appropriate in the context of conserving overall genetic diversity. Thus species that are present in multiple gene pools within the gene sea would be given the highest priority for conservation because they would better represent the breadth of the plant genetic diversity in a limited number of populations or accessions.

This utilitarian and anthropocentric view of the selection of conservation priorities may offend some conservationists who believe all species to have equal value (Naess, 1984). Sylvan (1985a,b) argues convincingly, however, that if all species are given equal value, they are by definition valueless and therefore none is worthy of special consideration for conservation. Thus the arguments for uniform designation of value appear mistaken. When resources for conservation are limited and are likely to remain so, and while humans are still dying of starvation, there appears no practical or ethical alternative to giving those plants that are of human use the highest 'value' and therefore conservation priority.

The actual value given to a species is unlikely to be a specific monetary value. Shands (1994) observes that there is no current way of estimating precise monetary value for species. Although precise values may be

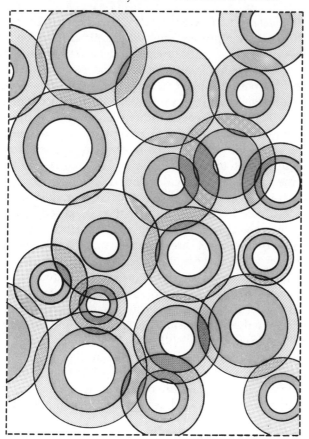

Figure 3.2 Schematic diagram of a segment of the gene sea, indicating multiple gene pools centred on several species.

difficult to ascribe, comparative values may be more easily designated. For example, major crops are likely to be given a higher priority than minor crops and plants of current human use will be given higher priority than those of possible use in the future. As with the other factors being considered below, the precise selection of target taxa will vary depending on the perspective of the agency commissioning the conservation. However, it seems likely that the selection of target taxa for conservation will commonly be correlated with their perceived monetary value.

3.2.3 Threat of genetic erosion

Estimates of the total number of flowering plants vary between 250 000 and 300 000 (Groombridge, 1992). It is difficult to estimate rates of species extinction, but the consensus view summarized by Lugo (1988) is that

15–20% of all species could become extinct by the year 2000. It is even more difficult to estimate precise levels of genetic erosion, but this must exceed the loss of species because there is likely to be some genetic erosion from the species than remain extant. So the loss of genetic diversity over the same time period is likely to be in the region of 25–30%. It is the loss of genetic diversity (within taxa) that is of prime importance for utilization, because it is largely genetic rather than taxonomic diversity that is used by humankind. It is difficult to quantify the loss of genetic diversity, though it seems likely that virtually all plant species are currently suffering genetic erosion to varying degrees.

Clearly, some species are in greater danger of genetic erosion (loss of genetic diversity) or even of complete extinction than others. These dangers must be evaluated carefully, so that those exposed to the highest risk are given increased priority for conservation. However, it must also be borne in mind that levels of threat often change rapidly and unexpectedly. Even within a single year the level of threat may be drastically altered. Thus, an area at low risk may suddenly come under the threat of urbanization, industrial development, tree felling, agricultural development, the building of dams and reservoirs, or natural disasters such as hurricanes, earthquakes or volcanic eruptions. Gomez-Campo *et al.* (1992), Groombridge (1992) and WRI *et al.* (1992) each discuss the kind of events that may lead to genetic erosion, but they may be broadly grouped under the general headings of overexploitation, habitat destruction or fragmentation, competition resulting from alien introduction, changes and intensification of land use, environmental pollution and climate change.

In order to compare the relative levels of genetic erosion faced by taxa, it is valuable to establish a system by which relative threat can be assessed objectively. Goodrich (1987) originally developed and Guarino (1995) modified and enhanced such a system for estimating the threat of genetic erosion that a taxon (wild or cultivated) faces in a particular region. The model may be operated without the necessity to visit the actual region, provided adequate background data are available on the taxon and the region. The model is based on scoring individual parameters, such as taxon rarity, local ecological conditions, local agricultural development, relationship to crop plants and local human activities. The higher the score, the greater is the risk of genetic erosion, and therefore the species is given higher priority for *in situ* or *ex situ* conservation. The scoring system is provided in Appendix 3A.

Rather than assessing the potential threat of genetic erosion to a taxon in a particular area, as Guarino's model attempts, IUCN have developed various categories of actual species threat based on such factors as individual numbers and changes in number, location of individuals, and analysis of the probability of extinction – the so called IUCN Red List

Categories (IUCN, 1994). The assessment in this case is based on the current status of the taxon, not the threat it faces; for example, whether the taxon is extinct, exists only as a result of human activities, exists in the wild, but faces various levels of threat or cannot be assessed because of a lack of background data. The eight IUCN Red List Categories are defined in Appendix 3B. If a species is considered to be Vulnerable, it would mean that the level of genetic erosion has been assessed, the species is suffering significant genetic erosion and so would be given a high priority for conservation.

3.2.4 Genetic distinction

When selecting conservation priorities, one approach would be to place the highest emphasis on the most genetically distinct groups. But how is this genetic distinction to be estimated? It is unlikely that molecular techniques will ever advance sufficiently to allow accurate estimates of genetic distinction to be made for all species, let alone populations. It is generally accepted that genetic diversity is reflected in taxonomic diversity, and that two distantly related species are less likely to share common genes that two closely taxonomically related species. In practice, this remains the most commonly applied method of measuring biological diversity: high levels of species differences are equated to high levels of biological diversity (Peet, 1974).

Heywood (1994) outlines a serious fault with this assumption associated with the lack of consistency applied by taxonomists when designating species. He states that all taxonomic species are not separated by the same, standard level of genetic distinction, and therefore use of taxonomic species distinction can be misleading. Heywood does not suggest an alternative strategy for assessing relative levels of biodiversity or genetic distinction, but does suggest that this problem may be avoided if higher levels of the hierarchy are used.

An example that is often taken to illustrate the problem with assigning equal weight between taxa is the comparison between two Floras. In the first Flora there are two species of dandelion, while in the second Flora there is one species of dandelion and one species of *Welwitschia mirabilis* (the monotypic taxon of a family from south-west Africa). If the conservationists have the resources to conserve only two species, logically they should conserve the most distantly related taxa (Vane-Wright *et al.*, 1991; Williams *et al.*, 1991; Williams and Humphries, 1994). The conservationist uses the taxonomic classification to deliberately select diverse target taxa. In this way the conservationist is considering not only the distinction of the taxa, but also the degree of taxonomic (= genetic) isolation between the taxa. Williams and Humphries (1994) suggest that it is only appropriate to use classifications based on cladistic analysis to assist in taxon selection because they are rigorously based on evolutionary, therefore genetic, differences between species. This is fundamental to the use of WORLDMAP (Williams, 1992), a program that is used to assess priority areas for conservation. However, there remain relatively few taxonomic groups for which cladistic classifications are currently available and so, at least in the short term, the accepted classification, however it was generated, needs to be used.

3.2.5 Ecogeographic distinction

It would appear reasonable to assume that plant species which are widespread, in terms of geographical and ecological range, are under less threat of extinction than those that are restricted to a particular country or localized in a distinct habitat. This assumption appears valid because increased extinction threat is unlikely to be applied consistently throughout the species' geographical or ecological range. Therefore, a species that is found on a wide range of soil types, for example, is less likely to suffer extinction than those naturally restricted to a single, rare soil type. Following this argument a taxon with severe ecogeographic limits would be given higher conservation priority than more generalist taxa.

It should be borne in mind, however, that a widespread species may well possess a range of ecological adaptation that may be reflected in the genotype. In this case the conservation objective will be to ensure the safety of the various ecotypes or geographically distinct populations. While the risk of extinction is greater in species, varieties or ecotypes which are very restricted, the loss of genetic diversity is arguably just as important in widespread taxa as in restricted ones; it is the loss of genetic material that is of significance. A single event, such as the building of a road, a house, changing agricultural practices or even the planting of a tree, may render a single valuable gene extinct.

Use of the term 'endemic' has been avoided in the context of identifying target taxa on the basis of ecogeographic evidence because, as pointed out by Williams (1992), endemism is not a true measure of diversity. Endemism is an arbitrary measure of geographical or ecological distribution: how can an endemic of Monaco be compared with one of Russia or an endemic of basaltic soils with one of redzinas in the Mediterranean? The use of the term 'endemic' ignores the concept of scale in defining the endemism and is therefore only practical when related to some form of standardized scale – for example, kilometre grid squares.

3.2.6 Conservation agency priorities

The conservation priority ascribed to a particular taxon will be affected by the internal priorities of the agency commissioning the conservation. Thus, the priorities established by the Ministries of Agriculture, Forestry and Environment, for instance, are likely to be quite different. Heywood (1994) points out that estimates of perceived value are not universal; different levels of value are likely to be given to the same taxon by different interest groups – community ecologists, park managers, plant breeders, population geneticists, agriculturalists, conservation biologists and taxonomists. Each of these professions will consider the conservation of a species in the light of its own priorities and therefore different advisors may ascribe different conservation priorities for the same group of species. Species priorities by the Ministry of Agriculture will be crops and crop relatives, while the Ministry of Forestry will give these species low priority and will give higher priority to tree species that grow rapidly or produce good quality timber.

The same logic applies to the different priorities of national, regional and international conservation agencies. A species may be abundant and not threatened internationally but is vulnerable within a particular country, perhaps on the edge of its natural distribution, and so warrants active conservation. An illustration of this point is provided by the occasionally cultivated *Vicia bithynica* (Leguminosae), which is native throughout central and southern Europe and the Mediterranean Basin (Maxted, 1995). The extreme north-western edge of its distribution is the southern coast of Britain. The species is found at a few locations in this area, all of which are subject to increased levels of tourism and natural coastal erosion. These populations are obviously threatened by increasing genetic erosion and the species may in the foreseeable future become extinct in Britain. Nevertheless, the species is thriving in its centre of diversity and is clearly not threatened at an international level. So should bithynican vetch be actively conserved in Britain? The answer will depend on the remit of the organization or person deciding conservation priorities. At an international level active conservation of the species in Britain would appear unwarranted. However, a British agency, based in a country with such a poor over-all flora, may decide that the potential loss of any species is sufficient to warrant an active conservation policy, at least instigating monitoring of the remaining populations. Thus priorities for selection of target taxa may vary depending on the perspective of the commissioning agency.

3.2.7 Biologically important species

In considering the conservation of species within their natural habitats, it is particularly true that when attempting to conserve a particular target taxon it may be necessary to conserve associated species. These associated species may have little value within themselves but as part of the natural ecosystem are required to ensure the sustained conservation of the priority target taxa. No target taxon exists in isolation, since it forms part of a community with other species with which it interacts to varying degrees.

One class of biologically important species with which the target taxon interacts has been referred to as 'keystone species' (Mills *et al.*, 1993; Tanner *et al.*, 1994). They are usually the dominant species within a habitat and tend to define the general habitat in which the other species exist. For example, the wild potato species (*Solanum jamesii* and *S. fendleri*) are native to the deserts of Arizona, New Mexico and Chihuahua in the United States and Mexico; these species rely on a whole series of tree species (*Juniperus deppeana*, *Pinus ponderosa* and *Cupressus arizonica*) as keystone species to provide the appropriate habitat for ensuring their survival. These tree species undoubtedly promote greater plant density of *Solanum* and other species than would be found in random plots in the same area that did not have the tree species present. A similar relationship is seen between the wild maize relative *Zea diploperennis* and the forest tree species of Manantlán, south-west Mexico. So in each case the local keystone species sustain and enhance a diverse mixture of species. If the target taxon is one of the species found in these habitats, then conservation efforts must also be devoted to conserving the non-target keystone species in order to secure the conservation of the target taxon. The destruction of forests of keystone species not only results in the direct genetic erosion of species and associated genetic diversity, but can also upset nutrient and water cycling, and the natural biological buffering within the environment.

The target taxon is likely to have a mutualistic relationship with animal pollinators and seed dispersers, herbivores, defending organisms, nurse trees and microbial symbionts. So, for plants, certain animal species may prove to be essential keystone species. The loss of any one of these associated species could lead to the loss of the target taxon and so the animal species may need to be actively conserved to ensure the conservation of the target plant species.

3.2.8 Culturally important species

Species may also be selected as target taxa because of their local or national cultural significance. For example, the cedar of Lebanon is an important national symbol in Lebanon and is shown on the nation's flag, currency and stamps. The area of native forest has declined extensively in recent years; the remaining specimens have little specific economic value, compared with other coniferous tree species, but because the tree is a national symbol this loss has been halted and now all native forests are actively conserved. Thus the species has cultural rather than economic value. Similar examples are provided by khezri and banyan trees in India, and the swamp cypress (*Taxodium mucronatum*) which was revered by the Aztecs in Mexico. Taxa may also be ascribed value because of their importance in religious ceremonies, such as *Bauhinia guianensis* among the Waimiri Indians of Brazil, and mistletoe (*Viscum album*) and yew (*Taxus baccata*) in pagan ceremonies in western Europe.

3.2.9 Relative cost of conservation

The relative cost (ease) of establishing reserves will affect the selection of target taxa. Faced with choosing between two target taxa and a limited conservation budget, if both have equal priority, the relative establishment costs for the two reserves would be a factor affecting the selection. The same logic is applied to the selection of reserve sites and their design (Chapter 8).

Another selection factor may be the number of target taxa it is possible to conserve in a reserve. If it is possible to conserve more that one, economics will dictate that those species that can be conserved together will be given a higher priority than those that each require a distinct reserve of their own.

3.2.10 Conservation sustainability

Conservation, whether in a genetic reserve or on-farm, is by definition long term and in the case of the former requires a large investment of resources. There would therefore be little value in establishing a genetic reserve or encouraging farmers to practise on-farm conservation unless the reserve or traditional farming system was sustainable over a long time period. This question of sustainability will obviously affect the selection of target taxa, as well as the selection of sites for reserve or on-farm conservation programmes.

3.2.11 Ethical and aesthetic considerations

Ethical and aesthetic justifications for species conservation are of increasing importance to professional conservationists. The growth of public awareness of conservation issues has undoubtedly resulted in a growth of conservation activities. It seems likely that the public demand for increased conservation activity is derived from ethical and aesthetic beliefs associated with the feeling that it is wrong for humans to eradicate species carelessly and that nature has intrinsic value and beauty. This is a view with which few would disagree. The ethical and aesthetic priorities of the public may vary from 'Humans have a responsibility (stewardship) to conserve all species,' to 'Orchids are pretty and therefore should be conserved,' to 'I was born and/or feel at home in this unspoilt valley and therefore it should be preserved'. These views are valid, but because they are difficult to quantify they are unhelpful when attempting to select target taxa for genetic conservation. However, as the majority of conservation activities are state funded, professional conservationists would be wise not to ignore the priorities set by the general public. This is illustrated by the large number of conservation programmes focused on 'flagship' species, such as pandas, whales, seals or orchids.

In conclusion it should be emphasized that when selecting target taxa the various factors discussed above are not considered independently. Rather, each of these factors should be considered and together form the basis of the selection of taxa to be actively conserved. This process results in particular species or habitats being ascribed value and priority for conservation. By definition these values are based on current, often incomplete knowledge of the species, realizing that we cannot predict today what may be a useful plant tomorrow or how much genetic diversity we shall require in the future (Given, 1994). The process is fallible, in that it is not thoroughly 'scientific', but remains necessary, because of the limited resources for conservation. Obviously the more detailed the consideration of the factors affecting the selection of target taxa and the more background data for the species available, the better is the attribution of value and the more appropriate the selection of taxa.

In order to demonstrate the points discussed above, two examples of the selection of target taxa are detailed below. Other case studies are provided in Chapters 14 to 21.

3.3 THE *IN SITU* CONSERVATION OF POTATOES (*SOLANUM*, SOLANACEAE)

Potatoes belong to the genus *Solanum* and are widespread in the Americas, and especially in the western regions. There are seven cultivated and over 230 wild species now recognized. Genes for resistance to a wide range of pests and diseases are found, particularly in the wild species but also to some extent in the cultivated ones. Thus, from an economic point of view their conservation is of considerable importance.

Cultivated potatoes are widely distributed throughout the South American Andes from Venezuela to southern Chile, extending for some 7000 km. Altitudinally, they range from sea level in southern Chile to over 4500 m in the high Andes of Peru and Bolivia. The greatest amount of diversity is found from central to southern Peru and northern Bolivia, generally from some 2500 to 4100 m. Five of the seven cultivated species (*S. ajanhuiri*, *S. chaucha*, *S. curtilobum*, *S. juzepczukii* and *S. tuberosum* subsp. *andigena*) occur in this central Andean zone. *S. tuberosum* subsp. *andigena* spreads southwards from Venezuela to northern Argentina, whilst subsp. *tuberosum* occurs in southern Argentina and southern

Chile. *S. phureja* is cultivated from Venezuela southwards to northern Bolivia.

Wild potato species are highly diverse and very widespread. They occur in the south-western states of the USA, throughout Mexico and Guatemala into Central America but always above the warm tropical zones. In South America they occur from Venezuela southwards to central Chile, and down on to the plains of Argentina, Paraguay, Uruguay and southern Brazil. Altitudinally, various species occur from sea level up to 4500 m in the high Andes from Venezuela to north-western Argentina. Wild potatoes thus extend for some 13 000 km in an approximately north–south direction.

3.3.1 *In situ* conservation

Many expeditions over the last 50–60 years have collected materials for *ex situ* conservation in gene banks in Europe and United States, though many are held in their countries of origin, and particularly in Peru (International Potato Centre) and Argentina (Balcarce, Buenos Aires province). However, *in situ* conservation has hardly been attempted for either cultivated or wild species. Most existing reserves in the Americas have been sited in tropical or subtropical forest regions (Costa Rica, Panamá, Peru, Bolivia, etc.), but given the very wide distribution noted above it would clearly be impossible or even inadvisable to consider establishing genetic reserves for every known wild potato species.

3.3.2 Cultivated potatoes

Species diversity lies largely (but not by any means entirely) in central to southern Peru and northern Bolivia, as mentioned above (Figure 3.3). Attempts have been made to establish reserves in southern Peru (Chapter 20) and in northern Bolivia (Rea, personal communication). Since potatoes are clonally propagated it is clearly necessary to conserve a wide range of clonal genotypes including as many as possible of the five species mentioned above. Discussions were held also at a recent conference of Proinpa (Programa de Investigación de la Papa) at Cochabamba, Bolivia, but so far no action to establish reserves for these species has been made, apart from that of Rea (see above).

Ideally, and as a matter of urgency, a chain of genetic reserves to conserve cultivated potatoes should be established throughout the Andes in areas with high levels of genetic diversity. These could be sited in western Venezuela (Merida), north-western Colombia (Boyacá to Santander), southern Colombia (Cauca, Nariño), northern Ecuador, northern Peru, central Peru, southern Peru, northern Bolivia (La Paz area), northern Bolivia (Cochabamba region) and central Bolivia (Oruro, Potosí). This

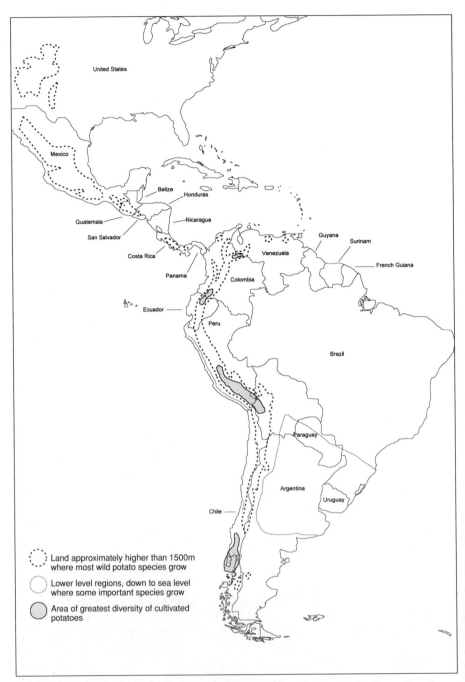

Figure 3.3 Distribution of wild and cultivated potatoes in North, Central and South America (excluding series *Etuberosa*).

does not, of course, preclude the establishment of other genetic reserves elsewhere in the region or the conservation of other indigenous tuber and grain crops in the reserves proposed.

3.3.3 Wild potatoes

It is clearly impossible to propose *in situ* conservation strategies for all 230 wild species occurring in different parts of the 13 000 km long area in which they occur. Unlike cultivated potatoes, which are adapted to temperate to very cold temperate regions with high rainfall during the growing season, wild potatoes exhibit a very wide range of adaptation as a whole, but with individual species restricted to their own ecogeographic regions. They can be found in arid to semi-arid habitats through cactus scrub, dry to semi-dry bush forest, temperate and humid mountain rain forest and alpine wastes. Many species are colonists of open ground and are ruderal ecological weeds, related closely to the cultivated species (Figure 3.4). Crop–weed complexes can often be seen round field borders, huts, backyard gardens, roadsides and path sides, and as weeds of other crops. The vast majority of wild potato species occur at altitudes of about 1500 m and above.

Some *Solanum* species are under much greater threat than others, according to their ecological adaptations. We can in fact attempt to

Figure 3.4 A weedy potato species (*Solanum sparsipulum*) growing in a maize field at Cuzco, Peru.

Table 3.1 Ecology adapted groups of *Solanum* species classified according to degree of threat

Group		Degree of threat
(i)	Groups of ruderal and adventive weeds	Little danger at present
(ii)	Those growing in seasonally dry scrub or semi-desert	Not in very much danger at present
(iii)	Those inhabiting ecotones between dry scrub forest and high mountain forest (these areas tend to be converted into arable fields)	Very endangered
(iv)	Species adapted only to tropical high mountain forests (these areas tend also to be converted into arable fields and grassy fields for cattle grazing, the trees used for firewood, fences, etc.)	Very great danger: many species now almost extinct
(v)	Species adapted to lowland southern forests (the national park 'Nahuel Huapi' in Argentina forms a reserve for *S. brevidens* and *S. tuberosum* subsp. *tuberosum*)	Considerable threat if not in reserves

classify them in groups according to the degree of threat (Table 3.1). A whole taxonomic series – *Conicibaccata* – is particularly adapted to regions (iii) and (iv) (Table 3.1) and is unable to grow elsewhere. Genetic diversity within many of these species has been decimated and they will become extinct if steps are not taken to preserve them.

3.3.4 Existing reserves

The species occurring in the south-western United States are not under threat in most existing reserve areas, but logging in the forested areas of Arizona and New Mexico is decimating populations of *S. fendleri*. In Mexico, there is a great diversity of wild potato species. Fortunately, many pine–*Abies* forests form part of the extensive national parks system in the Mexico City surroundings. More high altitude reserves are needed in the states of Puebla, Michoacán, Jalisco, Guerrero, Colima, Oaxaca, Chiapas and perhaps elsewhere. (The WWF Sierra Norte and adjacent regions project in Oaxaca State is an excellent example of the conservation of cloud forest, where wild potatoes and many other plants of potential economic value occur; unpublished report.) Forest conservation is urgently needed in Guatemala and Costa Rica where the high altitude rain forests (2700–3000 m) are largely converted into cattle-grazing pastures; thus, *S. longiconicum* is largely extinct, only surviving in one or two places. Similar problems exist in high altitude forest and páramo regions in Venezuela, Colombia and Ecuador, and in the high and middle altitude mountain rainforests of Peru and Bolivia.

A list of Protected Areas in the region has recently been produced by IUCN (1994b). Many of these coincide spatially with the area and altitudes mentioned above. At present, however, comprehensive botanical inventories do not generally exist for protected areas within the region. It remains important to find out the extent to which wild potato species do occur within areas set aside for conservation. It is also essential to make sure that the ecosystems and even the ecological niches inhabited by these species are cared for and frequently monitored. With this in mind it would be most valuable to establish a special task force to ensure that such ecosystems are being conserved within existing protected areas.

In Argentina many reserves already exist at medium to high altitudes and the problems in general do not seem to be so acute. We have very little knowledge of central and southern Chile. The species on the plains of Argentina, Paraguay, Uruguay and southern Brazil (*S. chacoense* and *S. commersonil*) are on the whole adventive and ruderal, managing to survive quite well as weeds of cultivation, road and tracksides, riversides, sea beaches, etc. Thus, quite clearly, more *in situ* conservation is urgently needed for the regions mentioned. In conclusion, the target species identified are those adapted to humid, semi-humid and arid forest and scrub areas, and reserves are urgently needed for their preservation.

3.4 THE *IN SITU* CONSERVATION OF VETCHES (*VICIA*, LEGUMINOSAE)

The genus *Vicia* L. (Leguminosae, Vicieae) comprises about 166 species (Allkin *et al.*, 1986) which were recently revised by Kupicha (1976) and Maxted (1993). The species of *Vicia* were divided into two subgenera, *Vicilla* (Schur) Rouy and *Vicia*. Subgenus *Vicia* contains fewer species than *Vicilla* but *Vicia* includes the majority of the agriculturally important species of the genus: the grain legume, faba bean (broad bean), *Vicia faba* L. and the minor forages, *V. narbonensis* L. (narbon vetch) and *V. sativa* L. subsp. *sativa* (common vetch), as well as their close wild relatives. The subgenus also contains taxa that have a high potential as forages of the future (Maxted *et al.*, 1990): *V. hyaeniscyamus* Mout., *V. noeana* (Reuter in Boiss.) Boiss., *V. sativa* subsp. *amphicarpa* (L.) Batt. and *V. sativa* subsp. *macrocarpa* (Moris) Arcang.

The genus is chiefly located in Europe, Asia and North America, with secondary centres of diversity in temperate South America and tropical Africa. However, the highest specific diversity is found in Turkey and north-west Asia. The majority of the species of subgenus *Vicia* occupy similar niches in dry, open areas of disturbed ground and are commonly considered weeds of cultivation or disturbed land on the margins of cultivation. While subgenus *Vicilla* species favour a broader range of habitats than those listed above for subgenus *Vicia*, they are found more often in moist, stable, closed communities in or on the margins of shaded woodland.

In recent years a series of forage legume conservation projects, discussed by Maxted (1995), have resulted in extensive *ex situ* collection from north-west Asia. These projects have also resulted in a broad understanding of the ecogeography and conservation biology of *Vicia* in the region and resulted in the publication of a detailed ecogeographic study of the group. Although changes in agricultural practices and industrialization have undoubtedly led to extensive genetic erosion in north-west Asia, the majority of the vetch species are weedy, colonizing species often found in habitats disturbed by humans and they seem unlikely to suffer serious genetic erosion in the immediate future. Although there has been a comprehensive *ex situ* collection programme for *Vicia* species over the past 10 years, there has been no systematic attempt to conserve the genetic diversity of the genus *in situ*, either in a genetic reserve or on-farm. However, the genus is particularly suitable for this form of conservation because it contains high levels of specific diversity and many of the species are used as minor forages (new lines are usually collected and selected rather than bred). This led Maxted (1995) to suggest the establishment of genetic reserves in Turkey, Syria, Lebanon, Israel, Iraq, Iran and the Caucasian Republics. Specifically, Maxted concluded that the species most seriously threatened by extinction are those restricted to Syria, Lebanon, Turkey and Israel; and that the highest concentration of potentially threatened taxa is located in Syria. Therefore he recommended the establishment of four reserves in Syria at: Ain Dinar, Al Hasakah (37°15′N, 42°20′E); Kessab town, Kessab (35°54′N, 35°56′E); Qal'at Al Hosn, Homs (34°46′N, 36°18′E); and Mimas, Djebel Druze (32°36′N, 36°43′E) (Figure 3.5). He also suggested that management practice be reviewed in the Olimpos National Park, Antalya, Turkey, which contains the rare endemic relative of the faba bean, *V. eristalioides*. Some of the specific reasons for selecting these sites for conservation in genetic reserves were current conservation status, potential utilization, threat of genetic erosion, genetic distinction, and ecogeographic distinction.

(a) Current conservation status

Until recently, *Vicia* genetic diversity had not been adequately or systematically conserved (IBPGR, 1985a). However, between 1985 and 1995 annual national and international expeditions collected germplasm from throughout the Mediterranean and it now seems likely that a high proportion of the gene pool is effectively conserved *ex situ*. The position

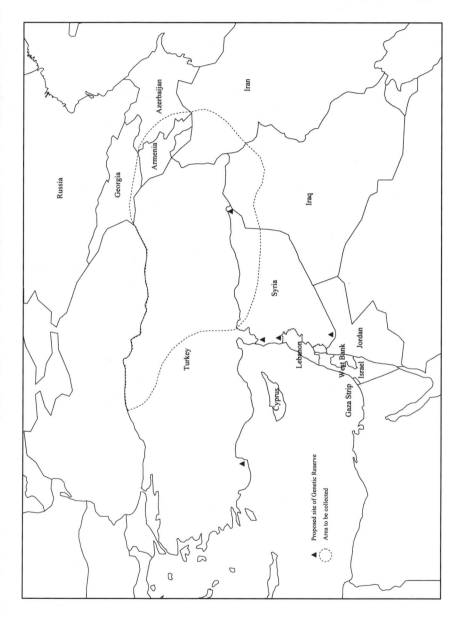

Figure 3.5 Near East distribution of genetic reserves and area to be collected for *Vicia* species.

regarding *in situ* conservation of *Vicia* is quite different: there is only one reserve (Chapter 15) and no on-farm conservation programmes for the vetches. Obviously, one reserve cannot effectively conserve the major crop, minor crop and wild species diversity in the genus *Vicia*.

(b) Potential utilization

The Middle East is the centre of diversity of the genus *Vicia* and the agriculture of the region has traditionally used a wide range of vetch species, *V. faba*, *V. narbonensis*, *V. sativa*, *V. bithynica*, *V. noeana*, *V. pannonica*, *V. ervilia*, *V. villosa* and *V. articulata* (Plitmann, 1967). However, the economic use of this range of species is undoubtedly decreasing with time (Ehrman and Maxted, 1990; Maxted *et al.*, 1990). The decrease in diversity was recognized by the Forage Working Group (IBPGR, 1985a), which concluded that *Vicia* species have a high commercial crop breeding potential and so should be given the highest priority for *in situ* and *ex situ* conservation. On-farm conservation is particularly appropriate in the Djebel Druse, Syria, because this is one of the few areas where these minor forage vetches are still extensively cultivated. As well as the traditional minor crop species, Ehrman and Maxted (1990) and Maxted *et al.* (1990) highlight the underutilized potential of several other *Vicia* species found in Turkey and Syria.

(c) Threat of genetic erosion

A search of the Vicieae database (Adey *et al.*, 1984) indicates that Turkey contains appreciably more *Vicia* species than any other country. However, due to the advanced agricultural practices, including extensive use of herbicides and widespread use of high yielding varieties, genetic erosion of crop land races and wild species is a major problem (GEF, 1993). The Turkish government is also implementing a major hydroelectric and irrigation project (Guneydogu Anadolu Projesi) in Urfa, Gaziantep, Adiyaman and Maras provinces (Özis, 1982, 1983), the provinces that have the highest concentrations of *Vicia* taxa (Davis and Plitmann, 1970). The project has led and will lead to the flooding of tens of thousands of hectares in the Euphrates and Tigris basins and will thus have far-reaching ecological effects. This real threat of genetic erosion in the centre of diversity for *Vicia* warrants the establishment of genetic reserves in these provinces. Therefore the Turkish agencies responsible for conservation are establishing reserves in this area (Chapter 15).

(d) Genetic distinction

The Vicieae database also shows that 59 *Vicia* taxa are found in Syria.

These taxa span Mediterranean, montane and steppe conditions, but the highest concentration of taxa is found on the plateau of Djebel Druse, in southern Syria. These taxa are not only high in number, but are also drawn from throughout the classification of *Vicia*.

(e) Ecogeographic distinction

V. dionysiensis is restricted to a small number of sites on the plateau of Djebel Druse, Syria. Recent surveys have shown the species to be restricted to moist (but not waterlogged) pastures of the valley bottom, with basaltic soil, at an altitude between 1200 and 1500 m. This combination of site characteristics is relatively uncommon in Syria and the obligate requirement for these characteristics has led to a very restricted species distribution, justifying the establishment of a reserve on these restricted edaphic enclaves. Another species, *V. eristalioides*, was thought to be endemic to the open, limestone hillside within the Olimpos National Park, Antalya, Turkey (a single plant was found at a second location in south-west Turkey in 1994). The park where the initial population was located is currently being extensively planted with conifers; the planting will undoubtedly threaten the small populations of *V. eristalioides*, and therefore the management policy within the Olimpos National Park should be reconsidered.

3.5 CONCLUSION

To conclude, conservation activities, whether *ex situ* or *in situ*, will always be limited by the financial, temporal and technical resources available to them. The conservationist is therefore forced to prioritize and select the taxa to be actively conserved. To be efficient and effective the selection process should be as objective and scientific as is possible based on the available data for the species. The selection of target taxa will be based on a consideration of some, if not all, of the factors that affect a taxon's comparative value. It is only once these factors have been considered that efficient and effective conservation can occur.

ACKNOWLEDGEMENTS

We would like to thank M. Sawkins for preparation of the maps.

APPENDIX 3A A MODEL FOR QUANTIFYING THE THREAT OF GENETIC EROSION (FROM GUARINO, 1995)

FACTOR	SCORE

1 General

1.1 Taxon distribution
- Rare — 10
- Locally common — 5
- Widespread or abundant — 0

1.2 Drought
- Known to have occurred in two or more consecutive years — 10
- Occurring on average one or more times every 10 years, but not in consecutive years — 5
- Occurring less than once every 10 years on average — 0

1.3 Flooding
- Area known to be very flood prone — 10
- Area not known to be flood prone — 0

1.4 Accidental fires
- Area known to be very prone to fires — 10
- Area not known to be prone to fires — 0

1.5 Potential risk from global warming
- Summit areas or low-lying coastal areas — 10

2 Crop species

2.1 Area under the crop
- Declining rapidly — 10
- Increasing or static — 0

2.2 Modern cultivars of the crop
- Available and used by > 70% of farmers — 15
- Available and used by 50–70% of farmers — 10
- Available and used by < 50% of farmers — 5
- Not yet available but introduction planned — 2
- Not available — 0

2.3 Performance of agricultural services
- Very strong, and biased towards modern varieties — 10
- No agricultural services — 0

2.4 Mechanization
- Tractors used by > 30% of farmers — 10
- Animal traction used by > 50% of farmers — 5
- Manual labour used by > 50% of farmers — 0

2.5 Herbicide and fertilizer use
- > 50% of farmers — 10
- 25% of farmers — 5
- None — 0

2.6 Farming population
- Declining rapidly 10
- Increasing or static 0

3 Wild species

3.1 Extent of wild habitat of target species within study area
- Very restricted (< 5%) 15
- Restricted (5–15%) 10
- 15–50% 5
- Extensive (> 50%) 0

3.2 Conservation status of target species
- Species not known to occur in any protected area 10
- Species known to occur within a protected area, but protection status poor or unknown 5
- Species known to occur within a protected area and protection status good 0

3.3 Extent of use of wild habitat of target species
- Industrial exploitation 15
- Exploitation by surrounding populations (e.g. fuelwood gathering from nearby towns 10
- Hunting and gathering by small local communities 2
- Completely protected 0

3.4 Extent of use of target species
- Industrial exploitation 15
- Exploitation by surrounding populations 10
- Local exploitation 5
- Protected or not used 0

3.5 Agricultural pressure on wild habitats
- Large-scale cultivation within habitat margins 15
- Subsistence cultivation areas within habitat margins 12
- Land suitable for cultivation, cultivated areas within 3 km of habitat margins 10
- Land suitable for cultivation, cultivated areas within 3–10 km of habitat margins 5
- Land unsuitable for cultivation 0

3.6 Human population growth per year
- > 3% 10
- 1–3% 5
- < 1% 0

3.7 Availability of agricultural land
- > 70 ha km^{-2} cultivated 10
- 30–70 ha km^{-2} cultivated 5
- < 30 ha km^{-2} cultivated 0

3.8 Species palatability
- High 10
- Medium 5
- Low 0

3.9 Ratio of present livestock density to estimated carrying capacity
  - \> 1.0   10
  - 0.5–1.0   5
  - \< 0.5   0

3.10 Average proximity to borehole or other all-year-round water supply
  - \< 10 km   10
  - 10–20 km   5
  - \> 20 km   0

3.11 Distance to major population centre
  - \< 20 km   10
  - 20–50 km   5
  - \> 50 km   0

3.12 Distance to major road
  - \< 10 km   10
  - 10–30 km   5
  - \> 30 km   0

3.13 Distance to development projects (irrigation scheme, tourism complex, mining site, hydroelectric power scheme, land reclamation scheme)
  - \< 20 km   10
  - 20–50 km   5
  - \> 50 km   0

APPENDIX 3B IUCN RED LIST CATEGORIES (FROM IUCN, 1994A)

Extinct (EX)

A taxon is Extinct when there is no reasonable doubt that the last individual has died.

Extinct in the Wild (EW)

A taxon is Extinct in the Wild when it is known only to survive in cultivation, in captivity or as a naturalized population (or populations) well outside the past range. A taxon is presumed extinct in the wild when exhaustive surveys in known and/or expected habitat, at appropriate times (diurnal, seasonal, annual) throughout its historic range have failed to record an individual. Surveys should be over a time frame appropriate to the taxon's life cycle and life form.

Critically Endangered (CR)

A taxon is Critically Endangered when it is facing an extremely high risk of extinction in the wild in the immediate future.

Endangered (EN)

A taxon is Endangered when it is not Critically Endangered but is facing a very high risk of extinction in the wild in the near future.

Vulnerable (VU)

A taxon is Vulnerable when it is not Critically Endangered or Endangered but is facing a high risk of extinction in the wild in the medium-term future.

Lower Risk (LR)

A taxon is Lower Risk when it has been evaluated, does not satisfy the criteria for any of the categories Critically Endangered, Endangered or Vulnerable. Taxa included in the Lower Risk category can be separated into three subcategories:

1. Conservation Dependent (cd)

Taxa which are the focus of a continuing taxon-specific or habitat-specific conservation programme targeted towards the taxon in question, the cessation of which would result in the taxon qualifying for one of the threatened categories above within a period of five years.

2. Near Threatened (nt)

Taxa which do not qualify for Conservation Dependent, but which are close to qualifying for Vulnerable.

3. Least Concern (lc)

Taxa which do not qualify for Conservation Dependent or Near Threatened.

Data Deficient (DD)

A taxon is Data Deficient when there is inadequate information to make a direct, or indirect, assessment of its risk of extinction based on its distribution and/or population status. A taxon in this category may be well studied, and its biology well known, but appropriate data on abundance and/or distribution are lacking. Data Deficient is therefore not a category of threat or Lower Risk.

Not Evaluated (NE)

A taxon is Not Evaluated when it has not been assessed against the criteria.

4

Ecogeographic surveys

N. Maxted and L. Guarino

4.1 INTRODUCTION

It is essential to understand the habitat preferences and geographical distribution of the target species if an effective conservation strategy is to be developed. Key sources of information on these are the passport data associated with existing herbarium and germplasm collections. The passport data for a particular species might indicate that it has previously been found only in mangrove swamps of south-east Asia. Clearly, these will be the areas that will have to be explored if germplasm is to be collected for *ex situ* conservation and/or to determine the best place(s) to site genetic reserve(s) of the species.

Ecological, geographical and taxonomic data are referred to collectively as ecogeographic data and the analysis of these data that can aid conservation. Maxted *et al.* (1995) provide the following definition of an ecogeographic study:

> An ecogeographic study is an ecological, geographical and taxonomic information gathering and synthesis process. The results are predictive and can be used to assist in the formulation of collection and conservation priorities.

The difference between a 'study' and a 'survey' is one of degree. An ecogeographic study involves a more detailed data analysis and interpretation phase than a survey. A study will involve detailed collation of fresh environmental data and multivariate analysis of the patterns of distribution and may take several years to complete. A survey will focus on collating data recorded by other plant collectors, rather than collecting fresh data, and may be restricted to a media search and collating passport

Plant Genetic Conservation.
Edited by N. Maxted, B.V. Ford-Lloyd and J.G. Hawkes.
Published in 1997 by Chapman & Hall. ISBN 0 412 63400 7 (Hb) and 0 412 63730 8 (Pb).

data from herbarium specimens or germplasm accessions. For this reason, as a routine prerequisite to locating and establishing a genetic reserve, ecogeographic surveys are more likely to be used.

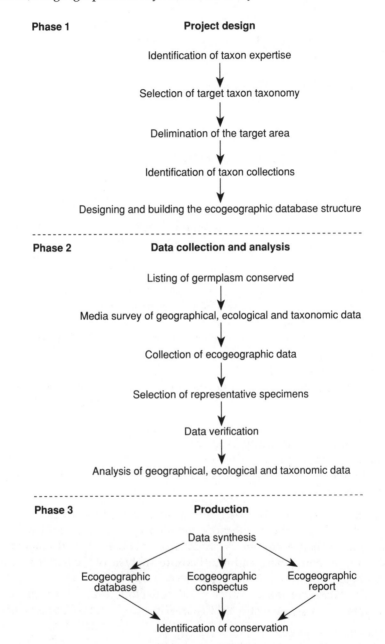

Figure 4.1 Ecogeographic paradigm. (From Maxted *et al.*, 1995.)

A possible procedure for undertaking an ecogeographic survey is outlined in the model or paradigm proposed by Maxted *et al.* (1995) (Figure 4.1). The model is divided into three phases: project design; data collation and analysis; and production.

4.2 PROJECT DESIGN

4.2.1 Identification of taxon expertise

The acquisition of ecogeographic data can be much easier if appropriate specialists are consulted at an early stage of the process. They will be able to suggest relevant grey literature, recommend Floras, monographs, taxonomic databases and ecological works, suggest which herbaria and/or gene banks should be visited and put the conservationist in contact with other specialists. The appropriate taxonomic and phytogeographical specialists to approach can be identified from the authorship of taxonomic treatments of target taxa, Floras and ecological and phytogeographical studies of the target region. *Index Herbariorum* (Holmgren *et al.*, 1990) often lists the specialist taxonomic group and geographical region of thousands of botanists working in herbaria worldwide. Increasingly, herbaria are acquiring electronic mail facilities, and there is a list of Plant Taxonomists Online (contact Jane Mygatt, jmygatt@bootes.unm.edu). A database of experts in botany and mycology worldwide is maintained at the University of Oulu, Finland (contact Anne Jäkäläniemi at the Department of Botany, anne.jakalaniemi@oulu.fi).

4.2.2 Selection of target taxon taxonomy

It is essential to have a good taxonomic understanding of the target group prior to undertaking an ecogeographic study (IBPGR, 1985b). This can be obtained from various sources: target taxon specialists, Floras, monographs, recent revisions of the group and, increasingly, taxonomic databases. These will help the conservationist to determine the generally accepted classification of the group. The classification will provide leads to other taxonomic literature: lists of accepted taxa, taxon descriptions, synonymized lists, distribution maps, identification aids (keys and illustrations), ecological studies, bibliographies and critical taxonomic notes. More obscure groups may lack a recent revision or monograph, but the researcher must collate whatever published taxonomic data are available to provide the taxonomic backbone to the study.

The most comprehensive guide to the botanical literature is the *Kew Record of Taxonomic Literature* (now *Kew Record*), which lists references to all publications relating to the taxonomy of flowering plants, gymnosperms and ferns. In addition to sections on taxonomic groups, there are references on phytogeography, nomenclature, chromosome surveys, chemotaxonomy, Floras and botanical institutions; also papers of taxonomic interest in the fields of anatomy and morphology, palynology, embryology, and reproductive biology, and relevant bibliographies and biographies. References are listed under various geographical subdivisions of the world and within these by country and/or administrative unit. *Kew Record* is computerized from 1982 onwards, but for internal (Kew) use only. Guides to which Floras cover which parts of the world are provided by Frodin (1984) and Davis *et al.* (1986). The standard international reference on plant nomenclature is *Index Kewensis* (now *Kew Index*). It lists the place of publication of new and changed names of seed-bearing plants from family level downwards (before 1972, only genera and species). It is available on microfiche (to 1975) and on CD-Rom.

4.2.3 Delimitation of the target area

The target area of the taxon being studied may be restricted by the terms of reference of the project commission, but if unspecified the taxon should be studied throughout its range. The commissioning agent may restrict the survey to a specific geographical area to save resources in the short term, but this may ultimately prove to be a false economy. Multiple studies of the same taxon, possibly with separate authors, are likely to form a less coherent whole and may unnecessarily limit the predictive value of the ecogeographic study. Having established which areas are to be included in the target region, additional ecogeographic information on the target taxon can be obtained from local Floras.

4.2.4 Identification of taxon collections

The researcher undertaking the study will need to visit the major herbarium and germplasm collections of the target taxa from the target region. The broader the sampling of ecogeographic data associated with herbarium specimens and germplasm, the more likely it is that the data will prove ecologically and geographically predictive. *Index Herbariorum* (Holmgren *et al.*, 1990) records where important dried plant collections are held. The important collections to be seen during the study fall into two categories: major international herbaria and local herbaria in the target area. The relative advantages and disadvantages of two categories of herbarium for the conservationist are set out in Table 4.1.

Ecogeographic data can equally well be obtained from the passport data recorded by previous germplasm collectors and maintained in catalogues and/or databases by gene banks and botanic gardens. IPGRI produces international directories of germplasm collections on a gene pool basis and also maintains a database, which may be queried on demand.

Table 4.1 Herbaria: advantages and disadvantages

Type of herbaria	Advantages	Disadvantages
Major international	Broad taxonomic coverage, possibly material used in the production of revisions and monograph Broad international geographical coverage, possibly material used in the production of local Floras Skilled researchers available to provide general advice Appropriate taxonomic and geographical specialists Type material of target taxa Good botanical library	Predominance of old collections, making extraction of passport data more difficult and likely predictive value lower Geographical names associated with older collections sites may have changed more recently
Local	Good local regional coverage of target area Better documented material, as the herbarium is likely to have been more recently established Regional specialists present, who can assist in deciphering local geographical names	Limited resources for herbarium mainenance Lack of target taxon specialists Limited botanical library

Information on botanical gardens (including taxa in cultivation, special and conservation collections and associated nature reserves and natural vegetation in the garden) is summarized in the *International Directory of Botanic Gardens* (Heywood *et al.*, 1991).

4.2.5 Designing and building the ecogeographic database structure

The ecogeographic information associated with herbarium and germ-plasm specimens can be collated on paper, but it is more efficient to enter the data directly into a computer database. A portable computer with relational database management system software can be used for data entry and editing. Time and effort would clearly be wasted by first copying the data on to paper and then transferring it to the database at a later stage. The process of data transfer is as time consuming as the original data collection and may introduce errors into the database. Generally, the fewer steps between data gathering and record entry into the database, the more efficient is the process.

It is important to standardize the data held in the database wherever possible. Thus, where possible, the database should include codified rather than textual data. This will save storage space, speed up retrieval and help to avoid errors. The codes used should be accepted standards and used consistently. Examples include standard codes for herbaria (Holmgren *et al.*, 1990), for authors of plant names (Brummitt and Powell, 1992) and for political units (International Standards Organization, 1981). The International Union of Biological Sciences Commission on Taxonomic Databases (TDWG) was established to facilitate data standardization and data exchange between botanical databases. It is producing sets of standard codes for botanical data, for example for Basic Recording Units (Hollis and Brummitt, 1992). Standards in preparation include ones for: economic use; habitat, soil and landscape; life-form; and plant occurrence and status. Information can be obtained from the TDWG Secretariat, based at the Missouri Botanical Garden. Since 1963 IUCN has been developing a system for describing the conservation status of species (IUCN, 1994a). Any published germplasm descriptor lists (which include passport, characterization and evaluation data) should be examined before beginning to develop a database.

4.3 DATA COLLATION AND ANALYSIS

4.3.1 Listing of germplasm conserved

Before embarking on the detailed data collation phase of the project, current conservation activities should be reviewed. If sufficient genetic diversity is already safely conserved either *in situ* or *ex situ*, further work may not be warranted. Details of the material currently being conserved can be obtained from the catalogues and databases of botanical gardens, gene banks and *in situ* conservation areas. However, care must be taken when interpreting information on current gene bank or botanical garden holdings. The material held may be incorrectly identified, though it should be possible to check the identification by consulting voucher material or identifying living material. The actual quantities of germplasm available could also be misleading: gene banks and botanical gardens are encouraged to duplicate their collection in other gene banks or botanical gardens and so duplicated accessions can give a false impression of the genetic diversity conserved. The conservationist should consider that, although accessions may be held in a gene bank, the material may for various reasons be unavailable to potential users and so create a false impression of a taxon's conservation status. Similarly, just because a species occurs within the confines of a reserve, it is not necessarily the case that it is adequately protected there.

4.3.2 Media survey of geographical, ecological and taxonomic data

The collation of these data could be undertaken while visiting the major herbaria, which often have good botanical libraries attached. The appropriate literature will include monographs, revisions, Floras, gazetteers, scientific papers, soil, vegetation and climatic maps, atlases, etc. Increasingly, however, information is becoming available in media other than the conventional printed literature. Abstracts of publications (and, in some cases, the full text) may be available on microfiche or in electronic bibliographical databases (on-line and/or on CD-Rom), for example. There may have been other attempts to survey herbarium label information, the results of which may or may not be formally published. For example, some herbaria and other organizations are developing floristic and indigenous knowledge databases. Herbaria may hold some label data in card catalogues. Many kinds of thematic maps are being made available in digital form. Prendergast (1995) has discussed in more detail the various published sources on taxonomic and ecological information and Auricht *et al.* (1995) on sources of environmental information.

Increasingly, computer networks, particularly the academic network known as the Internet, are being used as sources of information. The Internet links together some one million computers worldwide, which means that there are probably tens of millions of users. Many scientific interest groups have been set up on the Internet. Software such as Listserv and Usenet support electronic discussion groups and distribute electronic newsletters and scientific papers. In addition, many important information resources, such as university libraries and public domain software and databases, are being made available on the Internet. Compilations of list servers, of Usenet news groups and of information archives of relevance to biologists are provided by Dr Una Smith's *A Biologist's Guide to Internet Resources* (smith-una@yale.edu). TAXACOM (taxacom@harvarda.harvard.edu) is perhaps the best-known mailing list on taxonomy and related subjects.

The data that might be obtained from different media sources will include:

(a) accepted taxon name*
(b) locally used taxon name*
(c) where in the target area the species is reported to grow*
(d) timing of local flowering and fruiting*
(e) habitat preference*
(f) topographical preference*
(g) soil preference*
(h) geological preferences*
(i) climate and micro-climatic preference*
(j) breeding system employed*

(k) genotypic and phenotypic variation (are local variants found and is this variation genetically or environmentally based?)
(l) biotic interactions
(m) archaeological evidence
(n) ethnobotanical evidence
(o) conservation status* (e.g. Red Data Book status).

(An asterisk in the above list, or in the one in section 4.3.3 below, indicates data that could be coded in the database.)

4.3.3 Collection of ecogeographic data

The kinds of information that the conservationist can obtain from the passport data associated with herbarium specimens and germplasm accessions are:

(a) identification (and, if appropriate, previous identifications)*
(b) herbarium, gene bank or botanical garden where specimen is deposited*
(c) collector's name and number
(d) collection date* (to derive flower and fruiting timing)
(e) phenological data* (does specimen have flower or fruit?)
(f) particular area of provenance*, latitude and longitude or more detail if possible
(g) altitude*
(h) habitat type*
(i) soil type*
(j) vegetation type*
(k) site slope and aspect*
(l) land use and/or agricultural practice*
(m) phenotypic variation
(n) biotic interactions
(o) competitive ability*
(p) palatability*
(q) ability to stand grazing*
(r) vernacular names
(s) plant uses.

The listing is extensive and it is unlikely that all of these data categories will be recorded for each specimen. However, there are certain data items that must be recorded for the study to yield predictive results: taxon identification, geographical provenance (the more detailed the better), collection date, habitat preference and altitude.

Early plant collectors could not have predicted the in-depth analysis that would subsequently be based on their provenance information. They were often not particularly careful about recording data in detail:

site location data may be ambiguous and ecological details missing. Older specimen labels are almost invariably hand-written, adding the problem of deciphering the script, which may be in a foreign language and unfamiliar alphabet. Botanists from local herbaria can provide invaluable help, not only in herbarium label translation, but also in the identification of specimen localities if a local gazetteer is not available. (Herbaria sometimes have unpublished local gazetteers, for example in the form of a card index.) There is no comprehensive world gazetteer, but the *Atlas of the World* (Times Books, 1988) includes an extensive gazetteer, and an *Official Standard Names Gazetteer* is being constructed and is available in country volumes from the US Board of Geographical Names.

Care must be taken in accepting scientific names written on herbarium sheets. The identification should always be checked. Similarly, the identification of germplasm samples should be verified with reference to herbarium voucher specimens, or at least discussed with the authority who made the identification.

The ecogeographic database will inevitably contain many gaps. In general, it is much easier to record curatorial or geographical than ecological data from herbarium specimens. The conservationist may, however, be able to infer various features (latitude and longitude, geology, soil, altitude, etc.) of collection sites from location data by reference to appropriate maps.

4.3.4 Selection of representative specimens

It will not be possible, or necessary, to include in the ecogeographic database data from all the specimens that are seen. Maxted (1995) found that data from about a third of the specimens seen during the study were finally included in the ecogeographic database. The researcher will need to impose some kind of selection before including specimens of the target taxon in the database. It is important that emphasis be placed on obtaining reliable specimen provenance data for those specimens to be included in the database. Recently collected specimens often have higher quality passport data which is easier to read, being often typewritten. These data are also more likely to have remained current. Specimens are more likely to be selected for inclusion in the ecogeographic database if they have detailed ecogeographic passport data or if they show features of particular taxonomic, ecological or geographical interest, i.e. they are odd or rare forms, come from unusual environments or are found on the edges of their natural distributional range. Ideally, only specimens that either have latitude and longitude data available or for which these data can be established should be selected for inclusion in the database. Specimens should be positively selected to represent the breadth of geographical and ecological conditions under

which the target taxon is found. It is desirable to collect detailed passport data from a broad range of representative specimens, rather than
duplicate data already included in the database from previous specimens. Entering data from more than one specimen sampled from the
same population is likely to add little to the predictive value of the data
set as a whole.

The conservationist will be faced with the need to decide how many
specimens should be entered into the database. The law of diminishing
returns is applicable, and the conservationist must decide when the
amount of extra information gained from each specimen does not
increase the predictive value of the data set. There are no specific rules;
however, the conservationist should be on the look-out for the point
when novel ecogeographic combinations no longer occur in the specimens being examined, which will then indicate that the ecogeographic
range of the taxon will probably have been recorded in the database.

4.3.5 Data verification

Before the database can be deemed complete, the conservationist must
search for and correct errors. Indexing the database (i.e. rearranging the
records in alphabetical or numerical order) on each field in turn may
highlight typing errors or invalid entries. Mapping latitude and longitude data may reveal errors if particular localities are shown up as obvious outliers in impossible places. Herbarium specimen and germplasm
collectors often send duplicate sets of their materials specimens to different international collections. The conservationist should search the database for these duplicates and be aware of their possible effect on data
analysis: multiple specimens from a single population collection duplicated in multiple herbaria are likely to bias the results giving a false picture of geographical distribution.

4.3.6 Geographical, ecological and taxonomic data analysis

The raw ecological and geographical data included in the database can be
analysed to help to identify the particular geographical locations and habitats favoured by the target taxa. One of the simplest means of ecogeographic data analysis is to calculate the number of specimens (which can
be expressed as percentages) collected from sites characterized by different biotic and abiotic features, e.g. climate, soil types, aspect, shading
characteristics, habitat types, etc. Data arranged in this fashion will identify the particular niche occupied by the target taxon and so can be used to
relocate previous collection areas and indicate other areas where the taxon
is likely to be found. Frequency distributions can often be compiled
directly by querying the database. Correlation of the abundance or fre-

quency of taxa along environmental gradients (such as altitude, latitude and soil pH) will give error terms and can therefore be used predictively. Association of morphological characters with particular environmental conditions will help indicate possible ecotypic adaptation, in both wild and cultivated material.

Ecogeographic data can also be mapped. Taxon distribution maps can be used in conjunction with topographical, vegetation, rainfall, geological, soil and other thematic maps to predict where else the target taxon might be found. Stace (1989) describes the different means of visually displaying plant distribution: shading or enclosing an area with a single line; and using various kinds of dot distribution maps. Enclosing lines are ambiguous, providing no indication of the concentration of the taxon within the region. A single outlying specimen might erroneously suggest that the taxon is present throughout an entire region. The occurrence of a species is often sparse at the periphery of its range and there is rarely a distinct cut-off line. Indicating presence in this manner also means that any variation due to local ecological and geomorphological factors within the overall geographical distribution cannot be indicated.

To represent distribution patterns in detail there is a general trend towards the use of dot distribution maps (Stace, 1989). Morphological or ecological information can be superimposed on to a dot distribution map, as Strid (1970) has done for the uppermost internode length of various populations of *Nigella arvensis* in the Aegean region. The position of a rectangle indicates the population location, while its height shows the relative length of the internode. Piecharts can be used to display the relative frequency of a character in different places. They are commonly used to show the distribution of allelic variation among different geographical locations.

Enclosing line maps can be used to indicate concentration of species. These maps, known as isoflor maps, do not show actual species distributions, but each line is a contour delimiting the area where a given number of taxa may be found. Various species distributions are superimposed on to a single map, then contour lines are drawn around areas of the map to indicate the number of species found at any one point. An example of an isoflor map for *Vicia* section *Hypechusa* is shown in Figure 4.2. This is a useful method for picking out areas of high species richness, which may be particularly high priorities for *in situ* conservation, but three points should be kept in mind.

Firstly, two areas may have equal numbers of species, but the species may be closely related in one case and taxonomically widely separated in the other. Isoflor maps of both species and sections within the genus might need to be compared to get an accurate impression of the distribution of diversity. Williams *et al.* (1993) describe a computer program (WORLDMAP) which can be used to identify areas of high biodiversity defined according to four different weightings of simple species richness

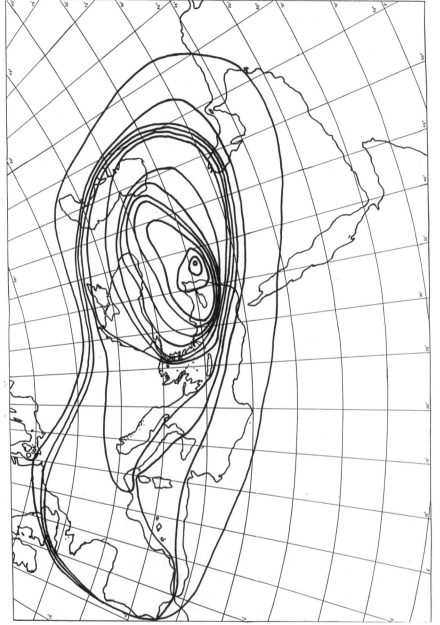

Figure 4.2 Example of an isoflor map for *Vicia* section *Hypechusa*. (From Maxted, 1995.)

derived from different measures of taxonomic relatedness.

Secondly, areas with equal numbers of species (or, indeed, land races) are not necessarily floristically similar, i.e. the species involved may be completely different. The WORLDMAP software has a facility for the exclusion of particular areas from the analysis. If the area with highest diversity is so excluded, and the diversity measurement recalculated for the remaining areas, and so on, a set of areas can be chosen which will optimally cover overall diversity. Rebelo (1994) also describes an iterative procedure for the optimal placing of nature reserves based on distribution information.

Thirdly, the number of species present in an area says nothing of the relative abundance of each. The distribution of plant records within classes, or of classes within higher-level classes, is an important aspect of diversity. Ten plant records may be distributed evenly among five species, or one species may be represented by six plant records and the other four by one each. Intuitively, the former case should score the higher diversity. The Shannon–Weaver diversity index and Simpson's index are two of the most commonly used measures of diversity which take into account the proportional abundances of different classes.

The analysis of ecogeographic data can be much facilitated by the use of a Geographic Information System (GIS). A GIS is a database management system dedicated to the simultaneous handling of spatial data in graphics form and of related, logically attached, non-spatial data (Burrough, 1986). For example, if the spatial data are the location of herbarium specimens in a country, the associated attributes could be the species name, the name of the herbarium storing the specimen, and the kinds of ecogeographic data listed above. The spatial processing system and database management system of a GIS allow one to bring together diverse data sets, make them compatible among themselves, analyse and combine them in different ways and display the results as a map or statistics on a computer screen or hard-copy. Some standard GIS capabilities, and their relevance in ecogeographic work, include:

- **Geometric correction** The scale, projection, etc. of different maps may be changed to make them comparable (and amenable to overlay analysis, see below). For example, a species distribution map may be compared with different thematic maps from varied sources to derive an ecoclimatic 'envelope' for the species.
- **Digital terrain model analysis** The altitude contours on a topographical map may be used to produce maps of slope, aspect, intervisibility, shaded relief, etc. These may constitute important additional provenance information.
- **Interpolation** Point data may be used to create isopleth (equal-value contour) maps, such as isoflor maps or maps showing areas where morphologically similar specimens have been collected.

- **Overlay analysis** Different maps of the same area may be combined to produce a new map. For example, maps of slope, soil and vegetation cover may be overlaid to synthesize a map of potential soil erosion or desertification, which may be useful in identifying areas at particular risk of genetic erosion. If some passport data are missing (e.g. the soil type or altitude at the collecting point), they may be derived by overlaying the species distribution map on different thematic base maps.
- **Proximity analysis** Buffers may be generated around features such as wells, villages and roads to determine the accessibility of potential reserves.
- **Computation of statistics** Means, counts, lengths and areas may be calculated for different features. Frequencies of the occurrence of different species in different habitats and ecological niches may be calculated, as well as the surface area of potential reserves and the percentage of the total occupied by different vegetation or soil types.
- **Location** Entities having defined sets of attributes (for example, all specimens recorded from particular soil types) may be located.

Attribute data can be entered into a GIS from a keyboard and maps by digitizing or scanning. Data entry is the major constraint on GIS implementation, as it is usually labour-intensive, time-consuming and therefore expensive. Thus, it pays to check whether relevant data sets are already available in digital form. A variety of data sets is available from organizations such as FAO, UNEP/GRID, the International Soil Reference and Information Centre (ISRIC) and the World Conservation Monitoring Centre (WCMC), though these are mostly (though not exclusively) at a regional or global scale (scales 1:1 million down to 1:100 million). There are also relevant regional sources of data sets, such as the Regional Centre for Services in Surveying, Mapping and Remote Sensing in Nairobi, the Regional Remote Sensing Programme in Bangkok and the Regional Remote Sensing Centre in Burkina Faso. Local-level data sets (up to scales of about 1:20 000) may be available from such centres and from national agriculture, environment, planning and cartography services, as well as from private companies. GRID's Meta-Database is an interactive electronic catalogue of spatial environmental data sets archived at GRID centres throughout the world. The most common data requirement will be for base layer data sets, showing country boundaries, internal administrative boundaries, rivers and lakes, altitude (contour and spot heights), and so on. The most commonly used medium-resolution data set of this type has been *World Data Bank II*, but the *Digital Chart of the World* (DCW), based on the 1:1 million Operational Navigational Charts and recently released on four CD-Roms, is of potentially greater use. However, the extraction of data subsets can be difficult and data quality is variable.

Remote sensing data already in digital form, such as satellite imagery, can also be entered into a GIS. GRID can give advice on how to obtain data sets from commercial and other sources, though it cannot procure these for users. FAO's Remote Sensing Centre has a comprehensive database of reference maps and imagery which is available to member nations and FAO programmes. Resolution varies from kilometres to tens of metres, depending on the system. The cost of obtaining these data sets can be very high. Also, analysis (which will involve image restoration or correction, image enhancement and information extraction) is complex, requiring specialized software, hardware and skills. Careful ground truthing is necessary for many applications. However, for inaccessible areas for which there are no detailed maps, for example, remote sensing may be the only source of some data. It is also often the only source of data on changes in vegetation and land use, whether from year to year or from week to week in a given growing season. Actual trends and developments in deforestation and desertification can be monitored using remote sensing data stretching back over many years, providing an indication of the threat of genetic erosion.

4.4 PRODUCTION

4.4.1 Data synthesis

The final production phase of the project commences with the synthesis of all the data collected during the study. The researcher should also be aware of the degree of completeness of the database, or the collections on which it was based, in terms of how fully the target area has been effectively covered. If a particular habitat is under-represented in the database, is it because the taxon is absent from that habitat, or because that type of habitat has not been sampled or is very rare, or even because the target taxon has not been recognized in such a habitat? This problem must be considered if the results of the analysis and the inferences drawn from them are not to be misleading.

4.4.2 The ecogeographic database, conspectus and report

The ecogeographic database, conspectus and report should be seen as the three essential products of an ecogeographic study. The ecogeographic database contains the raw data of the project. The conspectus summarizes the available ecological, geographical and taxonomic information for the target taxon through part or the whole of its range. The report interprets the data held in the other products and will aid the conservationist in selecting conservation priorities.

The conspectus is arranged by plant species name, listed either alphabetically or systematically. An example of the level of detail that might be included in an ecogeographic conspectus is given in Appendix 4A. The following information should be included:

(a) accepted taxon name, author(s), date of publication, where published;
(b) reference to published descriptions and iconography;
(c) short morphological descriptions or keys for important taxa or those that may be difficult to identify;
(d) phenology, flowering season;
(e) vernacular names;
(f) geographical distribution, countries from which the taxon is recorded, including reliable records from the literature – presence indicated by listing the country Basic Recording Unit, ISO-codes units or other standard codes;
(g) distribution maps, preferably dot distribution map produced direct from the latitude and longitude data held in the database;
(h) ecological notes, including altitude (minimum and maximum elevation), habitat, topographical, soil, geological, climate and microclimatic preference, biotic interactions and other habitat details;
(i) geographical notes, containing general data and interpretation of the taxon's geographical distribution;
(j) taxonomic notes, containing general data and interpretation of the taxonomic data, noting any distinct genotypic and phenotypic variation within the taxon;
(k) conservation notes, containing an assessment of the potential genetic erosion faced by and the conservation status of the taxon.

The report discusses the contents of the database and conspectus and must draw general conclusions concerning the group's ecogeography and present a concise list of conservation priorities. If possible, the following points should be covered:

(a) delimitation of target taxon;
(b) classification of target taxon that has been used;
(c) mode of selection of representative specimens;
(d) choice of hardware and software;
(e) ecogeographic database file structures and interrelationships;
(f) discussion of database content;
(g) discussion of target taxon ecology;
(h) discussion of target taxon phytogeography, discussion of distribution patterns and summary of distribution in tabular form;
(i) discussion of any interesting taxonomic variants encountered during study;
(j) discussion of current and potential uses of target taxon
(k) discussion of relationship between crop species and their wild relatives;
(l) discussion of any particular identification problems associated with group; presentation of identification aids to vegetative, floral and fruiting specimens;

(m) discussion of *in situ* and *ex situ* conservation activities associated with target taxon, including extent of diversity already conserved;

(n) discussion of genetic erosion threat facing taxon or group;

(o) discussion of priorities and suggested strategy for future conservation of target taxon.

The ecogeographic conspectus may be included within the report as an appendix or as a separate entity.

4.4.3 Identification of conservation priorities

The three products of the ecogeographic survey assist the conservationist in formulating future conservation priorities and strategies. Within the total distributional range of the species, zones of particular interest may be identified, such as areas with high concentrations of diverse taxa or low rainfall or high frequency of saline soils or extremes of exposure, etc. If a taxon is found throughout a particular region, the researcher can use the ecogeographic data to select actively a series of diverse habitats in which to site genetic reserves.

The ecogeographic survey or study should conclude with a clear, concise statement of the proposed conservation strategy for the target taxon and proposed conservation priorities. This will answer questions such as: where should the genetic reserves be sited? What part can local people play in conservation activities? How should population levels be monitored to assess the threat of genetic erosion? How large and what shape should the reserve be? If the primary motive of the survey or study was to designate reserves, what other complementary strategies should also be adopted? Or is a more detailed study required before these questions can be answered?

4.5 CONCLUSION

Herbaria, gene banks and botanical gardens, as well as containing plants and plant propagules, are storehouses of botanical data that can be used to facilitate plant conservation. The analysis of a taxon's geography, ecology and taxonomy allows its conservation status to be assessed and permits the prediction of areas and habitats that are likely to contain genetic diversity of the target taxon. Once specific areas of genetic variation are known, possibly associated with ecogeographic diversity, then sites for *in situ* reserves can be suggested and species effectively conserved.

ACKNOWLEDGEMENTS

The authors wish to acknowledge the support of the International Plant Genetic Resources Institute in production of this chapter. However, the views expressed here are solely those of the authors.

APPENDIX 4A EXAMPLE OF ECOGEOGRAPHIC CONSPECTUS (FROM MAXTED, 1995)

Vicia Section *Hypechusa* (Alef.) Aschers. & Graebner, Syn. Mitteleur. Fl., 6,2: 957 (1909).

Ref. Pub. Description: Kupicha, *Notes Roy. Bot. Gard.*, **34**: 320 (1976).

Annual; climbing; stem slender. Stipules entire or semi-hastate; 1–5.5 × 0.5–4 mm; edge entire or with 1–2 teeth. Leaf 14–105(–115) mm; apex tendrilous; 2–20 leaflets per leaf; leaflet 5–35(–40) × 1–15 mm; symmetric; margins entire. Peduncle 1–10(–28); with 1–4 flowers. Calyx mouth oblique; lower tooth longer than upper; base gibbous; pedicel 1–4. Flowers 12–35 mm; all petals approximately equal length; standard cream or yellow, rarely blue or purple; shape platonychioid or stenonychioid; claw bowing absent; upper standard surface glabrous or pubescent. Wing marking absent or present; wing limb with slight or strong basal fold. Legume 14–40 × 6–12 mm; oblong; round in cross-section; sutures curved; valves glabrous or pubescent; hairs simple or tuberculate (with swollen base); septa absent; 1–6 seeds per legume. Seeds 2–5.5 × 2–6.5 mm; round or oblong; not laterally flattened; hilum less than quarter of seed circumference; lens positioned opposite to hilum; testa surface smooth.

Number of taxa: eighteen.   **Chromosome number**: 10, 12, 14.

Geographical distribution: West, Central and Southern Europe, Mediterranean Basin and Transcaspia.

Geographical notes: The centre of diversity of sect. *Hypechusa* is focused on the fertile crescent countries of South West Asia. The distributional pattern of the two series within sect. *Hypechusa* is also centred on South West Asia, although ser. *Hypechusa* stretches further westerly to encompass Southern Europe.

Ecological notes: The species of this section are weeds of semi-arid areas, except for the rare species *V. esdraelonensis* which is reported to prefer moist areas. They are rarely found in shade, though I have collected *V. melanops* from shaded areas of Stone pine forest in Western Turkey. These species are most commonly found on the edges of cultivation through a broad range of soil types. The ser. *Hyrcanicae* species are in general larger plants and can stand more competition from surrounding plants than ser. *Hypechusa*.

Taxonomic notes: The current conception of sect. *Hypechusa* is similar to that used by Alefeld (1860) who first used the name for the genus *Hypechusa*. *V. mollis* was considered by Kupicha to belong to sect.

Peregrinae; this grouping of *V. mollis* with *V. peregrina* and its allies was originally suggested by Boissier (1872). However, the fact that it possesses a short peduncle suggests that this species is more naturally allied to sect. *Hypechusa*, to which it has been transferred (Maxted, 1991). This position was adopted by Townsend (1967) and is supported by Plitmann (pers. comm.). The sect. *Hypechusa* taxa are split into two series, *Hyrcanicae* and *Hypechusa*, on the basis of peduncle length, corolla shape, corolla size and standard pubescence.

V. anatolica Turrill, Kew Bull., 1: 8 (1927).

Ref. Pub. Description: Fl. Iran., 43; Fl. Tur., 3: 313; Fl. USSR., 13: 470; Illust. Fl. Iran., Tab. 32, fig. 2.

Phenology: April–July.   **Chromosome number**: 10.

Geographical distribution: CIS, IRN, TUR.

Ecology: Alt. 800–2000 m; Hab. disturbed land, orchards and mountain pasture.

Conservation notes: Not threatened.

5

Technical and political factors constraining reserve placements

J.T. Williams

5.1 INTRODUCTION

There are a number of constraints to the effective placement of reserves for the *in situ* conservation of wild species related to crops, and particularly for those sufficiently related that they have foreseeable and potential use in the genetic enhancement of the crop itself. These constraints are scientific, practical and political. Additionally, there are few operational blueprints for fully comprehensive systems of genetic conservation of crop gene pools ranging from long-term *ex situ* conservation of cultivars to long-term *in situ* conservation of wild species. Much of the needed action remains conceptual and broadly based in terms of implementation and methodologies.

This chapter highlights several existing constraints to reserve placement. To a large degree they are the same constraints hampering *in situ* conservation in its widest sense. There are therefore general priorities for programme development. Accordingly the discussion at some point must involve three major areas: firstly, the principles which underlie *in situ* conservation; secondly, the economic and political constraints to implementing the principles in practice; and thirdly, related gaps which need to be addressed to rectify the past and current piecemeal approach. A full understanding of the principles which guide action can go a long way to setting up new coordinated approaches. The principles are broadly dealt with in other chapters and the discussion here, in the main, deals with economic and political constraints to implementation; it is limited to *in situ* conservation in natural and semi-natural ecosystems

Plant Genetic Conservation.
Edited by N. Maxted, B.V. Ford-Lloyd and J.G. Hawkes.
Published in 1997 by Chapman & Hall. ISBN 0 412 63400 7 (Hb) and 0 412 63730 8 (Pb).

and does not deal with more recent interest in on-farm conservation of land races of crops.

5.2 CONSTRAINTS

Since spatial units for *in situ* genetic conservation involve the permanent appropriation of land, as well as management of the units, this must be a responsibility of sovereign states and the organizations dedicated to the purchase and management of land units. A series of constraints ranging from economic to scientific and organizational ones are apparent and are discussed below.

5.2.1 National planning and designation of reserve areas

Clearly many of the crop relatives are found in economically poorer parts of the world. It is in such areas that there is a need for special genetic reserves as well as for target species to be maintained within ecosystems protected for nature conservation. It is far easier for countries to understand the need for and to take action on establishing a series of national parks or heritage sites, since in this case there is likely to be some benefit to the people of the countries concerned. In many instances the benefits may be more apparent when there is a strong forestry sector, but these areas are often managed in a minimal way, using normal forestry practices. An interesting example of where the legal instrument for conservation specifically addresses genetic conservation management is the 1992 National Integrated Protected Areas System Law of the Philippines, and this is continually supplemented with Administrative Orders and Memorandum Circulars in which guidelines are laid out (Catibog-Sinha, 1993).

These protected areas clearly act to conserve genetic materials. In many cases these areas can also include genetic reserves even when the prime designation of the area is for broad ecosystem conservation (Ingram and Williams, 1993). The fact that populations of useful plant relatives occur within them is of interest, but unless they are known to represent particular segments of the gene pool this is not a reason to establish special reserves automatically within the broader reserve area. None the less this can be done if basic data are available and point to that end.

Although there are major costs involved in delineating and appropriating broad reserve areas, much progress has been made through the efforts of UNESCO, UNEP and IUCN. It has to be appreciated that many reserve area networks are voluntary; for example, the Man and the Biosphere Program network of biosphere reserves has no treaty or legally binding obligations. Other networks which are far less comprehensive do have a legal obligation, such as the convention covering the Protection of the World's Cultural and Natural Heritage. Whatever the legal background, there is a strong recognition that natural resources are critical in the economies of countries. This results from recognition that the world population increased from 1.6 billion at the start of this century to 5.2 billion by 1990, and environmental degradation has progressed as world economic activity has quadrupled over the same time period. Global concerns are now well to the fore and result in appreciation of the need to conserve biodiversity, using a range of methods from the designation of reserve areas to efforts such as genetic conservation of specific taxa and linking these activities to sustainable development.

Even though moral and legal obligations may have been recognized, government policies often need adjustment because many such policies provide incentives to mismanage the environment – especially in the agriculture and forestry sectors. International institutions such as UNEP, World Bank, IMF and OECD are currently considering new ways of national accounting so that some portion of the income from natural resources extraction can be reinvested for resource replacement. In this way a better balance can be struck between development and genetic conservation (Burley, 1993). Countries both of the north and those of the south recognize the need for joint interests, and the Global Environment Facility (GEF) is available through UNEP, World Bank and UNDP to assist developing countries to deal with such environmental concerns.

At present, with widespread debate on trade protection, external debt and other international problems, developing countries are challenged to produce national development strategies incorporating economic and environmental objectives, and all countries are being asked to promote sustainable development. These are not easy tasks and it is hardly surprising that many developing countries faced with huge problems related to education, health, development of infrastructure and the economy look to short-term benefits. Thus, the more esoteric needs for genetic resources conservation, with little immediate benefit, are considered to be of minor importance. Many such countries rely heavily on World Bank, UNEP and donor state-of-the-environment reports and on non-governmental organizations such as IUCN to help prepare country studies and national conservation strategies. Action in this area is recent: by 1992, state-of-the-environment reports had been completed in seven countries (with 40 in progress), and national conservation strategies were under way but by no means implemented in over 65 countries. Agenda 21, which resulted from UNCED, now calls for yet more action in the form of national sustainable development strategies, whilst the recent Convention on Biodiversity (CBD) calls for parties to identify the components of biodiversity that are important for conservation and sustainable use.

Faced with all these mechanisms to organize change and the new obligations of the parties of the CBD, it becomes clear that soundly based *in situ* conservation of crop gene pools with intensive management will show only slow progress. Two actions are clearly helpful: firstly, the availability of outside funding for the initial planning, whether from donors or the GEF; and secondly, the provision by an international organization of blueprints and organizational needs to national governments, and for these to be part of the national plan. Calls to national governments to do all this themselves will only result in a few actually doing it for political and economic reasons – unless the scientific community is strong enough to make it feasible, as is the case in a few countries like India and Brazil. It is much easier for developing countries to set up gene banks for *ex situ* conservation and this has been the trend over the past decade. In part this results from the dialogues of the FAO Commission on Plant Genetic Resources and the International Understanding on Plant Genetic Resources, although the latter is largely superseded by the CBD.

The overall conclusion is that the initial cost is not necessarily the prime constraint, but that rather it is the lack of succinct global and regional action plans in which relevant nations can participate, whether alone or cooperatively. Costs become a constraint when they recur in perpetuity, but until the orders of magnitude of these are readily available to nations they are unlikely to make commitments other than in principle. This is confounded by the fact that a nation may be expected to conserve a segment of a gene pool with probably no future benefit to the country itself but only of benefit to other nations. The development of mutual agreements to sort this out and result in tangible benefits is needed and this will rarely be done through technical assistance in a short-term project mode.

5.2.2 Availability of materials

If genetic material is going to be used it should be available to some extent from *in situ* reserve areas. Many reserve areas currently lack even basic inventory data and usually there are no voucher specimens, thus complicating access to suitable wild populations.

Under the CBD, Article 15 states that access is determined by sovereign governments but contracting parties should 'facilitate access to genetic resources for environmentally sound uses by other contracting parties'. This access is to be subject to the prior informed consent of the providing party and on mutually agreed terms.

Any scientific research on *in situ* genetic resources should also include participation of the provider and be conducted, where possible, in the providing state. Furthermore, users of the resources are to 'share in a fair and equitable way the results of research and development and the benefits arising from the commercial and other utilization of genetic resources with the contracting party providing such resources'.

These relevant articles of the CBD place a heavy onus on the sovereign state and its 'ownership' of the germplasm. An immediate constraint is apparent, and that is a scientific issue. Take, for instance, avocado. There are many potentially useful wild populations in several countries of Central America, South America and the Caribbean. The only way the full spirit of the convention could be accorded for avocado improvement would be through a specific cooperative agreement between many countries, most with no expertise in breeding avocado; and when diverse forms from diverse countries were actually used it would be almost impossible to ascertain what genes ended up in the product and from which country. A recent project of the World Conservation Monitoring Centre in Central American countries and Mexico discussed the needs for action on avocado but only within the context of that geographical area. Although an ecogeographic survey and *in situ* conservation in that area will be valuable it could well be conducted irrespective of cooperation with states elsewhere. Even taking action within the framework of the project the question to be asked is: will the location, study of population variation and designation of reserves be delayed due to the need to accord with the convention, or will nations be so enlightened that they will willingly invest funds and personnel to do the necessary research for the wider 'common good'? Fortunately, the political decision in October 1994 to create a Central American Council for Sustainable Development through an alliance to take specific political, economic, social, cultural and environmental actions to enhance peace, democracy, quality of life and sustainable development opens the door to promoting conservation of biodiversity on protected areas, and agreement might well flow when action is needed on avocado. There is thus an onus on scientists to be aware of the framework within which they can work effectively.

Even if the work becomes ongoing, there is a need for adequate curation of the resources in the *in situ* reserves. Internationally, the lack of trained human resources to conserve and manage effectively genetic materials *in situ* will constrain availability, utilization and flow of benefits.

Past systems for plant genetic resources have been, in principle and in effect, dealing with common property of potential value. This has now changed. As Witmeyer (1994) puts it:

> The CBD establishes regime norms for the biodiversity regime, which will apply to much of the world's plant genetic resources, that States have sovereign rights over their national genetic resources, and only need provide them to other States when mutually agreed terms can be arranged. This will, for plant genetic resources, replace the norm of free access.

The CBD aims, of course, to help to maintain the supply of genetic resources (increasing the economic incentives for conservation), to foster technology development in developing countries and to strengthen intellectual property rights in developing countries.

Widely quoted examples of mutually agreed sharing of biodiversity include the agreement between Costa Rica and a pharmaceutical company for biodiversity prospecting, and there is debate on the use of instruments such as 'material transfer agreements'. How these relate to provision of a particular sample from an *in situ* reserve, and its screening and possible discard as not actually useful in the breeding objective at hand, remains to be seen; the dialogue concerning interpretation of principles, agreements and other matters is likely to continue for a number of years. Whilst the FAO non-legally binding International Undertaking on Plant Genetic Resources is developed to harmonize with the CBD we are likely to see continued disagreements. These might well add to the 'wait and see' policies, with consequent delay of much-needed field screening and scientific action, pending satisfactory political solutions.

5.2.3 Valuing the resources and paying for them

Maybe the most contentious aspect of the negotiations for the CBD has been the principle that if the resource is used there should be some benefit payment. The discussions have become entangled with intellectual property rights, biotechnology and inequities involving transnational seed companies selling improved seed varieties to countries where the crops were domesticated or diversified. Almost certainly a number of countries will see the exchange of their germplasm as a source of more immediate financial benefits.

Shands (1994) has pointed out that it is difficult to find finance to evaluate genetic resources and there are no current ways of valuing them. There is a need for efficient seed industries in many poorer countries (Plucknett *et al.*, 1987), and Shands argues that if too many market disincentives exist for seed companies, the lack of private-sector interest will bring neither the best nor the cheapest seed to the farmer. The success of the biggest plant-breeding enterprise in the world, the Consultative Group on International Agricultural Research, has been based on the historic free availability of plant genetic resources. The products of the International Centres supported by the Group have been, and continue to be, of untold benefit to countries of the south. This work, now heavily focused on the resource-poor farmers, is likely to continue using materials readily available in *ex situ* collections and any constraints related to valuing a sample from an *in situ* reserve are likely to affect greatly this plant-breeding enterprise.

However, methods for valuation will need to be developed quickly.

Fair collecting fees, future royalties and other aspects such as paying for indigenous knowledge or mobilizing training or scientific work in a country all need to be considered (Gray, 1991; Lesser and Kratiger, 1994). Thus, the effect of international property rights and other exclusionary practices on the use of plant genetic resources is likely to dominate the discussions on valuation of a sample supplied and the practical constraints cannot be fully visualized at the present time.

5.2.4 Collaboration between states

Many plant materials which need to be conserved *in situ* will exist in reserve areas that transcend national boundaries of sovereign states. At present, inadequate attention is being given to this in development of national conservation plans and in the formulation of projects for support from donors or GEF.

The principles related to environments and ecologies, and linkages between *in situ* and *ex situ* conservation, discussed above, are particularly relevant and should be built into intercountry collaboration. However, constraints are obvious because organizational aspects will vary commodity by commodity. The surest way of overcoming the constraints is to promote regional activities, even though there may be many attendant problems. Regional activities have, over the past 20 years, been promoted time and time again, with varying results – not all positive. For *in situ* conservation the process would be helped by organizations representing states to develop the planning across national boundaries. This is possible within the rules of the CBD, but political entities will need to be very enlightened to do this and not be influenced unduly by politics. An illustrative example of how this can be done relates to the biodiversity of the Mediterranean aimed at crop genetic resources in protected areas under the auspices of the European Union as a 'Peace Campus Program' agreed in June 1994. Partners in this will be Israel, Italy, Germany, West Bank of Palestine territories, Egypt and Jordan.

The interdependency of nations in crop improvement makes it imperative that constraints are dealt with quickly. However, plans for *in situ* conservation with uncosted and only potential long-term benefits will not form a basis. More than ever the partnership of scientists and international organizations is needed to provide leadership in this area.

5.2.5 Development of common terminology

The need for common ground in relation to terminology, especially for the designation and description of spatial units, has already been referred to. Clarity is needed for administrators, politicians and scientists.

The proposed genetic resources management units (GRMUs) developed by some foresters (Riggs, 1982; Krugman, 1984; Ledig, 1988) is a most useful scheme which could be modified for *in situ* conservation of crop relatives and other categories and would not involve a network separate from the IUCN categories of reserves. These could be species specific and often having the protection of the IUCN categories – but they would require additional management. GRMUs can be at several levels, depending on knowledge of broad ecogeography, population structure and other factors; they can easily be designated for information retrieval as types GRMU 1, GRMU 2, etc. Finalization of such categories and their names is ideally a task for an international organization and is a logical way to link *ex situ* and *in situ* activities.

5.3 GAPS TO BE FILLED TO ENHANCE ACTION

Although it is possible to list a number of strategic areas of research which would aid *in situ* conservation development, such as stochastic loss of genetic variation from small populations, methodologies are available for broad ecogeographic surveys and delimitation of patterns of variation. These need to be vigorously pursued even as methodologies are refined. There appear to be four areas which need attention and would therefore enhance action.

5.3.1 Definition of priorities

The first step is the definition of priorities and broad assessment of which taxa are under threat and which need *in situ* action. There appears to be no current world list for reference. Each gene pool needs justification, and efforts to define priorities certainly appear to be piecemeal. This is a great pity after the needs have been expressed for so long.

For instance, we still rely on the scientific work of those interested in crop evolution (e.g. Debouck *et al.*, 1993) or international genetic resources efforts (e.g. IRRI's interests in wild rice species or CIMMYT's in maize), or on a few strong national programmes. The gaps are confounded by the fact that nobody has issued a list of the economic plants of interest along with the experts on their taxonomy in specific countries, or international experts, and moreover by the fact that assessments of variation based on ecogeographic survey and taxonomic, morphometric, biochemical and molecular analyses (in various combinations differing for each group of species) are still too geared to research interests rather than being mission-oriented to *in situ* conservation.

If the human resources are not available for this action, and if the work can only be done if development assistance funds are available, then the wider scientific community must be mobilized to help, since the crop genetic resources community has hardly responded to the challenges in the past decade.

5.3.2 Legal instruments

The second step will be the clarification of legal instruments to support scientific attempts to conserve the target wild species related to crop gene pools. Much more attention must be given to the basic legal instruments for area-based conservation, particularly since restrictions on specific land use in the area would affect people (either owners and occupiers or the public). Thus regulation of human activity in the public interest is required. Regulations may or may not be voluntary and much depends on the legal structure of an individual country as to how the regulating is done. The two extremes are 'public property countries' versus 'police power countries'. In a detailed discussion of these, De Klemm and Shine (1993) conclude that neither system is entirely satisfactory and that combinations of the two with incentives for voluntary measures might often lead more to land becoming an asset and a source of income.

The complexities of the legal instruments are such that each has to be tailored to fit the individual nation. There are useful examples to be followed and De Klemm and Shine cite combined regulatory/voluntary instruments in respect of South Australia's Native Vegetation Act of 1991 and the UK's Sites of Special Scientific Interest. However, surveys of existing instruments show that the management of areas such as national parks, and hence their regulation, may be vested in diverse bodies – a central ministry of forestry or environment, autonomous communities or local government bodies – established for the purpose. Interestingly, De Klemm and Shine state:

> The key to the success of nature parks seems to be the designation of a Government-agency or the establishment of an administrative body as specifically responsible for the area with the jurisdiction to prepare ... plans ... regulations ... and ... permits ... Where these matters are left to the Government agencies ... experience has shown that these lead to many implementation and enforcement difficulties.

For genetic resources of crop gene pools conserved *in situ* much more attention will need to be given to developing transnational boundary cooperation and interstate cooperation in the coordinated management of such areas. Whereas the instruments are available for this in Europe through the Council of Europe and bilateral agreements, and through treaties such as the Berne Convention of 1979, the Benelux Convention of 1982 and an EC directive 1992/43, there are not too many similar things elsewhere. The Asian countries concluded an agreement in 1985 which provides a framework for legal action. A Convention on Biological Diversity and the Protection of Priority Wild Areas in Central America was signed in 1992, the African Convention on the Conservation of Nature and Natural Resources 1968, and a Convention on Nature in the South Pacific 1976, are further possibilities but have either been slow to come into effect or require amendments.

Even though some frameworks exist they would apply to transnational areas rather than to transnational species. The CBD places greater emphasis on conservation of ecosystems than on conservation of particular species. In fact the key issue related to *in situ* conservation of wild crop relatives, i.e. the establishment of protected areas, is subject to follow-up conferences of the parties concerned. Technical and political viewpoints are likely to be divergent and progress as far as legal instruments relating to specific crop gene pools *in situ* is unlikely to be rapid. The follow-up and the development of protocols and annexes are still likely to focus on needs to counterbalance conservation obligations of states having rich biological diversity against the recognition of rights over their genetic materials, since equitable sharing of the benefits from utilizing such materials is a major objective of the CBD.

5.3.3 Access to germplasm

Much targeted effort for conserving segments of crop gene pools *in situ* can be negated if there is not ready access to the materials for utilization. It is a paradox that many have come to believe that all the resources will indeed be valuable. Undue restriction on access or undue payment for materials can readily result in a much lowered interest in screening by breeders who would intensify their efforts on materials already in gene banks.

In this area of concern, the ground-swell of opinion on *in situ* conservation of primitive forms at the farm level might represent a new trend for breeders. The whole question of on-farm conservation and community collections is hampered by over-generalizations across diverse agroecologies, and lack of attention to the amount of security needed.

5.3.4 Planning and execution

Finally, many national programmes require help in planning and executing relevant and effective work on *in situ* conservation. Many scientists at the genetic resources programme level will be overwhelmed by the complexities of anything other than their practical work, especially if they become drawn into discussions on legalities, ownership, payments for germplasm and the need for bilateral agreements. Many curators of national gene banks have never tried to forge bilateral agreements with colleagues in other countries to rationalize tasks and to get more work accomplished by sharing activities (Williams, 1985). Given also that the scientists rarely hold dialogue with their country representatives at international conventions or organizations, they are unlikely to help draft complex agreements on *in situ* conservation. International organizations must respond quickly with sets of easily understandable guidelines and offer help in developing strategies, planning, and introducing technically feasible methodologies.

5.4 CONCLUSIONS

As genetic conservation of crop gene pools has progressed over the past 25 years, significant action has occurred but *in situ* conservation has lagged behind despite its promotion by the international community.

In the past the constraints were largely technical and financial and many were overcome by the collaborations of the wider scientific community. The past constraints have now become overlaid with more political and legal problems than hitherto. On the positive side, the diversity of plants has become a common concern of mankind, even though genetic resources are no longer regarded as common property. In the long run the international community will surely approach the natural justice problems to encourage conservation and change the current perceptions of many scientists, politicians and administrators.

Promotion of *in situ* conservation in a scientific manner is not an easy task in the light of the practical and other constraints outlined in this chapter. It is clear that international organizations need to take a much stronger role as facilitators since reserve placements are so urgent for many crop gene pools.

6

Plant population genetics

M.J. Lawrence and D.F. Marshall

6.1 INTRODUCTION

Over the last decade, there has been an exponential growth in interest
and investigation of the problem of how best to conserve populations of
endangered species. As a result, there is now a very extensive and widely
scattered literature on this subject. Among recent papers which cite a
large number of references are those of Boyce (1992), Ellstrand and Elam
(1993) and Nunney and Campbell (1993). Soulé's (1987) book gives a
valuable survey of the subject, in which the chapter by Lande and
Barrowclough (1987) deals with the population genetics of the problem.
Much of this literature is concerned with the conservation of species
where *ex situ* conservation is not a realistic alternative. The demographic
and population genetics theory on which discussion of this broader issue
is based, however, is also relevant to the special case of those species
where the conservation of material in gene banks is possible. The purpose of this chapter is to give a brief outline of the population genetics
theory on which discussion of *in situ* conservation is based and to introduce some of the terms and concepts used in the literature. We also indicate the kinds of experimental investigation of populations which are
necessary to produce better recommendations about minimum viable
population size than are possible at present.

6.2 FACTORS AFFECTING CHANGES OF GENE FREQUENCY IN POPULATIONS

In an infinitely large population, in which the effects of mutation, migration and selection are negligibly small, allele and genotype frequencies

Plant Genetic Conservation.
Edited by N. Maxted, B.V. Ford-Lloyd and J.G. Hawkes.
Published in 1997 by Chapman & Hall. ISBN 0 412 63400 7 (Hb) and 0 412 63730 8 (Pb).

remain constant from one generation to the next. Thus, a fundamental consequence of Mendelian inheritance is that, in the absence of these effects, the hereditary mechanism conserves allele frequencies. A population in which allele and genotype frequencies remain constant from one generation to the next is said to be in Hardy–Weinberg equilibrium. Though this result is often combined with random mating, it holds (eventually) for all mating systems, including mixed selfing and random mating.

While the Hardy–Weinberg equilibrium serves as a convenient starting point in a discussion of the genetics of populations, it cannot be regarded as providing more than a very approximate description of an actual population of individuals. First, actual populations have a finite size, N. Unless N is very large, the frequency of the alleles in the sample of $2N$ gametes which unite to form the zygotes of the next generation will differ by chance from their frequency in the parental zygotes. In small populations of the kind that are likely to concern conservationists, the magnitude of this sampling variation can be very much greater than any other factor affecting allele frequencies. Second, while it may be possible to ignore the effect of mutation in small populations in the short term, it is not prudent to do so in large populations or in the long term, because mutation, though rare, plays a fundamental role in replenishing genetic variation that has been depleted by other factors. Third, few natural populations are so isolated from others of their species that the possibility of migration of genes between them can be discounted. Yet only a small amount of migration is sufficient to prevent adjacent populations undergoing independent evolution, a result which could be of considerable interest to the conservationist. Fourth, the individuals of all populations are subject to natural selection. The effect of this selection on genetic variation depends on its type, some kinds maintaining it, while others are expected to exhaust it. Lastly, the effects of these four factors of sampling variation, mutation, migration and natural selection on allele frequencies are not independent. It follows, therefore, that it is necessary to consider their joint effect in any comprehensive discussion of the evolutionary dynamics of a population.

6.3 TYPES OF VARIATION

The great majority of the phenotypic variation that is observed in populations is of the quantitative type, which is determined by many genes of individual small effect whose expression is modifiable by the environment. It is this type of variation on which natural selection acts and which is involved in adaptive response to changes of the environment. It follows, therefore, that it is overwhelmingly this type of variation which needs to be conserved. Most of the theory of population genetics, on the other hand, is concerned with the frequencies of alleles whose substitution has a discontinuous effect on the phenotype; these are the so-called major genes which determine single-locus Mendelian polymorphisms. It might appear, therefore, that much of population genetics theory cannot be applied to this quantitative variation. The genes determining quantitative characters, however, behave in the same way as major genes in that they are carried on the chromosomes and can display linkage in their inheritance, allelic variation, dominance and epistasis; that is, they differ from major genes only in respect of the magnitude of their individual effect on the phenotype. Hence, contrary to first impressions, the theory of population genetics applies as much to these polygenes as it does to major genes.

6.4 SAMPLING VARIATION

6.4.1 The idealized population

Suppose a very large population of diploid individuals which mate at random becomes subdivided into a large number, n, of subpopulations. Suppose, further, that each of these subpopulations contains the same number of breeding individuals, N, which mate at random in each generation and that each is completely isolated from the others (no migration). We shall also suppose that there is no mutation or selection and that each generation is distinct (no overlap) and of the same size as every other (N is constant over generations, each individual contributing, on average, one offspring to the next generation). Lastly, suppose that the frequencies of two alleles, A_1 and A_2, of a single gene in the initial or base population are p_0 and q_0, respectively.

Now, assuming (as will nearly always be the case) that the number of male and female gametes produced by individuals is large, the frequencies of A_1 and A_2 in the gametes produced by the individuals of the base population will be the same as in these adults, namely, p_0 and q_0, respectively. The same argument holds for the average frequency of these alleles in the subpopulations considered as a whole, not only in the first generation, but also in all generations thereafter, because the product nN is a very large number, so that the effect of sampling variation is negligible. This is not true, however, of the subpopulations considered individually. Thus, while the number of gametes produced in any one generation may be large, the number uniting at random to form the N zygotes of the next generation is $2N$. If $2N$ is small, sampling effects will cause random changes in allele frequency from one subpopulation to another, such that the frequency of A_1 in the first generation of some subpopulations will be higher than p_0 and in others lower. This dispersive effect of sampling variation, known as random genetic drift, is cumulative over time, so that the subpopulations become genetically differentiated by chance alone. Eventually a proportion of the subpopulations will contain only the A_1 allele, the other having been lost (A_1 is fixed and all individuals are A_1A_1 homozygotes), and the remainder will contain A_2 only (A_2 is fixed; only A_2A_2 homozygotes are present), so that each subpopulation will have lost all genetical variation with respect to this locus. Drift, therefore, not only causes subpopulations to diverge in terms of their allele frequency, but also causes them to become progressively more and more inbred, despite the fact that mating within them is random throughout their descent.

6.4.2 The quantification of drift

Though it is not possible to predict which or when an allele will be lost in any one subpopulation, because drift is a random process, it is possible to say which of a pair of alleles is most likely to be lost and when this will happen on average. Thus, the probability that A_1 is eventually fixed in a subpopulation is equal to its initial frequency, p_0, and the probability that A_2 is fixed eventually is q_0 (Fisher, 1930; Wright, 1931). It follows, therefore, that eventually a proportion, p_0, of the subpopulations will contain the A_1 allele only, A_2 having been lost, and that the remainder, q_0, will contain A_2 only. Hence, if one of these alleles occurs at only a low frequency in the base population ($p_0 << q_0$), it is much more likely to be lost from a subpopulation than the other; that is, it is difficult to conserve rare alleles.

The average time to fixation, $\bar{t}$, known as the mean absorption time (Watterson, 1962; Ewens, 1963), is approximately:

$$\bar{t} = -4N(p_0\log_e p_0 + q_0\log_e q_0) \tag{6.1}$$

This equation shows that mean absorption time depends on the initial frequency of an allele and on subpopulation size, N. If, for example, $p_0 = q_0 = 0.5$ and $N = 10$, then $\bar{t} = 28$ generations; whereas if $N = 50$, then $\bar{t} = 139$ generations. But if $p_0 = 0.05$ and $q_0 = 0.95$, then the mean absorption time becomes only 8 and 40 generations, respectively. Thus, rare alleles are not only much more likely to be lost than common ones, but they will be lost quite soon. Furthermore, it must be emphasized that the rate of loss of alleles over time is negatively exponential, a few persisting for a long time, but most being lost quite rapidly. For example, though the mean absorption time is 8 generations when $p_0 = 0.05$ and $N = 10$, the A_1 allele will have been lost in more than half of the subpopulations well before this time.

Another way of looking at the effect of drift is in terms of the increase of the variance of allele frequency over time between subpopulations and the reduction in their heterozygosity as a result of inbreeding. Thus, after one generation, the variance of allele frequency over subpopulations is:

$$\sigma^2_q = p_0 q_0 / 2N \qquad (6.2)$$

which becomes, after t generations:

$$\sigma^2_q = p_0 q_0 [1 - (1 - 1/2N)^t] \qquad (6.3)$$

This equation shows that the variance of allele frequency is expected to increase by a factor of $(1 - 1/2N)$ per generation. It follows that the average proportion of heterozygotes, H, at the locus must decline by the same factor; that is:

$$H_t = (1 - 1/2N)H_{t-1} \qquad (6.4)$$

Provided that N is not very small, heterozygosity declines at an approximately exponential rate over generations, so that:

$$H_t = H_0 e^{-t/2N} \qquad (6.5)$$

For example, after N generations, heterozygosity is only 0.61 of the initial heterozygosity, H_0.

An alternative and, perhaps, a more useful way of examining the effect of drift in these circumstances is to suppose that the gene, A, determines a quantitative character and to describe its effect in terms of changes in the distribution of additive genetical variation (that due to differences between homozygotes) within and between the subpopulations. Then, after t generations, the average additive genetical variance within subpopulations is approximately:

$$V_w = V_0 e^{-t/2N} \qquad (6.6)$$

and that between subpopulations is:

$$V_b = 2V_0(1 - e^{-t/2N}) \qquad (6.7)$$

where V_0 is the additive genetical variance in the base population from which the subpopulations were derived (Crow and Kimura, 1970). These equations show that the variance within subpopulations decreases, on average, at the rate of $1/2N$ per generation and eventually becomes 0, while the variance between subpopulations increases with time at this rate until it eventually becomes $2V_0$.

6.5 MINIMUM VIABLE POPULATION SIZE

Populations of species that are candidates for *in situ* conservation are unlikely to be very large in either their size or their number. We need to consider, therefore, the minimum size of a population, commonly referred to as minimum viable population (MVP) size (Franklin, 1980; Soulé, 1980), that is required to reduce, to an acceptable level, the loss of genetical variation and heterozygosity by the inbreeding caused by drift.

The loss of heterozygosity by inbreeding causes inbreeding depression of any character controlled by genes that display dominance in the increasing direction. Mather (1960, 1966, 1973) has argued that the genes controlling characters that have been subject to directional selection – those where greater expression of the character is unconditionally favourable – are expected to display unidirectional dominance. Most, if not all, components of fitness, such as viability and fecundity, must be subject to directional selection; such evidence as is available shows that the genes controlling fitness characters do in fact display dominance in the increasing direction, as Mather's theory of genetical architecture predicts (Kearsey and Kojima, 1967; Mather, 1982; Lawrence, 1984; Lane and Lawrence, 1995). Characters which are components of fitness are, therefore, expected to display inbreeding depression. Since significant reductions in viability and fecundity are obviously undesirable, we need to consider the minimum size of a population which reduces inbreeding depression to an acceptable level. Animal breeders have found that an increase of inbreeding of 2–3% per generation can be sustained without obvious detrimental effects. This suggests that a population size that would limit inbreeding to 1% per generation should be safe. Equation (6.4) shows that heterozygosity is expected to decline by an average of $1/2N$ per generation in the idealized population. Hence, putting $1/2N = 0.01$ gives an MVP of 50 individuals.

Hitherto we have ignored mutation, largely because mutations occur only rarely, but in the long run the generation of new genetic variation by mutation will replace that lost by drift. Mutation rates of major genes are generally very low, being of the order of 10^{-5} to 10^{-6} per locus per generation. Quantitative characters, however, are controlled by many loci. Though estimates of the number of genes determining such characters are unavoidably rather imprecise, it is likely that most components of fitness are controlled by a hundred or so genes. Hence, the total spontaneous rate at all loci controlling a particular quantitative character could be a hundred times larger than that of a single gene, giving, say, a rate of 10^{-3} per trait per generation. Equation (6.7) shows that the additive genetical variance of a quantitative character is expected to decrease at a rate of $1/2N$ per generation within the subpopulations of the idealized population. Thus, equating $1/2N$ to 10^{-3} suggests that an MVP of 500 individuals is of sufficient size for the new variation arising from mutation to replace that lost by drift (Franklin, 1980). The outcome of these two simple calculations is referred to as the '50/500 rule'.

6.6 EFFECTIVE POPULATION SIZE

6.6.1 Theory

Natural populations are very unlikely to conform to the idealized population we have been considering so far, because the individuals of many species of flowering plant set at least some of their seed by self-pollination; because populations are likely to vary in size over time; and, in particular, because the contribution that individuals make to the next generation is unlikely to be random, some producing much more seed than others. All of these factors make populations behave as if they are smaller than their census size, N. Wright (1931) called this smaller size the effective population size, N_e, which is defined as the size of the idealized population that would undergo the same amount of random genetic drift as the actual population. It is the effective size of a population, rather than its census size, which determines the magnitude of the effect of drift on the dispersion of allele frequencies, and on the rate of loss of heterozygosity and genetical variation. It follows, therefore, that it is necessary to replace N by N_e, in all of the formulae given previously.

In general, the prediction of effective population size is a far from easy task, not only because it is difficult to obtain good estimates of the vital statistics of a population (see later), but also because of the complex way in which these factors combine in determining the relationship between N and N_e (see Caballero, 1994, for a recent review of the theory). Some insight into this relationship can be obtained by considering some simple cases.

As before, we start by considering the idealized population in which individuals mate strictly at random. Suppose that the number of alleles contributed by the ith individual of the population to the next generation is k_i. Let k vary from one parent to the next with a variance of V_k. Suppose, further, that this variation in reproductive output is caused solely by chance and differences in the environment affecting their fecundity. If N remains constant from one generation to the next, the average number of alleles contributed by each parent to the following generation must be $k = 2$. Then in these circumstances, Wright (1938) showed that:

$$N_e = (4N - 2)/(2 + V_k) \tag{6.8}$$

If differences in fecundity are purely random, the k_i follow the binomial distribution, so that $V_k = 2(1 - 1/N)$ and:

$$N_e = N \tag{6.9}$$

which is the situation that obtains in the idealized population. We may note in passing that if we intervene to ensure that every parent contributes exactly two alleles to the next generation, $V_k = 0$, so that:

$$N_e = 2N - 1 \qquad (6.10)$$

That is, in these circumstances, the effective size of the population is approximately twice its census size, a result which is of very considerable practical interest for those planning the *ex situ* regeneration of seed (Gale and Lawrence, 1984). In nature, V_k is always larger than $2(1 - 1/N)$, so that $N_e < N$, as we shall see.

When individuals that are hermaphrodites mate at random, a random proportion, $1/N$, of matings are expected to involve self-fertilization. However, if a species is self-incompatible, such matings cannot occur. In these circumstances (Wright, 1931):

$$N_e = N + 1/2 \qquad (6.11)$$

For populations of the size we are interested in, the difference between (6.11) and (6.9) is, of course, trivial. If, on the other hand, the amount of selfing is greater than $1/N$, then N_e will be less than N. Suppose individuals set a proportion, s, of their seed by self-pollination, the remainder, $(1 - s)$, being set by random mating. Assume that the population is in equilibrium, such that the proportion of heterozygotes lost by selfing in each generation is exactly balanced by the proportion of new heterozygotes arising from random mating. Then, even if $V_k = 2(1 - 1/N)$:

$$N_e = N(1 - s/2) \qquad (6.12)$$

Thus, for a predominantly self-pollinating species, s is near 1 and N_e will be close to half N.

We have assumed so far that variation in reproductive output is caused solely by chance and environmental effects. If, however, this variation is partly genetic in origin, Nei and Murata (1966) have shown that:

$$N_e = 4N/[2 + (1 + 3h^2)V_k] \qquad (6.13)$$

where h^2 is the narrow-sense heritability of reproductive output. We note that this formula is similar to (6.8) except that it has an extra term, $3h^2V_k$, in its denominator. Components of fitness, being subject to directional selection, are typically found to have low heritabilities. Suppose $h^2 = 0.15$, $V_k = 4$ and $N = 500$. Then when reproductive variation is non-genetic in origin we find from (6.8) that $N_e = 333$; whereas with $h^2 = 0.15$, $N_e = 256$.

We have assumed so far that population size remains constant. What happens if this size varies over generations? The formulae given above apply to each generation considered on its own, so that N_e can be found for each. Suppose N_e in successive generations is $N_{e1}, N_{e2}, N_{e3}, \ldots, N_{et}$. Then in the short term at least the population behaves as if it had a constant size of N_e^* where:

$$1/N_e^* = [1/N_{e1} + 1/N_{e2} + 1/N_{e3} + \ldots + 1/N_{et}]/t = \Sigma(1/N_{ei})/t \quad (6.14)$$

That is, N_e^* is the harmonic mean of the N_es in each generation (Wright, 1939). The harmonic mean of a set of different numbers is always less than their arithmetic mean because it is particularly sensitive to the smallest number of the set. Thus, suppose in five successive generations N_e is 500, 500, 10, 500 and 500. Then $N_e^* = 46.3$, whereas the arithmetic mean of these numbers is 402. It is clear, therefore, that it is highly desirable to ensure that the number of individuals in a population remains constant over generations and that care must be taken in managed populations (e.g. an *in situ* reserve) to avoid catastrophic reductions in their size.

6.6.2 Evidence

For convenience, we have examined the effects of variable progeny size, mating system and variation in population size over generations on effective population size independently. In practice, all of these factors need to be considered because their effects will reinforce one another in reducing N_e relative to the census size of the population, N; that is, we expect the ratio N_e/N to be less than one.

Most of the estimates of this ratio are derived from animal populations and many are calculated by considering only one of these factors at a time. Nevertheless, recent evidence suggests that N_e/N is less than 0.1 for small, highly fecund organisms (Briscoe et al., 1992; Frankham, 1994). One of the most instructive estimates comes from a study of a small, semi-natural population of the poppy, *Papaver dubium* (Mackay, 1980; Crawford, 1984), in which the numbers of seed produced by each of the 2316 plants it contained were scored. These data revealed that 50% of all seed came from the 2% of the most fecund plants; indeed 4.6% of this seed was produced by the largest plant. This markedly non-random variation in fecundity gives a ratio of $N_e/N = 0.07$. If, in addition, allowance is made for the fact that individuals of this species set, on average, 75% of their seed by self-pollination (Humphreys and Gale, 1974), the ratio is reduced to 0.024 (Gale, 1990). If this estimate from *P. dubium* is typical of species of flowering plant that are candidates for *in situ* conservation, the 50/500 rule becomes, approximately, the 2000/20 000 rule. This calculation, however, makes no allowance for the fact that populations of this species persist, between flowering episodes, in the form of a very large number of long-lived seeds in the soil, allowance for which could make quite a difference to the outcome of this calculation.

6.7 SUBDIVIDED POPULATIONS

Our discussion, so far, has been confined to a population in which each individual has an equal chance of cross-pollinating with every other; that is, a population which is panmictic. In practice, neither pollen nor seed travels, on average, very far from its source, so that a population is likely to consist of a series of what Wright (1943, 1946) called neighbourhoods, within which mating is random, but between which gene flow is progressively restricted by distance. In addition, the majority of species of interest to conservationists are likely to be represented by at least several more or less isolated populations, rather than just one. In short, both species and their constituent populations are likely to have geographical structure. What implications does this have for planning the *in situ* conservation of a species?

We saw earlier, when discussing the idealized population, that subpopulations of finite size founded from a very large base population, which are completely isolated from one another, are expected, in time, to become genetically differentiated by chance alone. In addition, it is extremely unlikely that two subpopulations occupy habitats that are identical. The effect of natural selection on one is unlikely to be the same as that on the other. Hence, we expect allele frequencies to vary over subpopulations, because of both drift and selection. Where resources allow, therefore, a case can be made for the *in situ* conservation of more than one subpopulation of a species at risk.

In practice, the strength of this case is less than appears at first sight, for two reasons. First, most of the genetical variation of cross-pollinating species occurs within rather than between their constituent subpopulations. It follows, therefore, that the well-designed *in situ* or *ex situ* conservation of the genetical variation of just one of these populations goes a long way towards achieving the objective of all conservation programmes, namely, the conservation of the genetical variation of the species. Second, theory shows that a migration rate of as low as one or two individuals (e.g. seed) per generation between subpopulations, provided that these individuals reproduce, is sufficient to prevent their divergence (Malecot, 1975; Crow and Aoki, 1984). The same argument holds for the neighbourhoods of one of these populations; that is, with a migration rate of this order, a subpopulation will behave as if it is panmictic. For both of these reasons, when resources are limited, it might be better to concentrate on the conservation of the genetical variation of one population, rather than to disperse effort in an inadequate attempt to conserve this variation in several.

6.8 ESTIMATING EFFECTIVE POPULATION SIZE

It is clear, from the foregoing discussion, that there is a substantial body of theory which can be used to plan programmes designed to conserve the genetical variation of species of interest. It is not possible to apply this theory, however, in the absence of a knowledge of the life history, reproductive biology and quantitative genetics of the species in question. Some preliminary information usually will be available, such as whether the species is perennial or annual; whether its individuals are dioecious, monoecious or hermaphrodite; and whether the seed of an annual remains viable sufficiently long in the soil for generations to overlap (a topic not discussed here; see Lande and Barrowclough, 1987). Most of the information required to estimate effective population size, however, can be obtained only by appropriate survey and experimental investigation of the population. Among the most important aspects of this information are the census size of the population, the breeding system of the species, the distribution of progeny size among individuals and the genetical architecture of components of reproductive output.

6.8.1 Census size

Unless a population is confined to an island or refuge, remote from all other populations of the species, and its individuals are large and easily found, considerable care has to be taken to determine or estimate the census size of the population, because of the difficulties in identifying the boundaries of the population and the risk of overlooking small individuals. It may also be difficult to carry out a complete census without damaging the habitat by trampling. In most circumstances it should be possible to estimate the average density of individuals, using a systematic procedure (for example, a line transect run through the middle of a population), and estimating its census size N by multiplying this estimate by the total area occupied by the population. If this census could be repeated annually, variation in population size could be monitored and its effect on N_e estimated (6.14).

6.8.2 The breeding system

Provided that the species is polymorphic for such loci, the outcrossing rate t can be estimated from data obtained by scoring the individuals of open-pollinated progenies for enzymes whose allelic variation can be detected by gel electrophoresis (Brown and Allard, 1970; Ritland and Jain, 1981; Ritland, 1986). The seed from which these progenies are raised can be obtained by collecting seed from a random sample of individuals and packeting the seed taken from each separately from that taken from every other. In these circumstances, the members of each progeny are known to have the same maternal parent and, hence, have at least a half-sib relationship. This seed could be collected from plants situated at set intervals along the transect employed to estimate the density of the population, the length of this interval being chosen so as to minimize the chance that individuals at adjacent sampling points are closely related.

Although this is not always done, it is highly desirable to determine the inheritance of the isozyme polymorphisms used to estimate outcrossing rate in a preliminary investigation involving the analysis of full-sib families produced by making controlled crosses between plants of known enzyme phenotype.

6.8.3 Distribution of progeny size

The variation of the female contribution to progeny size can be estimated by determining the seed output of each of a random sample of, say, 500 plants. The plants used for this purpose could be the same as those recorded in the census survey. It will be necessary to ensure that all seed produced by these plants is retained and that none is lost. The best way of enclosing fruits to ensure this will vary from one species to another. In the investigation of distribution of seed output in *P. dubium* discussed earlier, capsules were sealed with small pieces of 'Parafilm' shortly before their pores opened (preliminary tests had shown that this had no detectable effect on seed quality).

6.8.4 Genetical architecture of quantitative characters

Unless a population is very small or has recently been founded from a very small number of plants, it is virtually certain that it possesses additive heritable variation for all or very nearly all quantitative characters. It is, however, necessary to demonstrate this in every case. The simplest way of accomplishing this is to raise 20–30 open-pollinated (i.e. natural) progenies of, say, size 10, in a completely randomized experimental field trial. An analysis of variance of the results obtained by scoring the individuals of this experiment for quantitative characters of interest will reveal whether these characters are heritable and, if so, can provide a rough estimate of the heritability of each character (Lawrence, 1984; Lawrence *et al.*, 1995b). A knowledge of the outcrossing rate of individuals in the population can be used to narrow the range of values which these heritability estimates can take. Estimates of the heritabilities of components of reproductive output can be used to determine the reduction in N_e relative to N (6.13). Better estimates of the heritabilities of quantitative characters can be obtained only from experiments involving families produced by controlled and systematic crosses between plants (see Lawrence, 1984, for a summary of experimental designs suitable for this purpose); such experiments could be carried out in the season following the preliminary investigation with natural progenies.

The detection and, particularly, the estimation of non-additive genetical variation, including dominance, is, in general, a much harder task than the detection and estimation of additive genetical variation. When the individuals of a species can be self-pollinated, the simplest and most direct way of detecting the presence of dominance variation is to demonstrate that a character displays inbreeding depression. If plants can be vegetatively propagated, a suitably designed experiment containing clonal replicates of 20–30 parents and their self-progenies should accomplish this purpose. In general, however, it will be necessary to replace parents by progenies produced by crossing them at random in pairs which, with their self-progenies, gives the augmented biparental (ABIPS) experiment (Lawrence, 1984). Consistent differences between the means of selfed offspring and either their parents or the means of their related biparental progenies indicate that the character in question shows inbreeding depression and, hence, is controlled by genes which display directional dominance. This experiment can also be used to detect and estimate additive genetical variation, to estimate heritability and to detect ambi-directional as well as directional dominance.

Quantitative genetics has, perhaps, the not wholly unjustified reputation of being a difficult and esoteric branch of genetics. There is no doubt that quite elaborate and large experiments are required to obtain detailed and precise information about the genetical architecture of quantitative characters. Nevertheless, as we have shown, it is possible to obtain some basic information about such characters from relatively simple experiments. It is necessary to take considerable care in the planning, conduct and the interpretation of the results obtained from even these simple experiments, if the conclusions drawn from an analysis of their results are to have any value. Investigations involving such characters should not be undertaken without expert advice.

6.9 CONCLUSIONS

Population genetics theory shows quite clearly that sampling variation is the chief cause of the loss of genetical variation in small populations; that it is their effective size, N_e, rather than their actual size, N, which determines the rate of this loss; and that in natural populations N_e is almost certainly appreciably less than N. Provided that information of the kind discussed in the previous section is available, it is possible to obtain an approximate estimate of the ratio N_e/N and, hence, the size of the population, N, required to reduce the loss of genetical variation by drift to an acceptable minimum. It may be, however, that the threat to an endangered species, for which *ex situ* conservation is not an alternative, is so immediate that it is necessary to decide whether to attempt its *in situ* conservation in the absence of this information. In these circumstances, it would be necessary to make a guess of the minimum population size necessary to avoid significant loss of genetical variation until at least some information about its reproductive biology and genetics became available. Reference to the *P. dubium* case mentioned earlier suggests that a population size of not less than $N = 5000$ is probably reasonably safe. This may, at first sight, appear to be a rather large number which is unlikely to be achieved in practice, but populations of this size of herbaceous species – such as the poppies *P. dubium* and *P rhoeas* (Lawrence *et al.*, 1994) – and those of wild wheat in Israel, in which the density of flowering plants is not uncommonly of the order 10 m^{-2}, occupy only about 500 m^2 of ground. At the other extreme, for a dipterocarpous species of the tropical rainforests of south-east Asia, whose density can be as low as only 2 km^{-2}, a population of this size would require a reserve of 2500 km^2. This calculation is of particular interest because Wallace (1910), in what must be one of the first discussions of the need to conserve botanical resources in these forests, recommended the establishment of a number of reserves of 3 km^2. Though he envisaged that these reserves would be situated within extensive tracts of forest, our calculations suggest that a reserve of this size is much too small to conserve the genetical variation of its constituent dipterocarpous species. Most of the species that are under consideration for *in situ* conservation will, of course, lie somewhere between these extremes.

It must be emphasized that our recommendation of a minimum population size of $N = 5000$ is tentative and should be regarded as no more than an attempt to get things started until such time that better estimates of the ratio N_e/N are available for the species in question. Indeed, because of the large number of factors that affect the relationship between N_e and N in natural populations, the difficulty of obtaining reliable estimates of the magnitude of their effects and the unknown impact of natural selection on the individuals of these populations, *in situ* conservation will almost always be more risky than the *ex situ* conservation of genetical variation. We have shown elsewhere that a random sample of 172 seeds, collected from the constituent populations of a species, is sufficient to conserve all or very nearly all of its genetical variation (Lawrence *et al.*, 1995a). If the viability of this seed could be maintained in perpetuity at low temperature in a gene bank, this genetical variation will have been completely conserved. Usually, however, it will be necessary to produce fresh seed from time to time. Provided that this rejuvenation is carried out in the systematic way recommended by Gale and Lawrence (1984), the effective size of the population of plants which produce fresh seed will be approximately double its actual size (equation 6.10) and no natural selection will take place. In short, it is possible to control or eliminate the processes which erode genetical variation in a well-designed *ex situ* conservation programme. This is, of course, much more difficult to accomplish in natural populations. It can be argued, therefore, that genetic diversity is more likely to be lost *in situ* than *ex situ*. Hence, wherever possible, genetical diversity conserved *in situ* should be conserved in a duplicate collection *ex situ* in a gene bank.

7

Plant population ecology

M. Gillman

7.1 INTRODUCTION

Identification of an endangered species is not a trivial task. Information on some species is so limited that it is impossible to tell if they are threatened: indeed, it is likely that there are many species, particularly in the tropics, whose existence will never be known. However, there are detailed data available for many species which allow rare and possibly endangered species to be identified. This is particularly true in Britain and parts of North America and mainland Europe, where there is a long tradition of ecology and natural history and a relatively high ratio of ecologists to species.

Many of the best sets of field data are for plant species. These include both distributions of plant species and censuses of plants over time. We can now say for many species, often with a high degree of accuracy, how their geographical range is changing and how their populations are behaving at particular localities. Quantifying population size and its fluctuations and trends in time and space, i.e. quantifying population dynamics, helps conservation biologists to identify candidate species for protection and management. Population genetics is also dependent on estimates of population size and dynamics, e.g. in estimating effective population size (Chapter 6). *In situ* conservation of a species requires accurate assessment of population size and dynamics because, as we shall see, both contribute to the likelihood of local population extinction and perhaps, ultimately, species extinction.

In this chapter some fundamental issues in plant population ecology are explored which are relevant to *in situ* plant conservation. The chapter begins with a consideration of geographical range estimation and con-

Plant Genetic Conservation.
Edited by N. Maxted, B.V. Ford-Lloyd and J.G. Hawkes.
Published in 1997 by Chapman & Hall. ISBN 0 412 63400 7 (Hb) and 0 412 63730 8 (Pb).

traction; it then discusses the link between local and metapopulation dynamics, the key results arising from this work and how population dynamics and extinction can be quantified. Finally, the chapter returns to plant species change within landscapes. Throughout the chapter the population dynamics of plant species are considered in context of the monitoring, management and protection of those species.

7.2 GEOGRAPHICAL RANGE

The most basic information available on an endangered plant species is its geographical range or spatial distribution within a given region. Such information is in the form of presence/absence data at localities or within grid squares of a certain size. In the tropics it is not unusual for a locality to be a whole mountain range or river system, perhaps covering thousands of square kilometres. In some temperate regions the resolution is much better. Figure 7.1 shows the change in the distribution of the early spider orchid (*Ophrys sphegodes*) in Britain, recorded over three periods (Hutchings, 1987). The marked contraction towards the south-east of Britain may reflect the largely southern distribution of the species in mainland Europe (Hutchings, 1987). The species is now very localized in Britain, being known from only 10 localities (Figure 7.1).

Whilst the type of information presented in Figure 7.1 is helpful in describing a contraction in part of the geographical range of the early spider orchid, it says nothing about the abundance of the species within any one grid square. A dot may represent just one plant or a large population of plants. Detailed population studies are needed to complement range data, as carried out by Hutchings and co-workers for the early spider orchid at a site in southern England (Figure 7.2). Other local population studies are described later in this chapter. Geographical range data are useful when embarking on detailed population studies. If the range of a species appears to be contracting, it may be informative to study a site at the edge of the species range where the factors reducing its range might be expected to be most pronounced. In fact, the choice of site for population studies is often determined by more pragmatic reasons: studies may need to be undertaken over long periods, so site security is important; sites may also be chosen for ease of access, or because the history of the site is well known.

Data such as those in Figure 7.1 have been used in two ways to quantify range (Gaston, 1991a). One is the extent of occurrence in which the outermost points are joined up with no internal angle greater than 180 degrees. The second measure is the area of occupancy, which is equal to the total area of grid squares in which the species occurs. Area of occupancy depends on the size of the grid square chosen, whilst extent of occurrence may be a gross overestimate of the range as it ignores habitat

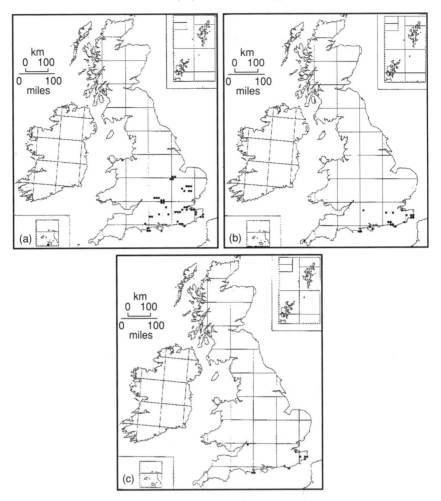

Figure 7.1 Changes in distribution of *Ophrys sphegodes* in the British Isles, recorded within 10 km × 10 km squares over three periods: (a) pre-1930; (b) between 1930 and 1974; (c) since 1975. Mapped by the Biological Records Centre, Institute of Terrestrial Ecology (Hutchings, 1987).

composition within the boundaries. With improving habitat data from remote sensing coupled with ground truthing we are now able, even for tropical species, to make moderately accurate predictions of the extent of potential habitat available. This is a more ecologically useful measure of range which can be produced by overlaying known locations of a species on to distributions of suitable habitat and estimating the amount of habitat available to a species using a GIS package (McGowan *et al.*, to be published).

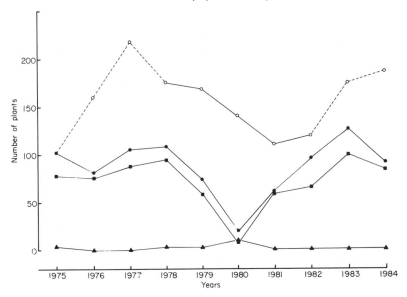

Figure 7.2 Early spider orchid (*Ophrys sphegodes*) observed at a site in Sussex each year from 1975 to 1984. Dotted lines are used for periods when the total population size may have been overestimated. ● = number of plants observed; ■ = number flowering; ▲ = number grazed; ○ = estimated population size. (From Hutchings, 1987.)

Finally, the spatial distribution of localities (occupied grid squares) within the range is also of interest. For 139 'scarce' plant species in Britain – those which occur in 16–50 grid squares of 10 km × 10 km – Quinn *et al.* (1994) have shown that the species have an aggregated distribution. This was related to their poor dispersal ability. One of their conclusions is that reintroduction may be necessary to supplement natural dispersal.

7.3 LOCAL AND METAPOPULATION DYNAMICS

Population size can be defined as the number of individuals of one species living in a particular area. In recent years a distinction has been made between local and metapopulations. A metapopulation is defined as a set of local populations linked by dispersal. In the original metapopulation model of Levins (1969, 1970a) it was assumed that all local populations were of equal size and that a local population could either become extinct or reach carrying capacity instantaneously following colonization. Therefore only two states of local population were envisaged: full (carrying capacity) or empty (extinct). The persistence of a metapopulation depends on the local population colonization rate not falling below the corresponding extinction rate. The relationship between metapopulation dynamics and local population dynamics provides a link between the geographical range analyses in the previous section and detailed local population analyses (below).

In reality the definition of a local population (and therefore a meta-population) is very difficult. Hanski and Gilpin (1991) define a local population as a set of individuals (of the same species) which all interact with each other with a high probability. But how high is that probability? Also 'local' may be different for different interactions; for example, two plants may show intraspecific competition over a scale of a few centimetres but be reproductively linked by pollination over hundreds of metres. It is also very difficult to say over what distance colonization of new areas (and therefore the 'birth' of new local populations) may occur. Typically the frequency of movements of propagules such as seeds over short distances is known but longer distance movement is not, because it is often a rare event. This is unfortunate because, as habitats become increasingly fragmented, long-distance colonization is likely to be central to the persistence of metapopulations.

Even when local populations can be identified, the pure Levins model of local populations with equal carrying capacity is unusual. More realistically, a spectrum of possibilities is envisaged, from mainland–island or core–satellite to pure Levins (Figure 7.3). These and other possibilities have been discussed by Harrison (1991, 1994), who considered the rarity of true Levins metapopulations in the field, and by Hanski and Gyllenberg (1993), who showed how to model both mainland–island and pure Levins with related equations.

Various processes will promote something close to a metapopulation structure for a particular plant species in the field, creating conditions under which local extinction and colonization are integral features of the population dynamics. These processes include:

1. gap creation or other disturbance within non-recruitment habitat, including forests and grasslands;
2. a mosaic of successional habitats – thus for (1) and (2) a pioneer plant species must move from one transient gap or early successional habitat to another;
3. increasing fragmentation of habitats;
4. localized resources such as canopy trees or decaying logs, which may be colonized by epiphytic plant species.

The plant metapopulation model of Carter and Prince (1981, 1988) combined the ideas of Levins with those about infectious diseases in an attempt to explain the geographical range limits of plant species. In particular they challenged the view that range was determined solely by correlation to climate variables – for example, that the northerly distribu-

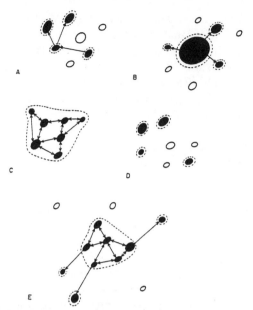

Figure 7.3 Different kinds of metapopulations: (a) Levins; (b) core–satellite (Boorman-Levitt, 1973); (c) 'patchy population'; (d) non-equilibrium (differs from (a) in that there is no recolonization); (e) intermediate case that combines (b) and (c). Closed circles = habitat patches (filled = occupied; unfilled = vacant); dashed lines = boundaries of 'populations'; arrows = migration (colonization). (From Harrison, 1991.)

tion limit of plant species in Europe was determined by physiological intolerance of cold winters (see examples in Carter and Prince, 1988; Beerling, 1993). Carter and Prince used the following equation:

$$\mathrm{d}y/\mathrm{d}t = bxy - cy \qquad (7.1)$$

where x = number of susceptible sites (sites available to be colonized), y = number of infective sites (occupied sites from which seed is produced and dispersed), b = infection rate and c = removal rate.

Essentially, b and c are local population birth (colonization) and death (extinction) rates and therefore equivalent to m and e in the Levins equation ($\mathrm{d}p/\mathrm{d}t = mp(1 - p) - ep$). Similarly x and y are related to $(1 - p)$ and p in the Levins equation, where p = proportion of local populations which are occupied and potentially 'infective' and $(1 - p)$ is the proportion of vacant (and therefore susceptible) sites.

The important conclusion of Carter and Prince was that, along a climatic gradient, a very small change in, say, temperature might tip the balance from metapopulation persistence to metapopulation extinction.

In their own words (Carter and Prince, 1988), 'a climatic factor might lead to distribution limits that are abrupt relative to the gradient in the factor, even though the physiological responses elicited might appear too small to explain such limits'. Thus climate (and physiological) factors are still important but their effects are amplified and made nonlinear by the threshold properties of the metapopulation system.

The predictions of this simple model are supported by the conclusions from a more complex, field-parameterized model. Herben *et al.* (1991) examined the dynamics of the moss *Orthodontium lineare* which occurs on temporary substrates such as rotting wood. The persistence of the metapopulation requires dispersal from one local rotting wood population to an uncolonized log. *Orthodontium lineare* is a native of the southern hemisphere but is now spreading rapidly through western and central Europe. The model of Herben *et al.* included a description of increase on occupied logs until carrying capacity was reached (unlike the Levins/Carter and Prince model, which did not have any local population dynamics other than zero to carrying capacity and vice versa), and the assumptions that dispersal by spores was in proportion to local population size and that spore dispersal distance declined exponentially from an occupied log. The results of this model supported the idea of Carter and Prince (and Levins) that there is a threshold for metapopulation persistence. In this case the percentage of logs occupied was a nonlinear function of probability of local population establishment (Figure 7.4). At a value of p_{est} of about 0.0002 the model predicted a sudden increase in the percentage of occupied sites. In other words there was a threshold value of p_{est} above which metapopulation persistence

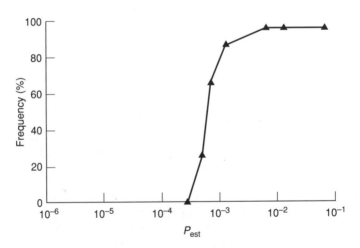

Figure 7.4 Frequency of *Orthodontium lineare* after 100 simulation steps at different values of p_{est}. (From Herben *et al.*, 1991.)

was likely to be high. If p_{est} depends on climate, this would produce exactly the type of sharp break in species range predicted by Carter and Prince.

An understanding of metapopulation processes shows how the extinction and colonization of local populations can tip the balance between persistence or extinction of a species at a larger spatial scale. Detailed local studies, at as many points throughout the range as possible, may therefore be important predictors of the regional fate of a species.

7.4 LOCAL POPULATION DYNAMICS AND EXTINCTION

Assessments of local population abundance are often based on samples of that population. Sampling is necessary because, first, the population may exist over a very large area, so there is the problem of effort and time required. Second, some of the individuals of the population may be very small and/or hidden. The detailed census of the early spider orchid by Hutchings (section 7.2), in which the positions of all plants within a 20 m × 20 m area were recorded so that they could be relocated the following year, showed that individuals may be dormant, remaining underground as a pair of rounded tubers with a few fleshy roots (Figure 7.5). Indeed, even in a population with very large individuals, such as a canopy tree species, there will also be small inconspicuous individuals, such as buried seed or newly emerged seedlings. Hutchings (1991) provides further information and case studies of plant population monitoring for conservation.

Once data are available on local population abundance over time we may proceed to analyse the likelihood of extinction. Extinction may occur due to deterministic (predictable) or stochastic (unpredictable, random)

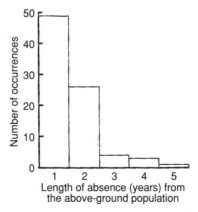

Figure 7.5 Frequency histogram of lengths of dormant periods observed in *Ophrys sphegodes* at study site from 1975 to 1984. (From Hutchings, 1987.)

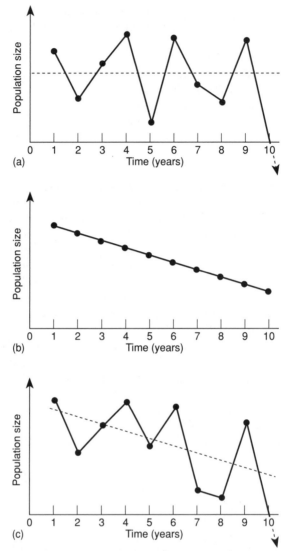

Figure 7.6 Types of population extinction, due to: (a) fluctuations around a constant mean population size (shown by dashed line); (b) declining mean population size; (c) fluctuations around a declining mean population size (shown by dashed line).

processes or a combination of the two (Figure 7.6). The source of the stochasticity is a combination of environmental and demographic factors. The former is external to the population, e.g. unpredictable weather patterns, whilst the latter is internal to the population caused by the unpredictability of seed production or mortality at different stages. The relative

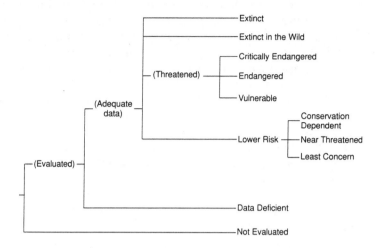

Figure 7.7 Structure of IUCN Red List Categories. (From Species Survival Commission; IUCN, 1994.)

importance of demographic stochasticity increases with decreasing population sizes.

If extinction is stochastic we need to calculate a probability of extinction or an expected (mean) time to extinction, given a certain initial population size. The smaller the initial population size, the higher is the probability of extinction or the faster the time to extinction (other factors being equal). In this chapter the focus is on assessment of stochastic extinction, as assessment of predictable change with time requires a standard application of regression statistics. Application of extinction analyses is now enshrined in the IUCN Red List categories and criteria (IUCN, 1994a; Figure 7.7). These criteria explicitly require analysis of change in abundance and distribution with time, including extent of occurrence and area of occupancy described above. Quantifying one or more of these criteria for a species enables it to be placed in appropriate categories of threat (Figure 7.7).

7.5 ASSESSMENT OF STOCHASTIC COMPONENT OF EXTINCTION

Let us assume that a population is changing in abundance according to the simple density-independent rule:

$$N_{t+1} = \lambda N_t \tag{7.2}$$

where N_{t+1} and N_t are the population sizes, densities or biomasses at times $t+1$ and t, respectively. Under deterministic conditions the value of λ describes the fate of populations. When λ = 1, the population does not change in size; when λ > 1, the population increases geometrically; when λ < 1, the population declines asymptotically towards extinction. Ignoring density dependence is clearly unrealistic: populations do not continue to increase *ad infinitum*; survival, fecundity and migration are affected by population density. However, for the analysis of stochastic extinction which follows, density dependence is unlikely to be important ; or, more truthfully, we do not know enough about the strength of density dependence in natural populations to know how important it is.

N can be interpreted either as all individuals in a population, or as the numbers of different stages of the plant (seed, seedling, etc. represented as a column vector of stages) or, as in most plant conservation applications, only the numbers of a particular stage, usually the number of flowering herbaceous individuals or trees above a particular size. The individuals within the latter broad category may have wide variation in mortality and fecundity, but the lumping is convenient for several reasons: the IUCN criteria refer to mature individuals; effective population size requires data on reproductives; and juveniles may be difficult to detect and/or identify, as noted above. If N does represent more than one stage and therefore (7.2) is a matrix equation, then λ is equivalent to the dominant eigenvalue of the population projection matrix which contains all the stage-dependent fecundity and survival terms (Caswell, 1989).

If λ varies from year to year we will get a population which fluctuates in abundance. A series of low values of λ could result in a population going locally extinct. Because N_t is multiplied by λ each year, we take the natural logs (ln) of (7.2) to give:

$$\ln N_{t+1} = \ln \lambda + \ln N_t \qquad (7.3)$$

in which ln λ (= r) is usually normally distributed and its mean and variance (v_r) can be found by standard statistical analysis of $\ln (N_{t+1}/N_t)$. The details of assessment of probability of extinction and the validity of this approach are reviewed in Foley (1994) for discrete models and Goodman (1987), who provides details of related continuous time models. Foley (1994) applied the above model and more complex ones to various animal species, calculating expected time to extinction (T_e) as a function of v_r, N_0 (the initial population size) and k (the maximum population size). N_0 and k are expressed as natural logs. For a population with a mean λ of 1 (r of 0), i.e. a population showing no overall increase or decrease over time, expected time to extinction is (Foley, 1994):

$$T_e = (2N_0/v_r)(k - N_0/2) \qquad (7.4)$$

In this model and the continuous time model (Goodman, 1987), k is not an average maximum value but a strict upper boundary with the population assumed to be proceeding on a random walk from N_0 bounded by k and 0 (extinction). If $N_0 = k$, then (7.4) simplifies to:

$$T_e = k^2/v_r \qquad (7.5)$$

It would be unwise to interpret these estimates of T_e too strictly. They are based on a very simple model with a series of assumptions. Perhaps their most useful function is to provide an estimate of the relative likelihood of extinction, as a contribution to the IUCN Red List process, allowing species to be prioritized for conservation.

The new IUCN criteria use the following probability of extinction levels:

- Critically endangered: 0.5 (50%) within 10 years or 3 generations, whichever is longer.
- Endangered: 0.2 within 20 years or 5 generations, whichever is longer.
- Vulnerable: 0.1 within 100 years.

To use the IUCN criteria with (7.4) or (7.5) we need to convert from probability of extinction to time to extinction. This can be done (following Foley, 1994) if extinction is viewed as a Poisson process and so T_e is approximated as:

$$T_e = -t/\ln(1 - P) \qquad (7.6)$$

where P is the probability of extinction and t is the number of years over which extinction is measured. If average generation time is longer than 3 to 4 years, T_e needs to be increased because t is effectively increased. In terms of T_e the IUCN thresholds for short-generation species (< 4 years) are therefore:

- Critically endangered: 14.4 years
- Endangered: 89.6 years
- Vulnerable: 949.1 years.

This gives approximately one order of magnitude difference in T_e between Critical and Endangered and between Endangered and Vulnerable.

To illustrate the calculation of T_e, consider a set of 23 year censuses of the green-winged orchid *Orchis morio* in Lincolnshire, England (Silvertown *et al.*, 1994). *Orchis morio* is found in lowland grasslands in Britain and throughout mainland Europe. Its abundance has declined markedly in the past 30 years, having once been one of the commonest British orchids (Summerhayes, 1968).

7.6 ESTIMATION OF TIME TO EXTINCTION FOR AN *ORCHIS MORIO* POPULATION

An *Orchis morio* population in Bratoft Meadow, Lincolnshire, has been subjected to eight fertilizer treatments (Table 7.1) in a blocked design with eight replicates. This unique experiment was set up in 1970 by Derek Wells and colleagues at the (then) Nature Conservancy Council and the Lincolnshire Trust for Nature Conservation (details in Silvertown *et al.*, 1994). The number of flowering plants has been counted annually in each of the 64 plots of 1.83 m × 9.14 m since 1970 (continuing in 1995). The dynamics of the population in one unfertilized plot, all eight unfertilized plots and all 64 plots are shown in Figure 7.8.

Using (7.4) and (7.5) the T_e for each unfertilized *Orchis morio* plot was calculated (Table 7.2). Even though these are samples from the same population, the eight plots differed by an order of magnitude in pre-

Table 7.1 Fertilizer treatments applied to experimental plots, expressed as a percentage of the same nutrients removed in the hay crop that was sampled in the two years before the start of fertilizer application

Treatment	N	P	Mg	K
1	50% (organic)	50% (organic)	50%	40%
2	100% (organic)	100% (organic)	100%	80%
3	200% (organic)	200% (organic)	200%	160%
4	100%	100%	100%	80%
5	0	1000%	0	0
6	0	0	0	123%
7	0	0	1000%	0
8	0	0	0	0

Treatments 1–4 and 6 were applied annually 1972–1977, Treatment 5 was applied only in 1972 and 1975. Nutrients were applied in inorganic form, unless otherwise indicated.

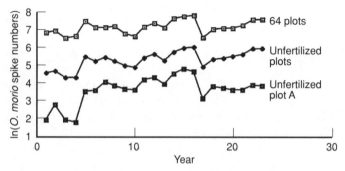

Figure 7.8 Flowering spike dynamics for *Orchis morio*.

Table 7.2 Estimates of variance in r (v_r), ln(maximum population size) (k) and time to extinction,(T_e, from k and from $N_0 = 4$) for *Orchis morio* in eight unfertilized plots (A–H), in plots A–H together and all plots in experiment

Plot	v_r	k	T_e (k)	T_e $(ln(N_0) = 4)$
A	0.398	4.718	55.9	54.6
B	0.485	4.394	39.8	39.5
C	0.667	4.060	24.7	24.7
D	0.366	4.060	45.0	45.0
E	0.752	3.738	18.6	18.5
F	0.284	4.564	73.3	72.2
G	0.483	1.609	5.36	$-N_0 > k$
H	0.614	3.367	18.51	17.8
Plots A–H	0.177	5.90	197	
All 64 plots	0.186	7.688	318	

dicted T_e (5.36 to 73.3 with $N_0 = k$). Increasing the sample size to the eight unfertilized plots combined and then all 64 plots increased the T_e because of the larger k, although v_r was very similar between the eight and 64 plots. Thus the spatial scale of sampling will affect the estimate of T_e and therefore potentially the IUCN category in which a species is placed. This and the effect of census duration is discussed in Gillman and Silvertown (in press).

Analysis of the effects of treatments on time to extinction are discussed in Gillman and Dodd (in press); effects of treatments on population density are dealt with in Silvertown *et al.* (1994). These analyses show that the fertilizer treatments, especially the 1000% phosphorus (treatment 5 in Table 7.1), produced a reduction in k and mean density, respectively. The variance in r was also much higher in the 1000% P treatment but the overall effect was not significant. The decrease in k and increase in v_r leads to a decrease in T_e for the 1000% P treatment, but again the overall treatment effect is not significant (Gillman and Dodd, in press).

Application of the IUCN criteria needs to be made for the species, not just local populations. Nevertheless, we have already seen that local population extinction may have metapopulation consequences. Local populations (occupied grid squares) of *Orchis morio* may be widely separated in space, as in Kent, England (Figure 7.9). This is in contrast to the general rule of aggregated distribution of occupied grid squares found by Quinn *et al.* (1994) although the clumps of occupied grid squares may themselves be widely separated in space. For a metapopulation to remain extant, colonizations need to balance extinctions. Wide separation of local populations is not necessarily a problem if suitable habitat

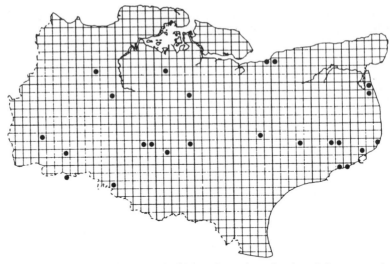

Figure 7.9 Distribution of native *Orchis morio* on downland and damp meadows in Kent, England. There has been a serious decline in the numbers of this plant in recent years, and whilst some localities have been destroyed by land drainage or by being ploughed up, it has also disappeared from other localities where the management has not changed. The reason for this decline is far from clear. Now very local and scarce. (From Philp, 1982.)

remains for colonization. Unfortunately we know this is not the case for this species and for many others (Fuller, 1987). In this light we may need to view many plant species as a set of separate local populations with the time to extinction of the species equal only to the time to extinction of the longest existing local population which will have a high N_0, high k, low v_r and a mean $\lambda > 1$.

7.7 LANDSCAPE PLANT DYNAMICS

Having begun with the wider picture of geographical range and linked this via metapopulation dynamics to local population dynamics, the chapter concludes by taking a step back again to view plant dynamics on a landscape scale. The metapopulation approach is a short, theoretical move in the direction of assessing local population dynamics in the context of real landscapes. Real landscape structure is obviously much more complex than can be accommodated with a Levins model, with species covering the full range of metapopulation structures discussed by Harrison (1991, 1994) (Figure 7.3). In addition there are a variety of possible vegetation types within a given landscape unit. One way of coping with the latter complexity has been to represent landscape change as a

Table 7.3 The percentage of vegetation type from aerial photographs in 1947 and 1989 in central coastal California (Callaway and Davis, 1993)

Year	Grassland	Coastal sage scrub	Chaparral	Oak woodland
1947	21.5	26.4	28	24.1
1989	23.3	25.9	24	26.8

Markovian replacement process (Horn, 1975, 1981). This involves determining the probability that a given plant species, or suite of species, will be replaced in a specified time by the same or different species within a landscape unit. It is assumed that these replacement probabilities do not change with time.

Callaway and Davis (1993) used aerial photographs to measure transition rates between grassland, coastal sage scrub, chaparral and oak woodland and their relationship to burning and grazing in central coastal California between 1947 and 1989. The percentage of vegetation types in 1947 and 1989 are given in Table 7.3, based on plots marked on aerial photographs. These plots corresponded to 0.25 ha on the ground.

Although the overall percentage cover was very similar there was considerable flux between the years within plots. Transition between vegetation type occurred in 71 out of 220 plots (32%). The transition probabilities were determined using these data (Figure 7.10).

Once the transition probabilities were known the current state (e.g. oak woodland) could be multiplied by the four transitions (including no change) in the 42-year period. This was then repeated in a Markov chain over time to predict the change in vegetation under particular environmental conditions. The predictions for three combinations of burning and grazing are shown in Figure 7.11. The greatest changes in vegetation were predicted to occur under conditions of no grazing and no fire.

7.8 CONCLUSIONS

The methods and results reported here can be applied across the plant (and indeed animal) kingdom in all terrestrial habitats. The likelihood of extinction of local populations may have regional, metapopulation consequences. To quantify the chance of extinction we require census data from local populations. Ideally the location of local population studies will be informed by geographical range data. In quantifying extinction it may be that assessments of reproductive individuals are sufficient to prioritize species for conservation action. We may therefore avoid time-consuming methods of identifying individual plants such as described for the early spider orchid and the construction of age/size/stage-

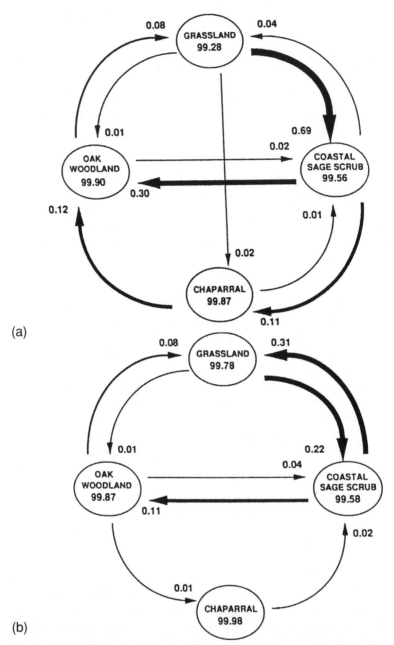

(a)

(b)

Figure 7.10 Annual transition rates among plant communities in unburned and burned plots, as determined from changes in vegetation between 1947 and 1989 shown on aerial photographs. (a) Unburned plots ($n = 78$): numbers in ovals estimate the probability, as a percentage, that a given community will remain the same; numbers on arrows estimate the probability that a community will change in the indicated direction (thickness of lines is proportional to the probability of that change). (b) Burned plots ($n = 53$). (From Callaway and Davis, 1993.)

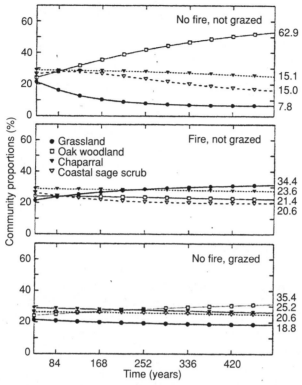

Figure 7.11 Markov chain model predictions of future change in proportions of plant communities. Final community proportions at stability (defined as < 0.1% change over 42 years) are presented at the right, opposite each curve. (From Callaway and Davis, 1993.)

structured models. If this cannot be avoided, and other assumptions such as density independence prove to be unreasonable, then application of population ecology to *in situ* conservation will prove to be too unwieldy and complex and ultimately too slow to cope with the accelerating pace of conservation needs. If the analytical and methodological problems can be overcome then population ecology will (and must) have an important role to play, once the robust elements of predictive extinction models have been identified.

ACKNOWLEDGEMENTS

This chapter has benefited from discussions with Mike Dodd, Phil McGowan, Nigel Maxted and Jonathan Silvertown.

8

Reserve design

J.G. Hawkes, N. Maxted and D. Zohary

8.1 INTRODUCTION

In recent years a considerable literature has grown, related to reserve design (Margules *et al.*, 1982; Margules and Nicholls, 1988; Shafer, 1990; Spellerberg, 1991a; Cox, 1993; Primack, 1993, 1995; Given, 1994; Meffe and Carroll, 1994). Most of it deals with the conservation of habitats for endangered wild animal species, and particularly for mammals and birds at or near the top of the food chain (Shafer, 1990). Much less has been written concerning plant species and even less specifically about *in situ* plant genetic resources conservation. This point having been made, the existing principles of reserve design can be equally well applied to plant genetic resources conservation.

Much discussion in the past has centred on two main controversial points. Firstly, what should be the effective size of a reserve in order to ensure the continued presence of viable population(s) of the target species with zero or minimal genetic erosion? Secondly, is there a greater advantage in establishing a number of small reserves, linked by immigration corridors on the one hand as distinct from single large reserves? These partly concern the geographical distribution of the species concerned. If they are naturally confined to one single area, is it due to extinction in other regions or because they are naturally adapted to that specific area? If they are widespread, a number of reserves in different regions would be required to cover as much as possible of the genetic diversity throughout their range. In this respect data on soil, climate, human activities such as cattle grazing, cutting and other features, natural predators and ecosystems together with the positive and negative

Plant Genetic Conservation.
Edited by N. Maxted, B.V. Ford-Lloyd and J.G. Hawkes.
Published in 1997 by Chapman & Hall. ISBN 0 412 63400 7 (Hb) and 0 412 63730 8 (Pb).

effects of commensural species must be taken into account.

The vast majority of wild relatives of crop plants are found in diverse habitats and geographical locations. In such a case a single reserve (or a single cluster of sites) would not suffice. A number of reserves, located in different segments of the distributional range and in different habitat types, would be required to cover its ecogeographic divergence and to deal adequately with the genetic changes which occur over its range.

The number of reserves needed, and where geographically it is best to locate them, could be decided only after an ecogeographic survey of the intraspecific differentiation in the target species, the range of habitats occupied by it, and the changes in polymorphic gene markers in its populations over its distribution and ecological ranges (also see IBPGR, 1985b).

There are three basic factors that require consideration when designing a reserve (Meffe and Carroll, 1994), of which only one is fundamentally scientific in nature. The first relates to the location, size, shape, etc. of the reserve itself. The second factor concerns the reserve's anthropological and cultural effects; where possible the establishment of a reserve should not disrupt the traditional, sustainable cultures of local indigenous people – the reserve should work with rather than against local peoples. The third factor is that the reserve will need to work within political and economic constraints at a local and global level. There may therefore be a requirement to trade-off scientifically based conservation against economic and political considerations when locating and designing a reserve. The existence of these three factors, each affecting reserve design, underlines an important pragmatic point: that the biological ideal often has to be balanced against what is expedient for society as a whole.

Within the biological realm, the planning of the conservation of plant genetic diversity in a reserve involves the marriage of several related disciplines: ecogeography (Chapter 4), political and economic factors (Chapter 5), population genetics (Chapter 6) and population ecology (Chapter 7), along with other practical aspects of reserve design. This chapter specifically addresses the latter, dealing primarily with the wild relatives of crop plants.

8.2 OPTIMAL RESERVE DESIGN

Although the theory related to the design of reserves has been largely developed by animal conservationists rather than botanists or plant geneticists, the basic principles do not vary a great deal. The current consensus view of an optimal reserve design is that based on the Man and the Biosphere programme (UNESCO), as discussed by Cox (1993, modified from Batisse, 1986). This establishes a central core area with a stable habitat, surrounded by a buffer zone and outside this, where possible, a

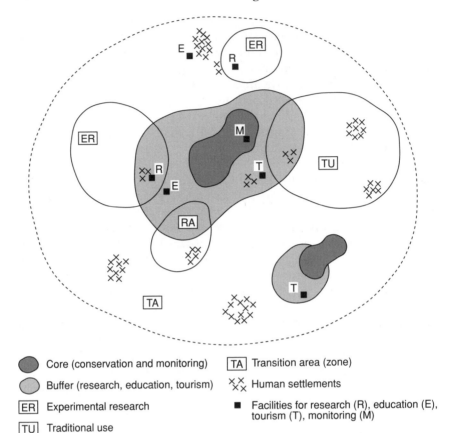

⬤ Core (conservation and monitoring)	TA Transition area (zone)
◯ Buffer (research, education, tourism)	×× Human settlements
ER Experimental research	■ Facilities for research (R), education (E), tourism (T), monitoring (M)
TU Traditional use	
RA Rehabilitation	

Figure 8.1 Model for reserve design (from Cox, 1993, modified from Batisse, 1986) including core, buffer and transition zones. The **core**, typically a protected area such as a national park, is designed to provide for conservation of biotic diversity, with research being limited to monitoring activities. The **buffer zone** is a region in which research, tourism, educational activities and traditional subsistence activities of indigenous peoples are emphasized. The **transition zone**, its outer limit not rigidly defined, constitutes an area where major manipulative research, ecosystem restoration and application of research to management of exploited ecosystems are carried out.

transition zone (Figure 8.1).

This plan assumes that the core area will be large enough to accumulate 1000–5000 potentially breeding individuals (MVP, minimum viable population) already growing as part of the natural ecosystem. This zone will be used for regular monitoring and will be open only to permit-holding scientists and officials. The zone outside this, the buffer zone, protects the core from edge effects and other factors that might threaten the viability of the target populations in the core. Given (1993) distinguishes two types of buffer zone. Extension buffering allows effective extension of the core area, thus permitting interbreeding populations that are much larger than theoretically required to survive. Extension buffering has a similar management regime to the reserve core and is open to *bona fide* non-destructive researchers, educationists and special eco-tourists. In contrast, socio-buffering involves the separate management of the core and the buffer, because some sustainable agriculture and forestry is permitted within the buffer zone. It may be advisable to establish a socio-buffer zone if the local people have lost traditional harvesting rights in the reserve core (section 8.9.1). However, it must be stressed that the activities of local people in the buffer zone must be closely monitored to ensure that they do not threaten the populations in the reserve core. The benefits of using buffer zones were summarized by Given (1993), modified from Oldfield (1988), and are given in Table 8.1.

Outside the buffer zone, the transition zone will be available for

Table 8.1 Biological and social benefits of using buffer zones (reproduced with permission from Given, 1993)

Biological benefits	Social benefits
Provides extra protection from human activities for the strictly protected core zone	Gives local people access to traditionally utilized species without depleting the core area
Protects the core of the reserve from biological change	Compensates people for loss of access to the strictly protected core zone
Provides extra protection from storm damage	Permits local people to participate in conservation of the protected area
Provides a large forest or other habitat unit for conservation, with less species loss through edge effects	Makes more land available for education, recreation, and tourism, which in turn helps in conservation advocacy
Extends habitat and thus population size for species requiring more space	Permits conservation of plants and animals to become part of local and regional rural development planning
Allows for a more natural boundary – one relating to movements of animal species that may be essential to some plants	Safeguards traditional land rights and conservation practices of local people
Provides a replenishment zone for core area species including animals essential to some plants	Increases conservation-related employment

research on ecosystem restoration and similar studies, limited human settlements and sustainable utilization, as well as general tourist visits. This optimal reserve design attempts to involve rather than exclude and alienate local people, while providing a practical buffer and transition zone between the core reserve and areas of intense human activity. It also provides a practical solution to the problem of the boundary between areas of conservation reserves and human exploitation. Wherever possible additional land should be acquired adjacent to the reserve to increase the overall reserve size, and also to act as additional buffer and transitional zones.

8.3 RESERVE SIZE

In practice, the size of a reserve is often dictated by the relative concentration of people and the suitability of the land for human exploitation (agriculture, urbanization, logging, etc.). Primack (1993) provides two examples of large reserves that fall into this category: the Greenland National Park, which is composed of a frozen land mass of 700 000 km^2, and the Bako National Park in Malaysia, which is set on nutrient-deficient soils. In contrast, lands with economic value for exploitation have few and small reserves. Given (1993) also illustrates this point by listing the 15 largest reserves found in the United States; all are found in agriculturally marginal areas. Although there may be a correlation between marginal lands and the land that governments are willing to allow become reserves, there is unlikely to be a natural correlation between marginal agricultural land and the distribution of species worthy of conservation.

Given the ideal situation where reserve sites are selected on the basis of scientific principles, the ideal size for a reserve has proved a continuing point of discussion (Soulé, 1987). Size is commonly related to theories of island biogeography (MacArthur and Wilson, 1967; Shafer, 1990; Spellerberg, 1991a; Cox, 1993) and relative rates of colonization and extinction per unit area. The debate is often centred on the relative advantage of single large versus several small reserves, the so-called SLOSS debate (Single Large Or Several Small). For example is it better to have one large reserve of 15 000 ha or a network of five each of 3000 ha? Large reserves obviously enable a more ecogeographically diverse environment to be included within a single location with minimal edge effect. Alternatively, if a network of smaller reserves is established, each reserve could be sited in a distinct environment, which would better enable the conservation of extreme ecotypes. So the conservation value of multiple small reserves can be greater than the sum of its individual components, especially if the reserves are closely juxtaposed or connected (section 8.5). However, as discussed above, if reserves are too small or too isolated the populations of the target taxon they contain will become inviable.

The current consensus is that the optimal number and size of reserves will depend on the characteristics of the target species (Soulé and Simberloff, 1986). Large reserves are better able to maintain species and population diversity because of their greater species and population numbers and internal range of habitats (Abele and Conner, 1979). They allow the physical integrity of the environment to be maintained (e.g. watersheds, drainage system) and are especially suited for low-density species, such as many forest trees, which are commonly found in disjoint populations. However, small multiple reserves may be more appropriate for annual plant species, which are naturally found in dense but restricted stands (Lesica and Allendorf, 1992). Practically, the vast majority of the wild relatives of crop plants are much more widely distributed. In such cases a single reserve (or a single cluster of sites) would not suffice. A number of reserves, located in different segments of the distribution area of the target species, would be required to cover its ecogeographic divergence and to deal adequately with the genetic changes which occur over its geographical range.

Within reason, the larger the reserve size the more diverse are the habitat variants included, which among other things should help to ameliorate short-term climatic fluctuations such as drought, heavy rainfall, fungal and insect attack and other variables (Chapter 15; Chapter 19; Dinoor *et al.*, 1991; Namkoong, 1991; Nevo *et al.*, 1991; Noy-Meir *et al.*, 1991a,b). The reserve should be sufficiently large to permit genetic divergence between populations in space and also in time (Brown, 1991) and promote gene flow where necessary (Golenberg, 1991).

It should be noted that the same level of deleterious human activity could decimate a small reserve, while a larger reserve may still remain viable. Smaller reserves will generally require more intensive management and monitoring to maintain the same population levels and diversity because of their inherently artificial nature. If multiple small reserves are established, each should have sufficient buffering to protect the core from catastrophes. Small reserves, which are often associated with urban areas, can serve an additional and vital purpose as an educational and nature study centre that will increase public awareness of the importance of conservation.

Having made these points, we are convinced that frequently the wrong question is being addressed.

8.4 POPULATION AND RESERVE SIZE

It is more objective and appropriate to target the ideal numbers of individuals that form a viable population, the effective population size (EPS), that will ensure the effective conservation of genetic diversity in the target species for an indefinite period. Shafer (1990) maintains that the minimum population size (MPS) for any given habitat is defined as the smallest population having a 99% chance of remaining extant for 100 years. This sort of decision is difficult to make with any degree of accuracy, of course, though Shafer mentions foreseeable events such as breeding success, predation, competition, disease, genetic drift, interbreeding, inbreeding, founder effect and natural catastrophes such as fire, drought and flooding that may effect population longevity. This is a pessimistic view, but if a reserve is established where natural populations of the target species already exist, this will probably eliminate at least some of the hazards mentioned and will continue the process of evolutionary changes related to pathogen diversity (Browning, 1991).

It is generally agreed that a severe reduction in population size will result in loss of alleles with possible negative effects on survival (Shafer, 1990). There are examples known of populations diminishing to a very small size and yet surviving and expanding later, but these 'phoenix' populations will undoubtedly have suffered extensive genetic drift during the process (Chapter 6). Precise estimates of the minimum viable population (MVP) size vary. Frankel and Soulé (1981) mention numbers of individual plants from 500 to 2000 . Hawkes (1991a) recommended a MVP of at least 1000 individuals, taking into account that not every individual may contribute genes to the next generation. There is an extensive discussion of minimum population size in Chapter 6, where Lawrence and Marshall tentatively conclude that a MVP of 5000 is appropriate in terms of *in situ* genetic conservation.

When considering the theoretical minimal size of a protected population, however, one should keep in mind, firstly, that catastrophes – although rare (or even very rare) – still do happen; and secondly, that different plant species may have different needs in terms of population sizes. Catastrophes such as continuous severe drought, the appearance of a destructive new pest, hurricane, fire, etc. can reduce the population in a given reserve to just a very small fraction of its original size. Also the life-form and the breeding system of the plant involved needs to be considered. In Mediterranean environments, climatic differences between years can be very wide and these factors strongly affect population sizes. Annuals seem to be much more vulnerable to such changes compared with perennial species. Two or three years of continuous, severe drought can reduce annual populations to only a small percentage of their average size. Native perennial bushes and trees show much smaller fluctuations. The risks for small annual populations (comprising only a few thousand individuals) is even greater when this life-form is coupled with self-pollination, which is the case in the majority of the wild relatives of Old World grain crops such as wheats, barley, peas or chickpea. In such species, where genetic variation is structured in true breeding lines, severe decimation in small populations of selfers is bound to result in loss of genetic diversity, particularly of the less frequent genotypes. As only a limited number of individuals can exist in a small reserve area, the genetic flexibility (the formation of new lines) will also be impaired. For these species larger population numbers and therefore reserve sizes will be required, as will be the case for species that are associated with disturbed rather than climax habitats.

All in all, when planning a reserve a minimal population size should be regarded only as a last resort and an extreme compromise (Chapter 6). For added safety, much larger populations (up to 5000 or even 10 000 are desirable) should constitute units of *in situ* genetic conservation. Frequently, this is also a spatial necessity. When considering habitat diversity (section 8.7) it is essential to include in the reserve a reasonable repetition of the ecological diversity found in the area. Very frequently it is impossible to achieve it in a small space. The unit area actually needed is usually much larger than that supporting (theoretically) only a few thousand individuals.

8.5 CORRIDORS

If multiple small reserves are selected, their potential conservation value can be enhanced by producing a coordinated management plan that attempts to facilitate gene flow and migration between the component reserves. In this way individual populations can be effectively managed at a metapopulation level. Gene flow may be further advanced by the use of habitat corridors linking individual reserves. These ideas were initially developed for animal conservation, but work equally well for plant species, especially those that rely on animal-based seed dispersal. In terms of animal conservation, the larger the animal the wider is the corridor required. So if a plant species uses a large animal for seed dispersal then the corridor will need to be of an appropriate width to permit migration of the animal dispersal agent. Habitat corridors do have some conservation drawbacks, because, by definition, they have a high edge-to-area ratio (section 8.6) and they also facilitate the rapid distribution of pests and disease between the reserves.

8.6 RESERVE SHAPE

It is important to consider shape when designing a genetic reserve. Ideally the edge should be kept to a minimum to avoid deleterious micro-environmental effects including changes in light, temperature, wind, the incidence of fire, introduction of alien species, and grazing, as well as deleterious anthropic effects. Therefore ideally the edges of the reserve must be kept to a minimum. A round reserve will have the minimum edge-to-area ratio; long linear reserves will have the highest edge-to-area ratio and should be avoided if possible. Fragmentation of the reserve by roads, fences, pipelines, dams, agriculture, intensive forestry and other human activities will necessarily fragment and limit the effective reserve size, multiply the edge effects and perhaps leave populations in each fragment unsustainable.

8.7 HABITAT DIVERSITY

When selecting sites for establishing a genetic reserve, sites with spatial or temporal heterogeneity should be given priority over homogeneous areas: the wider the range of habitat diversity and juxtaposition of different habitats in a site to be considered for a reserve, the better. This will ensure that the target species will preserve the various genes and genetic combinations associated with ecotypic differentiation, leading to more effective conservation of genetic diversity. Should the target species be restricted to a few soil types, certain levels of soil humidity, or certain natural or semi-natural habitats, then these must obviously be taken into account when the reserve is being planned and appropriate areas or habitats included (Yahner, 1988). This has led to theories of landscape ecology (Forman, 1987; Urban *et al.*, 1987; Hansson *et al.*, 1995).

Genetic reserves are by definition artificial in a biological sense. When a particular site is selected and a management regime instigated, the habitat may lose diversity; therefore the management regime should include habitat disturbance, which results in the desired patchwork of diverse habitat types. Natural causes of disturbance include fires, storm damage, pest and disease epidemics, herbivory, floods and droughts. All of these factors are non-uniform, in terms of coverage, and create habitat patches of an earlier successional stage, which will in turn promote species and genetic diversity.

Pickett and Thompson (1978) discuss the selection of reserve sites in conjunction with habitat disturbance and heterogeneity. They refer to minimum dynamic area, which is the smallest area with a complete, natural disturbance regime. This area would maintain internal recolonization to balance natural extinctions. As the majority of species are not exclusive to one habitat, the maintenance of reserve heterogeneity will promote the health (genetic diversity) of the full gene pool as represented in multiple populations or metapopulations.

The need for continued habitat heterogeneity is a factor that will need to be considered when formulating the reserve management plan. If fire, for example, was a natural causal agent of habitat disturbance and heterogeneity which promoted the target taxa, then the reserve design would have to permit continued use of fire, under the instigation of the reserve manager. If a reserve site is designated that includes human habitation, then regular fires may be undesirable or even dangerous. Thus the ultimate management regime will often affect the reserve design.

8.8 POLITICAL AND ECONOMIC FACTORS AFFECTING RESERVE DESIGN

The location and design of any reserve is rarely decided solely on the basis of biological expedience; political and economic factors will play a part (Chapter 5). Reserves, for economic reasons, will often be sited on publicly owned land and, as such, there may be conflicting or multiple uses of the same land. This will often mean that the idealized reserve design discussed above is pragmatically applied to allow complementary use as an agricultural, industrial or recreational resource. During the planning phase of the reserve, consultations should also be held with regional and national governments, thus ensuring that establishment and management agreements are in place before the reserve is functional.

As discussed in Chapter 3, the relative cost and ease of establishing reserves will affect the selection of reserve sites and their design. Faced with choosing between two equally appropriate sites and with a limited conservation budget, the relative establishment costs (cost of site purchase, current usage, displacement of local people, etc.) for the two reserves would be a factor affecting their selection and design. Following the same logic, if a reserve can be designed in such a way that multiple target taxa can be conserved in the same reserve, economics will dictate that this is a superior overall design.

8.9 COOPERATION WITH OTHER ORGANIZATIONS, COMMUNITIES AND INDIVIDUALS

Four distinct categories of people may use the reserve: the local indigenous population, the general public, reserve visitors and the scientific community. If the reserve is not to be unnecessarily limited, the use each group makes of the reserve must be considered when designing and managing the reserve.

8.9.1 Local indigenous population

Few reserves will be established without taking account of local communities, local farmers, land-owners and other members of the local population who may use the proposed reserve site. In areas where the reserve would be sited in a region of woodland, desert scrub and bushland which is used by local communities but not actually owned by them, detailed consultations and agreements must also be undertaken with them. These preliminary measures should apply also to hunter–gatherer community areas or regions of shifting cultivations and wild-plant 'harvesting'. In all such regions or areas a sympathetic and cooperative attitude should be adopted to ensure that the local communities are fully aware of the importance of the proposed reserve(s). Ideally reserve staff should be recruited locally and the whole community should be encouraged to take pride in local conservation work. Meffe and Carroll (1994) discuss in detail the rehabilitation of the Guanacaste National Park in Costa Rica and refer to the project being based on a philosophical approach called biocultural restoration, which incorporates local people in all aspects of the reserve's development and protection. The project has not only resulted in the restoration of degraded habitats, but has also restored to the local people a biological and intellectual understanding of the environment in which they live and has achieved the goal of creating a 'user-friendly' reserve that contributes to the quality of life of the local people.

8.9.2 General public

The second user group is the population at large, whether local, national or international, and its support may be essential to the long-term political and financial viability of the reserve. As discussed in Chapter 3, ethical and aesthetic justifications for species conservation are of increasing importance to professional conservationists. The growth of public awareness of conservation issues has undoubtedly stimulated a growth in conservation activities in recent years. If we as conservationists are to retain public support for our work we should ensure that the public is kept adequately informed of our activities, and where possible encouraged to visit the reserve.

8.9.3 Reserve visitors

The third user group is related to the second. Reserve visitors are members of the general public who wish to visit the actual reserve. If local people who traditionally used the reserve site are opposed to the reserve, they may be converted to supporters if they can see direct financial benefits to their community resulting from eco-tourist, school and other visitors to the reserve. The reserve design must take into account the needs of visitors, such as visitors centres, nature trails, lectures, etc. They are also likely to bring additional income to the reserve through guided tours and the sale of various media reserve information packs.

8.9.4 Scientific community

Along with the necessary monitoring and mapping of all species and habitats in the reserve that will form part of the management plan, the reserve will provide a research platform. One of the assumed values of *in situ* conservation is that the target species will be slowly evolving with general environmental changes and in particular with the constantly changing biotypes of pests and diseases. This will not happen with material which remains genetically static under *ex situ* regimes, apart from when seeds are regenerated in field plots or under glass, or when the explants are converted now and again from their culture tubes into whole plants, in contact with the outside environment.

No one can say how far the concept of plant pathogen evolution is justified, since evolutionary changes are slow, and perhaps no obvious changes may be observable within 50 to 100 years or more. On the other hand, this may not be so: from present experience with the malaria organism, for instance, changes can be remarkably rapid. It is therefore wise to assume that rapid changes in fungal, insect and other pathogens may well take place. For this reason, if for no other, *in situ* reserves of wild crop relatives will be needed to provide resistance genes able to combat the changes in virulence of the pests and pathogens as they occur.

The reserve may also permit some characterization and initial evaluation of the target taxon, leading towards a detection of desirable genetic traits (Qualset and McGuire, 1991). It would be valuable, also, to promote visits from national and international experts as and when possible, to encourage interest and gather suggestions as to management regimes and research initiatives. This is particularly relevant with wild crop relatives, which could be sampled for screening *ex situ* for resistance to disease or pest pathotypes that have appeared in the crops to which the *in situ* wild species are related (Dinoor and Eshed, 1991; Felsenberg *et al.*, 1991).

9

Management and monitoring

N. Maxted, L. Guarino and M.E. Dulloo

9.1 INTRODUCTION

It is usually possible to define, at least in rough terms, the sorts of places where a particular plant species might be expected to occur: the rare gymnosperm *Welwitschia mirabilis* is only found in the mists of the Namib desert of south-western Africa, or willow trees (*Salix* species) along temperate waterways, for example. This is because each species can only successfully grow and reproduce within a certain, more or less limited, range of conditions, which define its so-called niche. *In situ* conservation in a genetic reserve requires that the quality and/or quantity of genetic diversity within a species at a site be maintained in the long term. For that to be possible, it is clearly necessary that the niche and adaptation of the species be understood as fully as possible, so that the conditions necessary for its success at the site can be maintained, by appropriate human intervention – 'management' – if necessary.

No species exists in isolation. Each forms part of a community. The constraints that define a niche are not just environmental (climate, soil, fire, etc.) but also biotic (pollinators, dispersal agents, competitors, herbivores, pests, pathogens, symbionts, etc.). Sutherland (1995) stresses that it may only be by maintaining the integrity of the community as a whole that it is likely that the complex and often largely unknown demands of specific target species can be met. The management of a reserve may need to be directed not just at the particular target species, but at maintaining as a whole the community present at the site.

A management plan for the genetic reserve must be developed that regulates human intervention in such a way as to ensure that genetic

Plant Genetic Conservation.
Edited by N. Maxted, B.V. Ford-Lloyd and J.G. Hawkes.
Published in 1997 by Chapman & Hall. ISBN 0 412 63400 7 (Hb) and 0 412 63730 8 (Pb).

diversity within the populations of the target species is maintained or enhanced. Having developed a management plan and implemented it, the conservationist will need to monitor the reserve – i.e. study the reserve over time – to ensure that the management interventions remain appropriate. It may be argued that if a site already has a significant population of the target species, there is no need to expend scarce resources in formulating and implementing a plan for its management. However, conditions may change, as a result of either internal or external factors. Indeed, the very act of designating an area as a reserve may cause undesirable changes – for example, if its 'protection' results in the exclusion of herbivores which would normally control competitors of the target species. A management plan for the reserve must be formulated at an early stage and population(s) of the target species (and possibly others) within the reserve followed over time to ensure that the plan remains appropriate.

9.2 CONSERVATION OBJECTIVES OF A GENETIC RESERVE

It is important that a genetic reserve has clearly defined conservation objectives. If a reserve contains the only population of the target taxon, or one of few surviving populations, the conservation goal is fairly straight-forward: population sizes must be maintained, and if possible encour-aged to increase, at least within set limits. Certainly, decreases in population size or genetic diversity must be avoided. The maintenance of numbers and diversity within the target species is likely to be a basic goal of all *in situ* genetic reserves, but that might not be sufficient. For exam-ple, a reserve might be established because it contains a particularly diverse population or set of populations of the target species, with active speciation, or perhaps gene flow occurring between it and sympatric related species. Alternatively, a reserve might be set up to protect a population with a particular genetic constitution – for example, an adap-tation to acid soils or resistance to a particular disease. Mere maintenance of population size, ignoring the genetic dimension, may result in the con-servationist having a false sense of the efficacy of the reserve.

9.3 MANAGEMENT PLAN FOR A GENETIC RESERVE

Once clear conservation objectives for a potential reserve have been for-mulated, a decision must be made as to what sort of management inter-ventions, if any, will be necessary to fulfil these objectives. That is, a management plan for the site must be drawn up. Such a document will do more than simply specify the interventions to be undertaken. A full list of the functions of the management plan is provided by Hirons *et al.* (1995):

- Describe the physical and biological environment of the reserve.
- Articulate the objectives and purpose of the reserve.
- Anticipate any conflict or problems associated with managing the reserve.
- Describe the management practices required to achieve the objectives.
- Monitor community dynamics within the reserve to assess management effectiveness.
- Organize human and financial resources.
- Act as a training guide for new staff.
- Ensure consistency between the reserve and national and regional conservation plans.
- Ensure site management objectives and management practices reflect the policies of parent organizations.
- Facilitate communication and collaboration among genetic reserves.

A management plan can only be sound if it is formulated on the basis of an ecogeographic survey (Chapter 4). The biology of the target species must be investigated and the site must be studied to find out exactly what habitats, species and communities are present and to understand something of its dynamics. It is only on the basis of such knowledge that the precise means of implementing conservation goals – the main purpose of the reserve management plan – can be developed.

Hellawell (1991) points out that communities are intrinsically dynamic and identifies three kinds of 'natural' change: stochastic, successional and cyclical. Stochastic changes are unpredictable. They result from natural catastrophes, such as drought, floods, fire, cyclones, hurricanes and epidemics. Populations and communities have different levels of resilience to, or ability to recover from, such shocks. In some cases a natural catastrophe may even be necessary for the maintenance of the population or community. Examples include fire as a breaker of seed-dormancy in many arid-zone communities and tropical cyclones triggering the flowering of some tropical forest trees.

Successional change is directional and passes through predictable stages. It naturally involves the local extinction of species. If the target species thrives only in early-succession conditions, successional change may have to be halted by management intervention. However, successional change may be extremely slow and consequently difficult to detect by population monitoring.

Cyclical changes are usually associated with predator–prey and density-dependent interactions. Though in the short term they may be quite dramatic, by their very nature their effects do not persist. Therefore, species extinctions are unlikely, but genetic drift and founder effect may be important factors if populations persist at low levels for lengthy periods. If populations of target species in the reserve are undergoing cyclical change of this kind, no intervention may be necessary, depending on the level of stability of population size and genetic diversity required in the management plan.

These three kinds of natural changes often work together. For example, a succession may have its origin in a stochastic event and the succession may contain cyclical changes. In stable communities – so-called climax communities – natural changes are limited. However, Jain (1975) has pointed that none of the major food crops or their progenitors are associated with climax vegetation. Therefore, the primary targets for *in situ* conservation of plant genetic resources will often be associated with more dynamic communities and so the management plan should include upper and lower limits for the population of the target taxa, beyond which management action is triggered.

Important and all-pervasive though 'natural' changes are in biological communities, it is undoubtedly true that changes due to human activity are usually the most dramatic. Examples of agents of such change include urban spread, industrialization, introduction of alien species and agricultural development. For example, on the island of Mauritius, as a result of human interventions such as land clearance for agriculture and the introduction of invasive exotic plants (guava, *Psidium cattleianum*; privet, *Ligustrum robustrum* var. *walkerii*; and many others), the indigenous flora is suffering extensive genetic erosion and indeed extinctions. Over 200 out of 685 species of native plants are classified as endangered (IUCN and WWF, 1994). A recent inventory of the most threatened plants of Mauritius revealed that 53 native species are known from 10 individuals or fewer and 13 of these are down to a single individual in the wild.

Human activity may create, as well as destroy, habitats and this may favour target species on occasion. For example, the disturbed land associated with agriculture or found along roadsides is the favoured habitat of the relatives of several important crops. It is important when formulating management prescriptions to understand both the anthropomorphic and the natural factors that may lead to changes in the abundance and distribution of the target species within the reserve and in the quality and quantity of the genetic diversity found there.

Though there is no standard format for management plans, Hirons *et al.* (1995) discuss the specific points that are usually included for an ecological reserve. As the reason for establishing a genetic reserve will be to conserve a particular species or group of related species, the management plan for such a site will require additional sections focusing more precisely on the target taxon, and these are included in the outline set out below.

1. **Preamble**: conservation objectives, reasons for siting of reserve, place of reserve in overall conservation strategy for target taxon.

2. **Taxon description** (Chapter 4): taxonomy (classification, delimitation, description, iconography, identification aids), wider distribution, habitat preferences, phenology, breeding system, genotypic and phenotypic variation, biotic interactions (e.g. pollinators, dispersal agents, herbivores, pests, pathogens, symbionts), local name(s) and uses, other uses, present conservation activities (*ex situ* and *in situ*), threat of genetic erosion.

3. **Site evaluation**: evaluation of populations of the target taxon, reserve sustainability, factors influencing management (legal, constraints of tenure and access), externalities (e.g. climate change, political considerations), obligations to local people (e.g. allowing sustainable harvesting) and anthropomorphic influences.

4. **Site description**: location (latitude, longitude, altitude), map coverage, photographs (including aerial), physical description (geology, geomorphology, climate, hydrology, soils), human population (both within reserve and around it), land use and land tenure (and history of both), vegetation and flora, fauna, cultural significance, public interest (including educational and recreational potential), bibliography and register of scientific research.

5. **Status of target taxon in the reserve**: distribution, abundance, demography, and genetic structure and diversity of the target taxon within the site, autecology within the reserve, interaction with associated fauna and flora, specific threats to population(s).

6. **Site objectives and policy**: site objectives, control of human intervention, allowable sustainable harvesting by local people and general genetic resource exploitation.

7. **Prescription**: details (timing, frequency, duration, etc.) of management interventions that will need to be carried out, schedule of ecological and genetic monitoring, population mapping, staffing requirements and budget, project register.

When the first six topics are thoroughly understood, an appropriate prescription – the details of the human intervention required – can be formulated to manage the population(s) of the target taxon within the reserve. Cropper (1993) describes the kind of practical interventions that may be necessary to ensure that a reserve meets its conservation objectives. Examples include: different levels of exclusion of people; burning or protection from fires; exclusion or encouragement of herbivores; slashing; selective removal of particular species (e.g. invasive exotic weeds); assisted propagation; reintroduction or translocation; hand-pollination.

As mentioned earlier, many native plant species of Mauritius are threatened due to invasive alien species. Among these are many species which have actual or potential economic value, such as ebony (*Diospyros* spp.), makakas (*Mimusops* spp.), *Labourdonnaissia* spp., the wild relatives of coffee (*Coffea macrocarpa* and *C. mauritiana*) and many other species of medicinal importance. In order to conserve this range of species with the limited resources available, *in situ* genetic reserves have been established which target several species at a time. These reserves have comprehensive management plans which balance the needs of the various target taxa. The sites have been delimited, fenced to exclude deer and pigs (there were originally no native terrestrial mammals on the island), exotic weeds removed and seed predators (e.g. rats) exterminated. The results of the application of this management plan and prescription have been a spectacular recovery of the indigenous vegetation within the reserve, including the various target taxa.

Each reserve is likely to require a unique management plan, because of different conservation objectives, taxa, locations, local abiotic, biotic and anthropogenic influences and levels of resources available. However, whatever the taxon and whatever the location, the genetic diversity within the reserve will not be secure unless an appropriate management plan and prescription are applied. It is important to make the prescription as detailed as possible, but Hirons *et al.* (1995) warn against the plan becoming too large and complex, and thus running the risk of being perceived as being of no more than academic interest. The initial management plan may be largely subjectively formulated, but through trial and error, combined with regular monitoring, an effective management strategy should evolve. Detailed examples of reserve management plans are included in the case studies in the final chapters of this volume.

9.4 MONITORING A GENETIC RESERVE

The management of a genetic reserve will involve an element of experimentation and evolution. As discussed above, it is not likely that the ideal management regime will be known in full *a priori*. Therefore, the population or populations of the target species in the reserve (and possibly of competitors and other associated species) will need to be assessed regularly in terms of their size, genotypic composition and/or overall genetic diversity in order to be able to detect changes. If change is indeed detected, the management prescription will need to be reviewed. Management may or may not be amended, depending on the nature of the change and the exact reason for the establishment of the reserve. This process is usually referred to as 'monitoring'. It can be illustrated by the simple model shown in Figure 9.1.

9.4.1 Monitoring objectives

Hellawell (1991) emphasizes that the objectives of the exercise should be clearly understood before commencing a monitoring programme. This

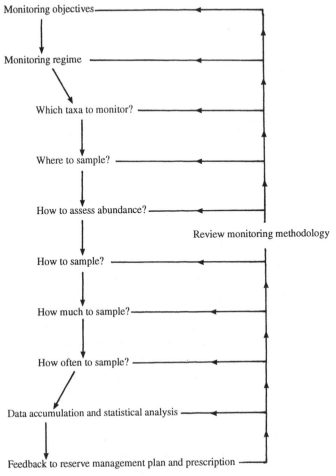

Figure 9.1 Schematic model of the monitoring procedures in a genetic reserve.

may appear trite, but he refers to several cases where it was not the case. For genetic conservation based within a reserve the objective is clear: to maintain or enhance the target populations representing the target species in the quality and quantity of their genetic diversity. It follows that the objective of the monitoring is to detect any detrimental changes in target population characteristics.

9.4.2 Monitoring regime

Even within a small reserve it would be impossible to record and monitor every plant of every species. The conservationist is forced to sample the reserve in such a way as to reflect accurately the overall picture. Sampling will have to be done at both the taxonomic and population levels. The conservationist must select key taxa to monitor and select sites within the reserve to monitor their population characteristics. How these sample locations are chosen and how the samples are taken is described in the following sections.

It is likely that a specific monitoring regime will be required for each reserve. It is critical to the conservation success of the reserve that the correct regime is established and to achieve this several specific questions need to be addressed, as outlined in the monitoring model (Figure 9.1).

(a) Which taxa to monitor?

The selection of which taxa to monitor may be problematic. The target taxon, which is the reason for establishing the reserve in the first place, will clearly need to be followed over time. It is likely that any taxonomically closely related species which may exchange genes with the target taxon will also be included in the monitoring programme. However, the abundance of other species may be directly related to or affect the abundance or diversity of the target taxon in the reserve. These include parasites, pollinators, epiphytes, symbionts and competitors, as well as indicator or keystone species. The final selection of what taxa to monitor will to a large extent depend on the resources available.

(b) Where to sample?

There are three main strategies that can be used to decide where to sample within the reserve: random, systematic and stratified random. Random sampling implies that every point in the reserve has an equal chance of being sampled. The sampling locations can be determined by drawing a box around a map of the reserve and using random number tables or the like to generate a set of coordinates. This is the most robust and statistically safe form of sampling. Indeed, most statistical analyses require that the data be derived from random sampling.

Systematic sampling means that samples are taken at regular intervals, for example in a grid pattern. Since many biological phenomena are spatially auto-correlated, this has the advantage over random methods of avoiding the over-sampling of 'uninteresting' areas at the expense of 'interesting' ones. However, the statistical tests that can be applied are more restricted, and care must be taken to avoid the sampling interval coinciding with phased ecological patterns in the area under study.

Stratified random sampling involves dividing the reserve into different but internally homogeneous zones and taking samples at random independently within each zone in proportion to the size of the zone. The zones could be areas of different vegetation or soil type, for example. This method has the advantage of ensuring that all habitats are sampled. It is perhaps the most common sampling strategy applied in ecological work.

Where there is a strong environmental gradient (e.g. decreasing mean temperature moving up a mountain, decreasing exposure to salt water with increasing distance from low-water mark in marshland, less water-logging with increasing distance from a lake), sampling is often done parallel to the gradient, along a transect. Sampling points may be taken at regular intervals or at random along the transect.

In practice, the strategy adopted may vary depending on the specific constraints. In a flat meadow, it will be difficult to find an argument against adopting the most scientifically appropriate strategy. In a more difficult site, such as a cliff face, the strategy may require modification. Precise decisions on where to sample will also depend on how individuals are distributed over the site and on overall abundance. For species which are dispersed by wind, the distribution pattern may be fairly uniform, but for clonal species which are likely to be clumped, random sampling may be unsuitable and stratified random sampling is a more appropriate option.

(c) How to assess abundance?

There are numerous methods of assessing species abundance or diversity (Figure 9.2) but the main non-destructive measures of plant abundance are density, frequency and cover. Density is the number of individuals per unit area. Frequency (which may be root or shoot frequency) is the proportion of samples within which the target species occurs. Cover is the percentage of the ground occupied by a perpendicular projection (shadow at noon) of the aerial parts of the species. Absolute measures of the density of a given taxon may be assessed in the form of number of individuals, demographic structure (particularly for long-lived perennials), distribution pattern and biomass or yield (this has the advantage that it is accurate, but the drawback that it is destructive). Frequency is the easiest measure to make, requiring only that the target species be recorded as present or absent within the sampling site. However, the frequency value obtained depends on the pattern of distribution of the species as well as its density, and also on the dimensions of the sampling site.

For most monitoring of genetic reserves, density and cover will be used. Density requires counts of individuals. This will be fairly straightforward for trees, shrubs, tussocks and single-stemmed herbs, but may be almost impossible for plants which spread vegetatively. Cover bypasses this problem, but is much more difficult to record, requiring either a sub-

Figure 9.2 Students assessing population abundance during a conservation field course to Almería, Spain.

jective visual estimate or, in short vegetation, the time-consuming use of frames of pins and the like (Goldsmith *et al.*, 1986). Cover values are often estimated using the Braun–Blanquet scale or Domin scales (Table 9.1).

It may well be that the abundance of different species will be recorded in different ways, depending on the accuracy required and the importance of the species to the conservation aims of the reserve. For example, the abundance of the species for which the reserve was

Table 9.1 The Domin scale of vegetation cover

Class	Cover value
+	1 individual
1	1–2 individuals
2	< 1%
3	1–4%
4	5–10%
5	11–25%
6	26–33%
7	34–50%
8	51–75%
9	76–90%
10	91–100%

established will probably be assessed as density. In addition, some estimate of age or vigour may be recorded for each individual counted, such as girth at chest height (a favourite of foresters), height or fertility. For other species, a fairly rapid visual assessment of cover may be sufficient. For others, their presence or absence at the sampling site may be all that is required.

However, abundance of the species within the reserve may not be enough information on which to base the management of the site. Fluxes in population numbers are due to the balance between recruitment and mortality. In turn, recruitment will depend on fecundity, germination rate, seedling survival, etc. To disentangle the causes of changes in population numbers (or the lack of change) which may be a necessary prerequisite to devising an appropriate management intervention, each plant of the target species within a sampling site, including seedlings, needs to be followed individually over time, with frequent observations. Such a detailed demographic study will include the preparation of life tables, fecundity schedules and the like (Harper, 1977).

In addition, as already mentioned, the reserve manager may well be more interested in the genetic diversity, structure and composition of the population of the target species than in its size. This will require isozyme and/or DNA studies (Chapter 12). In such cases, a decision will have to be made about how many individuals to sample and how to choose individuals for sampling. There are various techniques for the sampling of vegetative material in the field for subsequent isozyme or DNA analysis in the laboratory (von Bothmer and Seberg, 1995).

(d) How to sample?

The two main ways of sampling vegetation and plant populations are plot methods and intercept methods. Plot sampling involves taking observations at the sampling point within areas of standard size, usually called quadrats. In the line-intercept method, a measuring tape is laid out in a random direction at the sampling point and observations taken on those individuals which intersect the tape. Most monitoring in genetic reserves will probably involve the sampling of temporary or permanent quadrats.

Quadrats are usually square or rectangular, though circular ones have smaller edge effects. If quadrat size is small, portable wood or metal frames may be used. Larger quadrats may be demarcated by pacing out or measuring the sides using a tape measure, placing pegs or stakes in each corner, and running string or coloured tape around the perimeter. The required observations are made by systematically going through the quadrat counting and perhaps measuring and even tagging each individual of the target species encountered. A large quadrat may be subdivided to make it easier to work through it without missing individuals. When carrying out genetic diversity studies, a simple strategy is for material from a certain number of individuals of the target species from each quadrat to be sampled for molecular analysis back at the laboratory. Individuals should be chosen at random from within the quadrat. If all individuals have been tagged and labelled, numbers could be picked at random, for example. The number of individuals sampled from the whole population should probably be at least 20–30 (von Bothmer and Seberg, 1995).

Fresh quadrat sites may be chosen on each sampling occasion or repeat visits made to the same permanent quadrats. Using new quadrat sites each time is statistically more manageable, as the assumption of observations being independent of each other is basic to most statistical procedures, and repeated observations of the same sampling unit (quadrat) clearly contravene this rule. However, ways do exist of analysing time-series data from a set of fixed sampling units (e.g. based on fitting models separately to the data from each sampling visit and then comparing the parameters of the models from all the sampling visits), and using permanent quadrats is certainly easier if the quadrats are large and numerous and the terrain difficult. North *et al.* (1994) monitored vegetation change by estimating cover percentage in permanent quadrats on Round Island, Mauritius. Precise instructions on how to reach each permanent quadrat should be documented. Permanent plots are often photographed on each sampling occasion, though species abundance and other data can be difficult to extract from such photographs. The exact position of the camera should be marked as well as the quadrat itself, and details of camera settings recorded. The pegs or stakes used to mark permanent quadrats should be wooden or plastic (rather than metal), easy to spot and firmly planted. If vandals may be a problem, existing landmarks (rocky outcrops, trees) can be used as reference points.

For detailed population demography studies, individual plants may be labelled with a tag bearing a unique identifier, or their position within the quadrat may be accurately mapped – for example, using a pantograph (Harper, 1977; Hutchings, 1986). In some cases, it may be possible to map quadrat location and even individual specimens using a Geographical Positioning System (GPS) receiver, though it should be remembered that GPS receivers vary in the accuracy of their results and may be difficult to use in dense forest and highly dissected landscapes.

(e) How much to sample?

Quadrats should be small enough to be searched easily and permit sufficient replicates in the time and with the resources available, but large enough to accommodate whole plants of the target species. Quadrat size is particularly important to get right if frequency is going to be measured. Various techniques have been developed by ecologists carrying out vegetation surveys to determine what quadrat size results in the most efficient sampling of a particular vegetation type, i.e. the recording of the most number of species for the least effort. For example, if number of species is plotted against quadrat area, the relationship is usually asymptotic, which can be used to determine the quadrat size above which any further increase leads to the recording of few (if any) additional species. As a rule of thumb, Goldsmith (1991b) states that most species in the vegetation should have a frequency of 20–70% in the quadrats sampled and if several are present in all quadrats, then quadrat size is probably too large. However, whether optimal quadrat size for vegetation surveys will also be optimal for an assessment of the abundance of individual target species is not clear. Clarke (1986) makes recommendations on quadrat size and shape for different plant habits (Table 9.2).

If the list of species that it has been decided to monitor within the reserve includes trees, shrubs and herbs, nested quadrats of different sizes may be used. Having begun monitoring, quadrat size should obviously be kept constant, in particular if frequency is being measured, to permit comparison of data from subsequent surveys.

There are no simple rules for the number of quadrats required, but generally the greater the number the more reliable is the result (Goldsmith, 1991b). However, as with quadrat size, the extra information gained from an extra quadrat will diminish as the total number of quadrats rises (Spellerberg, 1991b). Therefore, the information each new quadrat provides must be balanced against the additional resources required to record that quadrat. Plotting data variance against the number of quadrats often yields an oscillating relationship; a useful guide to the minimum number of quadrats is the point where the oscillations damp down (Goldsmith *et al.*, 1986). Poole (1974) provides another method for

Table 9.2 Recommended quadrat size and shape for different plant habitats (from Clarke, 1986)

Target species	Plot size (m^2)	Ratio of sides
Herb	0.5–1.0	1:2
Shrub	50–100	1:5 or 1:10
Tree	200–1000	1:5 or 1:10

determining the appropriate number of samples. He suggests that variance (s^2), resolution (L) and number of samples (n) are related as follows:

$$n = 4s^2/L^2$$

If numbers of a target species are being monitored, L would refer to the maximum percentage difference (e.g. among sites) or change (e.g. between years at a given site) in the numbers of individuals of the target species which the project is willing to consider unimportant.

In most cases the size and number of samples taken from the reserve will often be a practical compromise between the number that are required to generate meaningful statistics and the resources available to the conservationist.

(f) How often to sample?

Perhaps the main determinant of the frequency of sampling will be the strength and nature of the perceived threat to the population(s). For rare or very threatened annuals, monitoring may occur as often as every week or fortnight during each of several growing seasons, in particular if the demography of the population needs to be studied in detail (Harper, 1977). For perennial species, the interval between observations of adult individuals may well be several years, though there may be frequent monitoring of seedlings and saplings during a growing season to assess recruitment. Once an effective management plan is established, the frequency of sampling may well decrease.

If the population is to be monitored once a year, it is important to sample at comparable stages in the life cycle of the target species. Early in the season there are likely to be large numbers of seedlings, but there is likely to be extensive mortality as the season progresses and so many fewer individuals will reach flowering and fruiting. In an adverse season some perennial species may not produce flowering bodies or even aerial parts, so it is important if recording at flowering or fruiting time to record all live plants, not just those flowering or fruiting.

In practice, it should not be surprising that the frequency of sampling is likely to be much higher in a newly established reserve than in a well-established one. As it is unlikely that the most appropriate management prescription will be known when the reserve is established, the initial monitoring frequency is likely to be high. However, as changes or adjustments to the management prescription become less necessary, so the interval between monitoring exercises may be extended.

9.4.3 Data accumulation and analysis

To ensure that the required monitoring data are recorded fully and in a uniform manner, standard hardcopy or softcopy forms should be used on each sampling occasion. It is important also to record the actual management that is undertaken between sampling events (Chapter 11), which may for practical reasons be different from the prescription. For instance, the prescription may say that a certain level of grazing is desirable over a certain time period. In practice, how often was the reserve actually grazed and at what intensity? Failure to record these details will make evaluation of the management plan difficult or even impossible.

Having collected the data for a particular monitoring exercise the conservationist will want to compare the population characteristics recorded on that occasion with previous surveys to see if any significant changes have occurred in the intervening period. It is important that the form of statistical analysis which such comparison will necessitate should be considered along with the choice of parameters to be followed before any data is actually recorded in the field. Otherwise, the statistical analysis may be unnecessarily restricted or irrelevant data may be recorded, so wasting recording effort. The major points to watch here have already been alluded to, in particular the requirement of standard statistical techniques for random, independent sampling.

When interpreting the results of monitoring it is important to distinguish between the 'normal' ranges for population parameters and those due to management. Monitoring may show a population is declining. Is this the result of normal, natural cycles, or an inappropriate management regime? The natural factors that may affect population parameters include seasonal effects (e.g. rainfall), long-term habitat directional changes or succession, natural biotic cycles (e.g. grazing cycles by herbivores) and longer-term climatic changes. It may well be necessary to monitor both populations being subjected to management within the reserve and nearby conspecific populations not subjected to management, which could be outside the reserve or within the reserve but excluded from human interventions by the management plan.

9.4.4 Feedback to reserve management plan and prescription

Analysis of the monitoring data should reveal trends in the selected population parameters for the target taxa. If these trends are interpreted as reflecting a deterioration of the conservation status of the target species in the reserve, then the management plan, and especially the prescription, will need to be altered. Shands (1991) cites the example of the establishment of a genetic reserve for the maize relative, *Zea diploperennis*, in the tropical forest of Sierra de Manantlan, Mexico. Routine monitoring of population sizes within the reserve indicated that if the original prescription was not amended the other forest plants would out-compete the wild maize. This is the crux of why active monitoring of conserved populations is vital. It is only through this monitoring process that the need for changes in the management plan can be recognized. The monitoring process acts as a feedback mechanism, triggering changes in the management of the reserve and ensuring that genetic resources are safely conserved.

9.4.5 Review monitoring methodology

The results of the analysis of monitoring data may also indicate that changes in the monitoring methodology are required. The analysis may, for example, indicate that the data recorded are not sufficiently detailed to permit valid statistical conclusions to be drawn. This may lead to more or larger quadrats being used in subsequent monitoring activities.

9.5 CONCLUSION

When the reserve is established it will most commonly be sited in a location with a 'healthy', genetically diverse population of the target taxon. The initial management plan will attempt to reflect the management regime of the site prior to establishment of the reserve. However, it may be difficult to clearly discern the prior regime and the manager may also wish to alter the regime to enhance the populations of the target taxon within the reserve. Therefore, the application of the management plan is likely to involve an element of experimentation and evolution. As the practical regime is changed, so it is necessary to monitor the size, genotypic composition, etc. of the populations of the target taxon to detect any population changes and ensure their 'health'. This process of management and monitoring is pivotal to the conservation of plant populations within a genetic reserve. It is expensive in time and resources but is the only way to ensure that the target taxon is conserved effectively.

10

Locally based crop plant conservation

C.O. Qualset, A.B. Damania, A.C.A. Zanatta and S.B. Brush

10.1 INTRODUCTION

Conservation of cultivated plants or domesticated animals follows the same principles and concepts for naturally occurring biological resources as outlined elsewhere in this book. This chapter focuses on crop plants, where a broad definition of a crop plant is applied. That is, we include plants grown under human cultivation and those exploited in natural stands, but under minimal management by humans. The latter includes those non-cultivated plants that are exploited extensively, such as for the extractive harvest of fruits, nuts and sap from trees or the destructive harvest of trees for timber. The primary distinction in methods for the conservation of crop plants from other biodiversity is that humans usually have a much stronger and essential role in conservation of crop plants. For example, a habitat reserve may be established for conservation of wild species, but once established there is minimal human activity in maintaining the *in situ* reserve. For cultivated crop plants the habitat is the farming unit itself or, occasionally, the local community, and humans are involved actively with the planting and harvest of each crop and have direct control of the fate of the crop and hence its biological diversity. Likewise, for natural stands of non-cultivated crop plants, humans have control over the intensity and frequency of harvest and, to some extent, regeneration of new plants.

The term 'locally based conservation' (LBC) has not been widely used,

Plant Genetic Conservation.
Edited by N. Maxted, B.V. Ford-Lloyd and J.G. Hawkes.
Published in 1997 by Chapman & Hall. ISBN 0 412 63400 7 (Hb) and 0 412 63730 8 (Pb).

but it has great utility in the practice of conservation of species used in agriculture. We define LBC as the conservation of biological entities at the farm, community or regional level. LBC is essential because it permits the evolution of crop genetic resources in their areas of origin. This involves such extensive areas of agricultural lands where conservation cannot be managed effectively or reliably by government programmes. It will be shown later where some incentives for LBC can be provided by government, but local issues stand foremost, such as preservation of traditional food and cultural practices, pride, ownership and economic value.

Unfortunately, the need for active conservation is derived from overgrazing of rangelands by sheep, goats and cattle and by human excesses in extraction of harvests of trees and shrubs for fuel, timber and industrial uses and of plants for ornamental use or extraction of chemicals. Human activities have resulted in the complete displacement of natural and cultivated vegetation in large areas, such as by the farming enterprise itself. Other significant destructive activities include the construction of artificial water reservoirs, open-pit mining, urban development, agro-forestry using exotic species and, in some cases, the invasion of exotic weedy species. It is because of these severely destructive activities that aggressive steps must be taken to implement biological conservation and restoration of habitat.

LBC involves people from both the farm and urban communities and has the potential to make great advances in the conservation of crop plants. For LBC to be effective it must be locally acceptable and locally practical. Incentives may be required to start the process. A primary goal will be to conserve crop genetic resources, but the conservation ethic follows for soil and water conservation as well. Two basic factors must be recognized in planning LBC:

- Compromises between ideal (full) conservation and demands upon natural and farmland habitat by human populations must be negotiated.
- Conservation can only be optimally planned and implemented if it is known what is to be conserved, where it is located, its abundance and, at least for a series of indicator species, the distribution of genetic diversity among and within populations.

Brush (1995) has emphasized the fact that there is unrelenting change in human populations with respect to social habits, economic incentives, reception of technology and response to government policies. These changes have affected the inventory of crops being managed and hence the genetic composition of agricultural species. This includes the wholesale displacement of certain crops and the introduction of new ones, as well as the introduction of modified forms of existing crop types. Both of these types of change result in significant genetic replacements and must be considered in developing strategies for LBC. For most cultivated and some non-cultivated crop plants, complementary *in situ* and *ex situ* conservation strategies are simultaneously applied. In Soulé's (1991) classification of global conservation strategies, *in situ* conservation is defined as conservation via protected natural habitats with little or no human intervention. His four classes of *ex situ* conservation strategies include: agroforestry and agro-ecosystems; zoos and botanical gardens; living collections; and suspended-living collections. These modes of conservation are all applicable to LBC of crop plants and will be applied appropriately, depending on the local conditions.

10.2 TARGETS FOR LOCALLY BASED CONSERVATION

The primary targets for LBC are land races of crop plants as they occur in primary and secondary centres of diversity. It is in these areas that genetic diversity is great within the crop plants; and wild related species are often extant. A secondary target is traditional (heirloom) varieties that have been largely replaced by new varieties and exist now in gene banks or are maintained by hobbyists. Heirloom varieties may be directly derived from land races or as the products of plant breeding. Seed crops have been collected for *ex situ* conservation for many years, while perennial crops such as fruits, nuts and berries have had limited attention for *ex situ* conservation. For all crops, little effort has been given toward LBC. Almost no efforts for conservation of non-cultivated crop plants have been undertaken and an emphasis on integrated natural resource management will be needed to conserve and restore those plant genetic resources. In view of the risks of habitat destruction, major useful species should be targeted for *ex situ* conservation as seed in seed banks or as plants in gardens and field gene banks. *Ex situ* conservation is an essential back-up to preserve populations that are at risk in their native habitats.

Targets for conservation, as summarized by Soulé (1991), are:

- entire ecosystems, including indigenous human populations;
- biogeographic assemblages, which include populations of species that are geographically or temporally isolated and present clear genetic differentiation;
- indigenous and endemic species;
- local populations;
- genetic variation within a species such as wild relatives, domesticates and non-economic genetic variation.

Locally based conservation of crop plants emphasizes the last three conservation targets, and all targets are relevant for wild crop relatives or species used for extractive harvest in their native habitats.

10.3 THE CONCEPT OF GENETIC EROSION

The process of replacement of old, indigenous germplasm by new, high yielding varieties was equated to loss of genes, and consequently the phrase 'genetic erosion' was coined (see also Chapter 3). However, the agricultural processes must be examined with respect to the loss of genes, gene combinations, or allelic forms. The concepts of gene replacement and **displacement** must be understood. Gene **replacement** occurs when indigenous varieties are replaced by introduced ones, resulting in substitution of alternative alleles within the same species; while gene **displacement** refers to the loss of whole genomes by substitution of one crop species for another or by elimination of the crop entirely. Genetic erosion can then be seen in two forms: **genic or allelic erosion** and **genomic erosion**. Replacement of old varieties by new varieties within a crop causes a change in allelic frequencies for all alleles differing in the old and new varieties from 0 to 1.0 or 1.0 to 0 if there is complete replacement of old by new varieties and at intermediate frequencies if there is partial replacement of old by new varieties. The replaced alleles are lost or eroded if not conserved elsewhere. Also lost are the specific combinations of genes that occur in the replaced varieties. With respect to *in situ* conservation of genetic resources, partial replacement or complete replacement of varieties are critical distinctions and must be evaluated at the individual farm level or at higher levels, such as communities or regions.

Genomic erosion is potentially more devastating to crop biodiversity because all of the genes for a crop would be lost when another crop is introduced or when agricultural habitat is lost due to urban and industrial development, flooding to create artificial lakes, or toxic invasions such as soil salinity. These events point to the essential need for *ex situ* conservation or relocation of critical populations to other sites.

The fundamental unit of conservation is the gene and its allelic forms. The concept of genetic erosion was primarily applied to the replacement of alleles within a crop species, as in the replacement of land races by improved varieties. Nevertheless, the simple act of replacing one variety by another is an evolutionary event in the process of agriculture and may clearly increase both Darwinian and agricultural fitness, i.e. higher reproductive rates and yields as measures of fitness. If a crop variety is replaced by another within the same species, there would be genetic erosion with replacement, that can ultimately lead to **genetic enhancement**. In the case of replacement of a few genes, leading to improved yield, stability or quality, there is clearly genetic enhancement instead of genetic erosion. The introduction of new alleles into an agricultural system that improves fitness is **genetic enrichment** and it would be legitimately called genetic erosion only if alleles are in fact lost. Therefore, genetic erosion must be evaluated on the basis of degree of replacement and conservation status.

In the years of the green revolution, adoption of modern varieties was rapid and independent of the farm size. In contrast, more recently the small-scale farmers have often been slower to adopt the newer modern varieties than those with large holdings (Byerlee and Moya, 1993). In most countries of West Asia and Africa, local durum wheat varieties and land races predominate, particularly under moderate to low rainfall conditions (Srivastava and Damania, 1989; Tesemma and Belay, 1991; Gebre-Mariam, 1993). Land races and diverse populations of wheat can still be found in these areas of the world (Byerlee and Moya, 1993) in an island pattern similar to the continued cultivation of land races of potatoes and maize in Latin America (Brush, 1995) and rice in Thailand, even though the main production and land area are devoted to modern varieties, which have been available to farmers for a long time (Srivastava and Damania, 1989; Zencirci, 1993; Damania, 1994). In the past, due to difficulties such as transport, the rate of exchange among these so-called islands is assumed to have been much lower than nowadays, resulting in reduced gene flow among these areas.

The theory of island biogeography states that immigration and extinction rates maintain species number in a dynamic equilibrium on islands, leading to the maintenance of species diversity, despite examples like high diversity in tropical forests and coral reefs being maintained in a non-equilibrium state due to the occurrence of intermediate disturbances. None the less, considering that the theory of island biogeography can be applied to a broad range of habitats, almost any patchy environment or spatially heterogeneous area may contain diversity–area or diversity–remoteness relationships. Applying the theory of island biogeography to the case of land races of some crops which remain in small islands encased in areas of cultivation of modern varieties, it would be expected that these islands of land races might become progressively genetically impoverished, i.e. declining genetic diversity with a lower number of genotypes and phenotypes, if there is no further migration from the 'type source'. If those genes are not maintained somewhere else, true genetic erosion would then take place, even though land races are still being cultivated. However, the fact that these populations are connected by the means of market exchange has to be considered (see also Chapter 20). Other issues such as minimum viable population size, and the effect of fragmented populations, need likewise to be better understood in crop populations

A high proportion of existent alleles of the major crop plants have been conserved *ex situ*, with the evolutionary processes abated. *Ex situ* conservation may have conserved this portion of the existent diversity, but it has basically curtailed such evolutionary changes as may occur in the usual crop husbandry practices. Several arguments have been put forward supporting the *in situ* conservation of land races of varieties:

- They possess adaptation for particular local conditions and selection from them can be used directly as new varieties.
- They can evolve freely under changing environmental conditions and benefit farmers who use them.
- *Ex situ* conservation at a remote gene bank could lead to shifts in population structure during storage and deterioration in genetic integrity due to regeneration in an environment different from the original.
- Local control of land races will ensure that benefits, if any, would accrue to the farmer and communities that developed them.

Therefore, LBC of agricultural species and their progenitors and relatives is best served by *in situ* conservation and supported by *ex situ* conservation.

10.4 *IN SITU* CONSERVATION OF LAND RACES

Land races are the most variable populations of cultivated plants (Frankel, 1971; Frankel and Brown, 1984; Chapman, 1986; Cuevas-Perez *et al.*, 1992). This extraordinary variability is present for many phenological, agronomical, morphological, physiological and qualitative characteristics. Besides being adapted to their environment, both natural and man-made, the genotypes are also co-adapted. Hence, genetic variation within a land race may be considerable, but it is far from random. The great genetic diversity present in land races make them good sources of genes for plant breeding. In reality, the term 'land races' is used by many authors to specify heterogeneous mixtures of genotypes or domesticated populations maintained by subsistence farmers. These populations are often highly variable in appearance, but they are each identifiable and usually have local names. They are not adapted to high soil fertility, high plant populations or high production (Harlan, 1975). Nevertheless, the example of extensive utilization of land races in breeding programmes of irrigated rice in Latin America and the Caribbean demonstrates that it is possible to broaden genetic variability while increasing yield potential by making use of land races (Cuevas-Perez *et al.*, 1992).

A very large quantity of germplasm accessions of cultivated, obsolete/primitive and wild forms of crop plants has been assembled and stored *ex situ* at various genetic resources conservation centres around the world in the form of seeds, plants in orchards, or tissue cultures. To date, *ex situ* conservation programmes are extensively used to preserve samples of crop genetic diversity in botanical gardens or gene banks. On the other hand, *in situ* conservation refers specifically to the maintenance of variable populations in their natural or farming environment, within the community of which they form a part, allowing the natural processes of evolution to take place. The germplasm so conserved will be adequate not only for current research needs, but also those for the future, such as responding to global climate change (Davis, 1989).

To overcome the limitations of the *ex situ* collections, preservation of crop plant genetic resources, as well as wild crop relatives populations in their natural habitats, is important for long-term benefits of national programmes and the international community as a whole. This form of conservation was largely ignored for land races until very recently, but indirect evidence obtained from studies on land races and wild cereal populations indicate that *in situ* methods are effective for the conservation of genetic diversity in cultivated species as well. During their long history of propagation in crop fields in centres of origin and primary diversity, land race populations of major crop species were in close association with their wild and weedy relatives. They occasionally exchanged genes, enriched genetic diversity and evolved together, being most certainly exposed to a multitude of biotic and abiotic stresses.

In the case of activities involving crop plants, LBC entails the continued cultivation of older varieties, through the retention of the ecosystem as a whole, including the cultivator. In due course, the needs and demands of local farmers must be identified and discussed in conjunction with preservation endeavours and developed as an integral part of land-use management. Human population pressures and the need for increased food production will restrict the success of strict conservation management practices when it is perceived that few or no economic benefits are forthcoming from these efforts (Cohen *et al.*, 1991a). Consequently, the objective of an applied conservation activity should be to demonstrate that the preferred economic enterprise of the people in a given area is consistent with the profits from the area's economically important biological diversity .

In situ conservation projects require a high degree of technical expertise and persuasion skills to monitor and manage the community structure and full cooperation from the local residents in protecting the targeted species, especially when the *in situ* sites are nearly all in remote areas (Chang, 1994). Jana (1993) has said that we should get away from the notion that *in situ* conservation of land races is for safeguarding breeding materials. He argues that it should be practised for its own sake; it enhances the quality of life and ensures the continuation of the agro-ecosystem.

Several authors have suggested the need for resorting to *in situ* conservation of land races in the communities in which they occur (Altieri and Merrick, 1987; Brush, 1995; Chapters 15–21). However, it is a challenge to undertake locally based conservation of cultivated crops without a return to or the preservation of traditional cultural systems, which may be unacceptable or impracticable under political systems in areas where diversity abounds. In the case of species which have recalcitrant seeds, or plant forms that cannot be preserved *ex situ*, there are few alternatives to *in situ* conservation.

The potential role that agro-ecosystems can play in conserving biological diversity in the context of using that diversity is beginning to be appreciated. Altieri and Merrick (1987) contend that although most traditional agro-ecosystems are under the process of modernization to varying degrees in different parts of the world, conservation of plant genetic resources can still be integrated with agricultural development, especially in regions where rural development activities preserve the vegetational diversity of traditional agro-ecosystems. Traditional farming systems preserve the interaction of plants with their environments. These systems often represent a strategy which promotes diversity of food intake, income and efficient use of available labour, and optimize inputs for production utilizing low levels of technological skills. Francis (1985) has estimated that as much as 15–20% of the world's per capita food supply still comes from traditional and multiple cropping systems, but these figures may be on the decline.

10.5 COMMUNITY INVOLVEMENT IN CONSERVATION

Up to now, there has been little regard at national level to preserving agriculturally important plants, as well as their progenitors and relatives, in their native state. In this matter, it is most important that local communities should become aware of the situation and become part of the conservation movement. Indigenous farmers have been known to retain folk varieties and continue to grow them even when they experiment with and adopt modern high yielding varieties (Brush, 1995). The reasons for this practice are as diverse as the crops themselves. Some of the most often mentioned are: storage properties, ease of cooking, nutritional and processing qualities, historical and cultural reasons such as dietary diversity and the use of folk varieties in traditional foods or religious ceremonies, and the filling of unique market niches (Cleveland *et al.*, 1994). There are agronomic reasons such as greater suitability to intercropping systems, early or late maturity, or more resistance to local biotic and abiotic stresses. Stability of yield in regions where seasons are unpredictable from year to year is another important factor in helping farmers to retain their land race varieties while planting improved germplasm.

Brush (1995) presents three cases of ongoing maintenance of land races by farmers who have also adopted high-input technology, including high yielding cultivars, in potatoes (*Solanum* spp.) in the Andes of Peru, in maize (*Zea mays* L.) in southern Mexico, and in wheat (*Triticum* spp.) in western Turkey. These examples show that land race cultivation persists in small areas where these crops were domesticated. Factors that promote *in situ* conservation include the fragmentation of land holdings, marginal agricultural conditions associated with hill lands, hetero-geneous soils, economic isolation, cultural values and preference for diversity.

The central Andes is the centre of origin and primary centre of diver-sity for the main cultivated potato (*Solanum tuberosum*). The highland Andean farmers grow seven species and subspecies according to Hawkes (1979) – evidence of a dynamic system of cultivation which could accommodate new varieties whenever available. Modern varieties of potatoes (with higher yields, smooth skin, even shape and light-coloured flesh) were introduced in Peru in the early 1950s and are found now in almost every village in the highlands (Brush, 1992). Nevertheless, native farmers in the Tulumayo and Paucartambo valleys have maintained different types of potatoes on their farms (Chapter 20). Indigenous varieties are appreciated for their culinary appeal and because they fetch higher market prices (Brush, 1991a). The modern varieties are cultivated for their yield potential and resistance to biotic and abiotic stresses. Andean farmers apparently do not intend to replace entirely the indigenous varieties with improved ones. Rather the general strategy has been to grow potatoes in a three-tier system where improved varieties are cultivated side by side with selections from the native varieties for sale in the market, and mixtures of native varieties for home consumption (Table 10.1). This pattern of cultivation should be encouraged elsewhere in South America and other such hot-spots of diversity around the world.

Mesoamerica is the centre of origin and diversity for maize (*Zea mays*). Some 32 races of maize were described from this region before improved varieties were introduced (Wellhausen *et al.*, 1952). Native farmers main-

Table 10.1. Potato farming systems in Tulumayo and Paucartambo valleys in the Peruvian Andes. Values in the table are given in percentages (adapted from Brush, 1992)

Valley	System	Area in potatoes %	Potatoes sold %	Potatoes consumed %
Tulumayo	Native improved	59	78	41
(*n* = 154)	Native commercial	30	20	37
	Mixed	11	2	22
Paucartambo	Native improved	31	62	30
(*n* = 240)	Native commercial	8	7	8
	Mixed	61	31	62

tain their own races separately, each of which is adapted to their own particular micro-environment. Farmers who adopt improved germplasm derive them from private seed companies or purchase them on credit from the state-run farming cooperatives. In both cases a dramatic impact on the diversity of local varieties can be expected. But, as with potatoes in the two valleys in Peru, the traditional maize farmers in the southern Mexican state of Chiapas are keenly aware of the desirability and short-comings of their native varieties and choose to maintain diversity in a mixture of seed management practices. This procedure allows the conser-vation of traditional varieties and different races at the same time as deriving economic benefits of improved germplasm (Brush, 1991a).

Some Italian farmers still grow *Triticum monococcum* and *T. dicoccum*, two obsolete forms of wheat known to be grown in Roman times as far back as 200 BC, but thought to have disappeared (Perrino and Hammer, 1982). In a classic case of farmer-based conservation of crop genetic resources, the two obsolete wheat forms were found to be cultivated in small fields in two mountain villages of Castelfranco in Miscano (Benevento province) and Monteleone di Puglia (Foggia). The glumes of the kernels of both species are persistent and the farmers utilize them to supplement the diet of swine and as a poultry feed. Straw from the plants, which has a high biomass, is used to provide roofing to the field huts. In recent years, *T. dicoccum* cultivation in Italy has received some boost because the consumption of 'farro' (*dicoccum* flour) is thought to prevent colon cancer due to its higher fibre content than normal durum or bread wheat flour.

Another example of community involvement in conservation can be found with the Hopi Native Indian farmers in Arizona in the United States. They primarily plant native varieties of field crops (Soleri and Cleveland, 1993). Their blue maize folk variety, for example, is retained because of its adaptation to drought, its short growing season and its use in tribal rituals. It is also becoming rather widely used in commercial food products.

Rice (*Oryza sativa*) farmers in the Chaing Mai Valley of Thailand have also demonstrated that adopting improved germplasm does not neces-sarily lead to the genetic erosion of the traditional varieties. As in Peru, the indigenous varieties have a special commercial value in addition to their value as a food source. Although Thai rice farmers change varieties every three or four years, they seem to adopt more varieties from the traditional gene pool than the introduced one.

The Vishnoi people, a basically farming community found on the west coast and central parts of northern India, have communal ownership of their land and they assign rights to the whole ecosystem in the land under their influence. Their benevolent and peaceful disposition towards all forms of life precludes the cutting of living trees for wood and they have conserved innumerable forms of indigenous crops, flora and fauna for many centuries (Tiwari, 1993).

Farmer community maintenance of land races *in situ* depicted by the examples cited above as part of the local farming systems may be difficult to sustain in the future. Little research has been done on indigenous seed supply networks or seed conservation. As local communities become more assimilated into mainstream industrial society, traditional markets or indigenous germplasm exchange networks may lose their importance, making it less easy to find land race seeds.

LBC includes, at one extreme, buffer-zone protection of parks and reserves and, at the other, conservation in rural areas. Community-based conservation shifts the focus from centre-driven conservation activities to the people who bear the costs, i.e. conservation of biodiversity by, for and with the local community, for whom the agenda usually is to regain control over natural resources and, through conservation practices, to maintain or improve their economic well-being. Community-based conservation and awareness is growing globally of its own accord despite several obstacles. At this stage what is most needed is recognition of a neglected set of participants and acknowledgment by more economically favoured urban dwellers of the rural landscape's significance in conservation of biodiversity (Western and Wright, 1994).

10.6 CONSERVATION IN HOME GARDENS AND ORCHARDS

Home gardens and orchards have been described as living gene banks where indigenous germplasm in the form of land races, obsolete cultivars and rare species are preserved and thrive side by side. They were first described as a conservation resource by Anderson (1952), Kimber (1973) and Hawkes (1983). This class of conservation sites is of great importance for the conservation of certain species whose land races may now be confined to home, orchard and kitchen gardens of families retaining traditional farming systems. Thus, for example, Hawkes counted 45 species cultivated in an orchard garden and 25 more wild ones growing locally that had medicinal value in a village in Central Java, Indonesia (Hawkes, 1983).

Home gardens are not only productive in developing countries. In the United States in 1944, for example, because of the world war, 40% of the total vegetable production of the country came from home gardens (Niñez, 1986). However, there is much greater diversity of forms and species in a tropical home garden than in one in the temperate zone.

It has been estimated that agro-forestry farming systems in the tropics typically comprise well over 100 plant species per site, the products of which are used for construction purposes, firewood, simple tools, medicine, livestock feed and human food (Altieri and Merrick, 1987). For example, in Mexico, the Huastec Indians manage several agricultural and fallow fields, complex home gardens and forestry which total over 300 species (Alcorn, 1984). Areas in the vicinity of their dwelling sites are known to contain 80–125 useful plant species, mostly of medicinal value (see also Hawkes, 1983:83–89).

The high potential of traditional agricultural systems as tools for *in situ* conservation has been recognized in recent years. These systems include the gardens of subsistence farming in tropical countries of the developing world. For example, the traditional habitation sites of the people of West Java usually contain 80–125 useful plants. Of these plants, about 42% provide material for fuel, building houses and other implements; 18% are fruit trees; 14% are used as vegetables; and the remainder constitute medicinal plants, ornamentals, spice, condiments and other cash crops (Christanty *et al.*, 1986).

Esquivel and Hammer (1992) give a detailed account of the role of home gardens called *conucos* in Cuba. These are relatively large gardens where Cuban farmers practise traditional agriculture based mainly on local cultivars. These cultivars have originated from almost all centres of diversity and were transported to Cuba during its colonial past. In the sixteenth century, European influence led to the introduction of plants from the Indo-Chinese, Mediterranean, Central Asian, Siberian and Near Eastern regions. Several crops, such as, mangoes, bananas and sugarcane, reached high economic importance. Plants of African origin and medicinal plants are also conserved in home gardens. Their cultivation has been encouraged by the rural doctors who freely prescribe them due to scarcity of allopathic medicines.

When traditional systems of *in situ* conservation of local varieties break down, new means of conservation are required (Cleveland *et al.*, 1994). Community gene banks, such as established by the Swaminathan Research Foundation in India for local conservation of land races of rice, millets and pulses, can be effective in meeting this need through *ex situ* conservation. Crops that do not readily produce seed and those with recalcitrant seeds should be conserved in permanent field gene banks and maintained by farmers, such as established for coffee in Ethiopia (Worede, 1993). Heirloom varieties of horticultural crops are being effectively conserved and used in home gardens through seed savers networks; the largest is the Seed Savers Exchange (Iowa, USA) which maintains catalogues and inventories of about 16 500 varieties. The fruit and nut tree crops are likewise networked through the North American Rare Fruit Growers, Inc. (California) and native plants by the Southwest Endangered Aridland Resource Clearing House (Arizona). Similar organizations have evolved in France and Canada (e.g. Heritage Seed Program).

10.7 INCENTIVES FOR LOCALLY BASED CONSERVATION

Locally based conservation seems powerful because it is based in the fundamental principle that individuals will take care of things in which they have a long-term, sustained interest. The major problem in LBC is the establishment of strong and lasting local recognition and support for the conservation of biodiversity important to the international community. The answer may be of a constellation of incentives to farmers and communities for the conservation of genetic resources. The removal of perverse incentives (McNeely, 1988) for crop conservation, such as giving credit only for the cultivation of modern varieties, should be emphasized. For important political or social reasons, governments often institute incentives that induce behaviour in the community which depletes biodiversity. The mechanized production of soybean in Brazil and Paraguay, for instance, promoted the displacement of many farmers, who ended up settling in previously forested areas (Schumann and Partridge, 1986). Another example of a perverse incentive can be found in a study conducted by the World Bank in the late 1980s, also in Brazil. It shows that the government offered a series of incentives to quicken the rate of settlement in the Amazon basin, thereupon leading to deforestation. Other measures adopted by governmental agencies likewise encourage the depletion of plant genetic resources. An illustration would be the conversion of forests to pasture and cropland to reduce tax liability, hence leading to extreme deforestation of marginal land on large farms. In cases when it is very difficult to remove an instituted perverse incentive to biodiversity conservation, as it often occurs with agricultural incentives, the governments, then, have to institute new conservation incentives to cancel the negative impact of perverse incentives on the environment (McNeely, 1988). All mechanisms, including the pressure of public opinion, would be applied to discourage local people, governments or international organizations from exhausting biological diversity.

Positive incentives are working rules used to divert resources towards conserving biodiversity by facilitating the participation of groups or agents which will benefit these biological resources. They are relevant not only at community level but also at national or international levels. Direct (not market-driven) incentives may be practised to accomplish specific objectives and they can be in the form of cash or in-kind payments. Indirect incentives, on the other hand, encourage the conservation of biological resources without any direct economic interventions from the government or other sources. They refer to services and social incentives that are designed to improve the quality of life of the community. These include strategies such as enhancing amenities in rural areas, the development of niche markets, and building facilities to expedite the commercialization of the products (better roads, for instance). It is also important to elevate the prestige of crop resources like land races and the support base for crop conservation by encouraging amateur conservationists, farmers and land owners, natural resources groups and scientists to conserve plant genetic resources. Government and other institutional interventions should be minimized. Another incentive for farmers to keep their plant heritages is to introduce on-farm plant selection techniques that will increase productivity or quality and maintain at least part of the basic genotype of the land races.

If LBC is based on farmers' needs and participation, it is more likely to prevail for many years. It should be tied to socio-cultural organization of the community and be decentralized so that the local people would have maximum control. Minimal institutional development and intervention should take place. Economic and social incentives may be used in a package of direct and indirect incentives to integrate conservation and rural development. Two examples are presented:

1. In Thailand's Khao Yai National Park, creative rural development techniques were used together with a conservation awareness programme and extension activities, for both children and adults, to promote local cooperation in protection of the park resources.
2. In Brazil, harvest rights were established by the National Council of Rubber Tappers to create protected areas to be managed by the people in forest communities as an incentive to conserve biological resources. The sustainability of extraction that does not destroy the forest makes this approach an important alternative to agriculture and cattle farm management (Allegretti and Scwartzman, 1986) and is offered as a model for *in situ* conservation. The harvest rights were achieved when the Upper Juruá Extractive Reserve was recognized by the Brazilian Federal Government in 1991 (Lleras, 1991).

Another aspect to be considered with respect to incentives for LBC is the fact that the sociological elements of crop plant conservation are poorly understood, as well as fundamental issues in population biology, such as the desirable allele frequency to be conserved and how population size and distribution affect allele frequency (Brush, 1995). The success of conservation strategies depends on the knowledge of the population genetics (dynamics) of the taxa in question. To ensure the success of LBC, research on population biology and ecology of land races must be undertaken together with economical, socio-cultural and political issues.

An important incentive to LBC of crop plants is the establishment of breeding programmes with the participation of the local community. Land races would be improved to increase crop production, resistance to biotic and abiotic factors, response to different inputs, nutritional value of plant products (protein and amino acid composition) and end-use quality of the crop. Locally based crop improvement can be seen as both an alternative to institutional plant breeding and an incentive to *in situ* conservation of crop plants.

10.8 PLANT BREEDING AND LOCALLY BASED CONSERVATION

To improve the genetic material of their crops over time, many farmers have themselves been selecting for superior phenotypes and therefore improved crop performance. Considering that land races have the advantage of being adapted to the local growing environment and being already accepted by farmers and consumers (Teverson *et al.*, 1994), breeding for LBC of crop plant genetic resources is, like breeding for sustainability, largely a process of suiting varieties to an environment instead of modifying the environment to suit the varieties (Coffman and Smith, 1991). It requires a method to introduce improvements, i.e. change in gene frequency, with minimal disruption of the local gene pool. This would be more easily accomplished with the direct involvement of the farmers working consciously as plant breeders. The farmers' criteria for selection are frequently wholly different from those used by researchers.

If farmers are to improve their crops by on-farm selection, then there are two options that must be explored with farmers, because their seed or plant selection methods no doubt give very little annual improvement and can best be characterized as stabilizing selection. The first option is to introduce new selection criteria so that improvement can be realized without changing the basic use character of the crop. The second is to introduce new seed selection techniques, such as single plant mass selection in the field using the grid system developed by Gardner (1961) for maize. In self-pollinated crops this method may reduce the intravarietal variation and some of the innate buffering capacity of diverse varieties may be lost. To guard against this loss of diversity some selection rules would be applied, such as a low selection intensity and deliberate selection for diverse phenotypes.

Breeding stations can contribute to the process by introducing single-gene traits, such as disease resistance, by back-crossing and making the derived lines available to the farmers. This would ensure improvement and retention of the basic population genotype. In the future, specific genes will be introduced by molecular gene transformation and this method will be valuable for conservation of genetic diversity. Other methods, such as Suneson's (1956) evolutionary plant breeding method for creating segregating populations using land race hybrids at research stations, permit selection in segregating populations by farmers directly on the station or from populations that they grow on their own farms. This method was apparently successful in a bean breeding programme in Rwanda (Sperling *et al.*, 1993). Examples of other on-farm selection techniques, with breeders' interventions, have been outlined for sorghum and other crops in Ethiopia (Worede, 1993) and rice in Sierra Leone (Longley and Richards, 1993; Monde and Richards, 1994). One suggestion presented by Lenné and Smithson (1994), similar to Suneson's method mentioned above, involves the improvement of land race mixtures, following the concept that mixtures may exhibit yield and resistance improvements, sometimes exceeding the better component.

10.9 CONCLUSION

There are many challenges in practising LBC of crop plants, but there are many approaches and methods that may be applied according to local conditions. It is a conservation strategy for agricultural crops that draws upon both *ex situ* and *in situ* conservation methods. It is especially valuable in retaining the evolutionary potential of the crops themselves and their associated species, such as insects and pathogens. Locally based conservation requires continuous involvement by farmers and other community members, which in turn promotes the conservation ethic and the retention of traditional cultural values.

11

Genetic conservation information management

B.V. Ford-Lloyd and N. Maxted

11.1 INTRODUCTION

It is widely accepted that the management of genetic resources which are conserved *ex situ* will involve firstly the location and collection of the plant material and then its effective conservation, usually in some form of long-term storage. Once actively conserved, the germplasm may be characterized, evaluated and possibly regenerated as part of a routine management programme. If it is agreed that the major reason for conserving any genetic resources is to promote their utilization, either now or in the future, then detailed documentation relating to each of these activities will always be required to enhance any utilization activities. Therefore, it follows that even if the genetic resource does not leave its natural location, but is conserved *in situ*, it is just as important to document passport and characterization data for the conserved material. Any genetic resource, whether conserved *ex situ* or *in situ*, is virtually useless without reliable information describing that germplasm. If germplasm conserved in an *in situ* genetic reserve or on-farm is to be utilized, then it must be thoroughly characterized and evaluated.

It is often the case that those involved in the conservation of biodiversity tend to work at the level of documenting species within habitats or ecosystems. For *in situ* conservation to be an effective way of conserving genetic diversity and making that diversity available for utilization, then extra levels of information need to be considered, in addition to those recorded for *ex situ* conserved germplasm. For example, much more

Plant Genetic Conservation.
Edited by N. Maxted, B.V. Ford-Lloyd and J.G. Hawkes.
Published in 1997 by Chapman & Hall. ISBN 0 412 63400 7 (Hb) and 0 412 63730 8 (Pb).

detailed site descriptive data will be required, such as the precise sizes and locations of populations of each target species within each reserve, as well as accurate subspecific taxonomic detail, their isolation from other adjacent populations and local edaphic and other geographical features. It is the acquisition of this sort of information, which is still often lacking for many *ex situ* collections of germplasm and for which there has been much criticism, that is required for germplasm to be conserved effectively *in situ*. Will such demands be too great for the *in situ* conservation of genetic resources to be fully effective?

Two basic issues require addressing in terms of conservation data management: identifying and accessing the sources of existing information that may assist in the development of appropriate conservation strategies, and how to manage the data associated with specific *in situ* conservation activities (Pellew, 1991). First there is a need to clarify some of the basic terminology associated with *in situ* conservation.

11.2 TERMINOLOGY

The discussion of data management for *in situ* conservation introduces the problem of appropriate terminology. *Ex situ* conservation is substantially based upon the **accession** – the population sample of seeds conserved within a gene bank. This cannot easily be accepted as a term for the population sample conserved by *in situ* conservation. Historically the accession has been defined as a sample of the target taxon transferred to a remote location for active conservation. It is only once the population sample leaves its original location that it is referred to as an accession. Therefore the *in situ* conserved material can not be an accession; firstly, because it is does not leave its original location and secondly, because it is not a sample in the same sense as an accession: it is a sample of the entire gene pool, but may contain an entire population, which would rarely be the case for an *ex situ* collection.

Recognizing this problem, Williams suggests in Chapter 5 that within spatial conservation units of varying size, **genetic resources management units** (GRMUs) should be recognized, each of which would be made up of one or more species populations. The clear arguments are that terminology needs to be defined which those involved in the science or politics of conservation can understand, and that GRMUs are already in use within forestry conservation circles. The problem with the use of GRMU is that it can refer to an entire reserve or a subunit within a reserve, or to several species or just the target taxon, and therefore cannot be equated to an *ex situ* 'accession'. It would be more appropriate to refer to the unit of *in situ* conservation as the 'population', even though this is likely to have a slightly different connotation to that applied in other spheres, such as ecology or genetics. In this context a **population** is an interbreeding group of individuals representing a particular taxon that is being actively conserved and managed as a single unit. It should be noted that a reserve may contain several isolated populations.

11.3 EXISTING INFORMATION SOURCES CONCERNING GENETIC DIVERSITY

Conservationists will always have limited resources for conservation, and this will necessitate a choice or selection of alternative target taxa (Chapter 3). It is important that this choice is as objective as possible and to assist this process it is important to have adequate background data on which to base the choice of target taxa.

11.3.1 Information sources at the ecosystems and species level

Even if purely at the species level, information on the conservation status of species and ecosystems will be crucial to planning and implementing conservation activities. The World Conservation Monitoring Centre (WCMC), a joint undertaking of the IUCN, UNEP and WWF, is a primary source of information on the conservation status of many species of plants and animals. It maintains a database of published literature, unpublished reports, government reports, and references to conservation organizations, contacts and correspondents throughout the world. The detailed biological information that is held covers the distribution, ecology and status of about 52 000 plant species, and data include population size, potential or actual threats and occurrence in cultivation. Also listed are key sites of high biodiversity and the locations and importance of about 16 000 protected areas world wide. Mapped digitized data also exist in the Global Biodiversity database, which is available for browsing.

For effective conservation at the species level, an understanding of the taxonomy of any target group will necessarily precede the interpretation of its conservation status. For example, if a target species has commonly used synonyms (nomenclaturally illegal names that refer to a taxon with a different accepted name) or the name being used is itself a synonym, then this may confuse interpretation of the conservation status for the accepted taxon. The importance of having a good taxonomic understanding of the target group prior to conservation is underlined by IBPGR (1985b) and Maxted *et al.* (1995). The primary sources of taxonomic information concerning any genetic resource are specialist publications (revisions, monographs, checklists, taxonomic journals, Floras and Faunas), educational materials (field guides, lectures, multimedia programmes) and taxonomic experts. Specialist publications are likely to be found in the libraries associated with various kinds of taxonomic collections (e.g. museums, herbaria, botanical gardens, germplasm collections and zoos), while general libraries will provide field guides and other media guides. The conservationist may gain access to taxonomic experts by contacting the taxonomic collections or educational establishments with which they are usually associated. When reviewing the taxonomic literature, it is worth consulting such publications as the *Kew Record of Taxonomic Literature* (now *Kew Record*), which is published annually by the Royal Botanic Gardens, Kew, and attempts to act as an abstracting source for the plant taxonomic literature.

Traditionally a conservationist studying a group would have undertaken a literature survey, but a contemporary review will include all media – microfiche, diskettes, multimedia CD-Roms and on-line services (BIDS, BIOSIS, CABI, etc.), as well as traditional printed text. The quantity of data associated with taxonomic information is vast and in recent years, to enhance the efficiency of data handling, computer database technology is increasingly being used to organize these data. Since the 1980s many multi-institutional database projects have begun to collate biodiversity information for either specific taxonomic groups or geographical regions. For example, ILDIS (International Legume Database and Information Service) has established a botanical diversity database for the 17 000 legume species (Zarucchi *et al.*, 1993). This project is managed as a cooperative, involving approximately 20 research groups from five continents, and the information system is available internationally. Similar projects have been established for several other groups (Table 11.1).

The long-term goal must be to combine these and other taxonomic databases to make a single biodiversity database, which would prove an invaluable tool to the conservation community. The examples provided in Table 11.1 are largely centred on specific taxonomic groups, but there are also several geographically restricted projects, such as ERIN (Environmental Resources Information Network), which is an Australian initiative to provide a geographically related environmental information system that will aid environmental impact assessment and monitoring of species, vegetation types and heritage sites within Australia (ERIN, 1991).

11.3.2 Information sources at the plant and population level

At the population and individual plant level, those involved in the *ex situ* conservation of plant genetic resources have been concerned with acquiring and storing information alongside germplasm for many years and have employed many documentation systems to assist with the task. The International Plant Genetic Resources Institute (IPGRI) has for some time been developing standards for germplasm documentation. The Institute has been producing descriptor lists for over 15 years, during which time there has been substantial evolution in their accepted format. Originally,

Table 11.1 Examples of database projects (from Bisby, 1994)

Group	Project	Location
Bacteria	DSM. List of valid bacterial names	Deutsche Sammlung von Mikroorganismen und Zellkulturen, Braunschweig
	BIOSIS TRF. BIOSIS Bacterial Taxonomic Reference File	BIOSIS Philadelphia
Protists	ETI. Linnaeus Protists	Expert Center for Taxonomic Identification, Amsterdam
Insects	ANI. Arthropod Name Index	CAB International, Wallingford
Molluscs	ETI. Linnaeus Molluscs	Expert Center for Taxonomic Identification, Amsterdam
Fish	Eschmeyer, W.N. Taxonomic Database for Fishes	California Academy of Sciences, San Francisco
Fungi	IMI. Species Fungorum Database	International Mycology Institute
Plants	ILDIS. International Legume Database and Information Service	ILDIS Coordinating Centre, Southampton
	CITES. CITES Cactaceae Checklist	Royal Botanic Gardens, Kew

minimum descriptors were defined on the basis of consensus amongst leading scientists as to what was the key information that was needed for any given crop. More recently the aim has been to compile the widest range of descriptors for each crop, resulting in the best possible standardization, and leaving the actual choice of which descriptors to be used for characterization of germplasm to the individual scientist (for example, Descriptors for Barley: IPGRI, 1994; Descriptors for *Capsicum*: IPGRI, 1995). Latterly emphasis has been placed upon recording more precisely environmental descriptors, a move that will obviously benefit the detailed description of reserves and sites for *in situ* conservation. All crop descriptor lists are freely available on request from IPGRI. Recently, IPGRI has produced a guidebook for germplasm documentation, which provides guidance on good documentation practice, rather than that concerned with the description of any specific crop or crop relative group (Painting *et al.*, 1993).

For passport, characterization and evaluation data many national and international crop databases now exist resulting from the formation of genetic resources crop networks (Frese and Doney, 1994). Participants generally provide specified data to a database host, where resides the responsibility of compiling, analysing and distributing the data to the participants. All available information can then be used to plan future conservation activities, characterization of germplasm or replenishment of seed stocks. At present these networks are almost exclusively focused on *ex situ* collections, but will in the future be complemented by the addition of information concerning *in situ* conserved germplasm. They will be able to provided useful background information on taxa to the genetic reserve manager.

There is little point in establishing a new genetic reserve if the target taxon is already adequately conserved using both *in situ* and *ex situ* techniques. Therefore it is important to know what material is currently being conserved and any associated characterization and evaluation data that can be obtained from botanical garden, gene bank and genetic reserve catalogues and databases. Searching these sources can prove time consuming. In an attempt to alleviate this problem national and international directories and databases of conserved material have been and are being established by a variety of national and international agencies, for example the European Cooperative Programme on Genetic Resources (ECP/GR), the International Plant Genetic Resources Institute (IPGRI) and the FAO Commission on Plant Genetic Resources.

Within the United States, the National Genetic Resources Program (NGRP) acquires, preserves and distributes genetic resources nationally and to the rest of the world. The programme incorporates a range of different activities focusing not only on plants but also on forest tree species, animals, aquatics, insects and microbes. The National Plant Germplasm System (NPGS) handles US plant genetic resources activities, while the National Plant Genome Research Program (PGRP) deals with all aspects of genome work. Importantly, AGIS (Agricultural Genome Information Server) on the World Wide Web consists of genome information for agriculturally important organisms, currently mostly plants, and allows for links to be made to GRIN (Genetic Resources Information Network) via germplasm accession identifiers (see below for further discussion).

11.4 MECHANISMS FOR ACCESSING EXISTING INFORMATION

One of IPGRI's main roles has always been as an information service, providing the plant genetic resources community with literature-based information, primary documents and bibliographical data. Published crop descriptor lists give important guidelines for the establishment of new databases throughout the world (IPGRI, 1994, 1995), while catalogues of germplasm holdings are a good starting point for anyone wishing to access germplasm in different parts of the world. Equally the WCMC provides documentary information about species distributions and conservation status. However, the way forward for worldwide access to information, whether it be about ecosystems, species, plants, genotypes, genes or gene sequences, is the use of the contemporary computing networks, such as the Internet.

The Internet consists of a global linkage between local computer networks which allows the transfer of mail, news, files and, most importantly, information. At the time of writing, about 50 countries contribute to the net with about 20 million users connected each day. Using the World Wide Web (WWW) interface to the net, it is possible to access the 'home page' of many universities and institutes which are concerned with genetic resources conservation. Via hypertext links to these home pages it is then possible to read a wide array of documents, including those describing the activities and mandate species of gene banks, storage conditions used, numbers of accessions stored and passport and characterization data related to individual accessions. In order to get at such important information it is often necessary only to follow menu-type instructions. With an appropriate 'web browser' such as Netscape or Mosaic, and a 'search engine' such as Lycos or Infoseek, it is simple enough to enter key words such as 'genetic resources' and then, for example, locate home pages for the UK National Plant Genetic Resources Working Group, the Nordic Genebank, the National Genetic Resources Program in the United States or IPGRI.

Developing forms of access that will allow the integration of information of widely differing kinds is now being undertaken by the US Department of Agriculture. Under the Plant Genome Research Programme, the RiceGenes Database is being developed to serve as a repository of all rice genetic information, which can be accessed freely using several different software interfaces. The database focused initially on genetic maps and molecular markers, and actually includes map information on maize as well, for comparative mapping purposes. It includes molecular profiles of genetically diverse material from germplasm collections, and GRIN (Genetic Resources Information Network) accession numbers are also included to allow for full access to the passport data for any material. While current references are largely to germplasm stored *ex situ*, as more and more scientific information is accumulated about genetic resources being maintained *in situ*, there is no reason to think that *in situ* reserve catalogues could not in the future be linked with a database such as RiceGenes – provided that the quality of the data being accumulated is sufficiently good and that documentation procedures are carried out correctly.

11.5 DATA GENERATION IN GENETIC RESOURCES

Much of the information generated by a functioning genetic resources collection, either *in situ* or *ex situ*, is of interest and value to some part of the scientific community. Information from specific studies will often be published in reports, catalogues or journals. The information produced during conservation activities will be of critical management importance for setting priorities, planning activities and managing resources in the reserve or gene bank. There is therefore an inherently central role for documentation of genetic resources within *in situ* conservation. Documentation also provides a basis for effective utilization of the genetic resources by the users of those resources. Perry *et al.* (1993) discuss the processes that generate *ex situ* genetic resources data and distinguish the following data types:

- **Passport data** are generated as a result of plant collection and introduction activities. Data are compiled when the accession is collected and when it enters the country and institution where it will be maintained.
- **Characterization and evaluation data** are used to describe the genetic resources. They assist in the maintenance and use of the genetic material and are, together with passport data, the basis for the rationalization of the genetic resources held in collections.
- **Management data** are essential for maintaining the viability of the genetic resources and their distribution to users. These data are used, among other things, to monitor the condition (e.g. viability) and assist with the safe management (e.g. amount of stock, location of stock, etc.) of the genetic resources.

As will be shown in the next section, these types of data are just as applicable for *in situ* genetic resources conservation.

11.6 THE DOCUMENTATION OF *IN SITU* CONSERVED GENETIC DIVERSITY

For *in situ* conservation of plant genetic resources to be effective, we need to consider the acquisition and use of information from a range of sources which can be conveniently described at different levels. We could, for instance, consider it to be necessary to monitor *in situ* reserves at the ecosystem level, and to be able to monitor global change by considering anthropogenic impacts on those ecosystems including the actual effects of management activities. We may wish to monitor very carefully, for example, the effects of any land use changes or, over a period of time, the effects of climate change upon the components of biodiversity. The purpose will be not just to make an inventory of biodiversity as it is now, but also to determine its sustainability: to make an assessment of its likely change with time – in other words, to monitor erosion.

The problem as far as 'levels of information' are concerned is that biodiversity can be measured at different levels and therefore give rise to different levels of information. Remote sensing, for instance, may allow us to acquire information about genetic erosion very quickly, and may tell us about changes in vegetation patterns, habitat sizes or even species occurrences in some cases. At the next level, biodiversity needs to be assessed in terms of numbers of populations and numbers of individuals within populations of target species. Without some sort of assessment of the genetic variation within and between these populations, the value of the biodiversity as a genetic resource is limited. The value can only be assessed if we know what sort of characteristics are possessed by plants within populations, and if any one population is substantially different to another adjacent or more distant population. Knowing whether plants possess particular disease resistance characteristics or are especially drought tolerant, or whether they are much more variable genetically than others, is very important information from the point of view of potential use and provides us with a useful yardstick with which to monitor biodiversity loss. Not actually at a different level of information, but adding to the difficulties of information gathering and management, is the need to be able to determine that the characteristics which we are observing, whether disease resistance, drought tolerance or simply amount of variation, are truly genetic components of biodiversity and not just manifestations of environmental variation. Ideally we need to be recording information on genes in order to be able to make best use of our information in the future, and if we are to be really serious about monitoring genetic erosion or change from the biodiversity point of view, then it is genes that once again should interest us as representing the finest level of detail.

The simple way forward is to consider the sort of descriptive information which is recorded for *ex situ* purposes. A substantial number of passport and environment/site descriptors (IPGRI, 1994, 1995) devised for *ex situ* conservation could be applied equally well to plants and populations growing in reserves. **Passport descriptors** are those which provide the basic information used for the general management of the accession and describe parameters that should be recorded when the accession is originally collected (discovered and observed in the case of *in situ*). **Environment or site descriptors** describe the conditions under which germplasm is collected, and are certainly applicable in the *in situ* sense. It is also possible that some characterization could be attempted under natural conditions in the same way as for *ex situ* purposes. **Descriptors for characterization** enable an easy and quick discrimination between phenotypes, and should represent highly heritable characters, easily seen by the naked eye. If they are equally expressed over a range of environments then they lend themselves to *in situ* information gathering. If characterization is the responsibility of the gene bank curator in the *ex situ* sense, then it should also be the responsibility of the *in situ* reserve manager.

One of the most important differences between the information requirements of the two different conservation strategies is related to the need to monitor changes which might take place within designated *in situ* reserves. Not only will species composition of the reserves need to be monitored, but also fairly accurate and regular measurement of individual population sizes and density may be required. These extra descriptors will need to be added to those used for *ex situ* conservation. Population size and density should be recorded for any germplasm collected and this will apply to germplasm conserved in a genetic reserve, but with the latter there will be a much greater need for accurate recording of associated taxa, such as keystone taxa, whose numbers and density may affect the number and density of the target taxon. The data collated from the reserve will, unlike *ex situ* passport data, be cumulative. To identify population trends, the accumulated data of many years will need to be analysed together to distinguish cyclical changes from population slumps.

In addition to the information at the gene level (characterization data), consideration needs to be given to variation expressed by **molecular data**. The objective of genetic reserves is to retain not only population numbers or density, but also diversity (Chapter 12). Molecular techniques can provide us with markers which may allow us to seek out the existence of genes of use to plant breeders, to identify plants at the taxonomic level, or to make very accurate assessments of the relative quantity of variation within populations and therefore to determine which populations to target for conservation. Consequently, it will then be possible to monitor the future existence of that population and whether genetic erosion is taking place. In the future, perhaps, one of the biggest challenges to those involved in the conservation of genetic resources, whether *in situ* or *ex situ*, will be the linking of data from the different levels such as populations, plants, genes and gene sequences (*cf* RiceGenes Database) to allow greater use of germplasm and understanding of patterns of variation and evolution in the future.

It is assumed that the efficient way to collate, manage and analyse these data will be to design and build a reserve or multiple-reserve database. The general advantages of using a database rather than a paper-based system include the avoidance of inconsistencies, the ability to share data between different programs, the ability to enforce standards, repeated editing, remote access, etc. The reserve database system will be no less important to the effective running of the reserve than is the gene bank management system to the effective *ex situ* conservation of germplasm. The development of a documentation system for plant genetic conservation is discussed in detail by Painting *et al.* (1993); it focuses on the development of an *ex situ* documentation system, but numerous sections are equally applicable to *in situ* documentation.

11.7 CATEGORIES OF DOCUMENTED INFORMATION

There are five basic kinds of data that need to be recorded for the 'populations' found in genetic reserves:

- Nomenclatural – information concerning the accepted taxonomic name of the particular population at any site.
- Curatorial – information used to manage and identify populations.
- Descriptive – information used to distinguish and describe taxa.
- Management – information related to the management plan for any particular site within the particular genetic reserve.
- Monitoring – information related to the monitoring policy for populations in a particular genetic reserve.

What follows is a general scheme for the organization of descriptive information within the *in situ* context. As stated above, these descriptors will overlap in a substantial way with *ex situ* passport descriptors falling under the heading of collection descriptors.

11.7.1 Nomenclatural information

This is the information concerning the accepted taxonomic name of the particular population – its scientific name (genus, species, subspecies, botanical variety) – and is essential passport data. It may also include other nomenclatural details, such as authorities, publication details, typification details and commonly used synonyms. The scientific name is an internationally recognized name for the crop and facilitates the collation of information on a species from many disciplines.

11.7.2 Curatorial information

This consists of information that is used to manage and identify particular populations within a reserve. The data basically relate to the practical maintenance of the germplasm and its distribution to any user. They include the population number, site number, initial date of location, precise location of population within the reserve and location of any duplicate *ex situ* samples. To conserve a population effectively, it is important that, once identified, it is registered and given a unique identifier, as is the case for *ex situ* conserved germplasm. This will be referred to as the

population number and is unique to that population; such numbers act as a tag to which all other relevant information is tied during the conservation process. In order to avoid the situation where the same number is accidentally assigned to more than one population, it is good documentation practice to use a single database file to register all populations, at least within one reserve.

11.7.3 Descriptive information

These data describe the reserve, site and population characteristics. Each reserve is likely to contain more than one site where the target taxon is located and therefore more than one population of the target taxon. Any particular site within a reserve will contain more than one species: the target taxon, and the non-target taxa which happen to be sympatric with the target taxon. It will be desirable to record the standard set of descriptors for the target taxon, but it may also be desirable to record some descriptors for some or all of the non-target taxa, if conservation of the non-target taxa at the particular site is integral to the conservation of the target taxon. These factors should be considered when designing the reserve database.

The descriptive data will include both passport and characterization data. These types of data have received considerable attention in the *ex situ* context and, in order to facilitate standardization of data, lists of descriptors have already been developed by IPGRI for many crops or crop complexes. These descriptors can be readily adapted for *in situ* use by the addition of some of the reserve and site descriptors discussed below.

The descriptive data fall into three categories: those which describe the general characteristics of the reserve, those which describe the specific characteristics of the site within the reserve where the population is found and those which describe the population itself. A distinction is made between the reserve and site descriptors because, for example, it would not generally be possible to record one latitude, longitude and altitude figure for the whole reserve unless the reserve was very small, but this is feasible and desirable for the actual site where each population is found within the reserve. In this context it may be more appropriate to think in terms of using a Geographical Information System rather than a standard database management system to document the spatial diversity within the reserve accurately.

(a) Reserve descriptors

These will describe the general characteristics of the reserve as a whole (Figure 11.1) and may be subdivided under the following headings.

Figure 11.1 Recording reserve descriptor on record sheets in the Caucasus mountains of Georgia.

- **Reserve location**, including details of the country, province, nearest settlement, location in relation to the nearest settlement, ranges of latitude, longitude and altitude.
- **Abiotic reserve description**, including general details of the physical description of the reserve: climate, rainfall, topography, parent rock, soil type, soil pH, depth of soil, aspect, slope, percentage ground cover, water relations, land use, percentage rock cover, type of rock cover.
- **Biotic reserve description**, including general details of the local ethnic group, dominant vegetation, dominant herbivores, grazing pressure, percentage tree cover, type of tree cover.

(b) Site descriptors

These will describe the specific characteristics of the site within the reserve where a particular population of the target taxon is located and may be subdivided under the following headings.

- **Site location**, including details of the precise location within the reserve, latitude, longitude and altitude.
- **Abiotic site description**, including specific details of the physical description of the site within the reserve: climate (temperature, rainfall, wind, frost, humidity), topography, parent rock, soil type, soil pH, depth of soil, aspect, slope, slope form, percentage ground cover, soil drainage, salinity, soil fertility, land use, rock type, percentage rock cover, type of rock cover.
- **Biotic site description**, including specific details of the biotic features of the site and their relationship with the target population: dominant and associated vegetation, dominant and associated herbivores, agricultural or forestry practice, grazing pressure, percentage tree cover, type of tree cover.

(c) Population descriptors

These will describe the specific characteristics of a population within the reserve and may be subdivided under the following headings.

- **Population characteristics**, including details of the number of individuals, area covered by the population, density, location of individuals and their mapping, causes of genetic erosion, whether voucher specimens have been taken.
- **Population genetic characteristics**, including details of any molecular variation studied.
- **Population descriptors**, largely conforming to those which, within the *ex situ* context, are used to describe germplasm accessions, and fall into the category of characterization and preliminary evaluation descriptors.

The recording of further characteristics of any populations within the reserve will largely be characterization rather than evaluation because of the basic difference between the two: **characterization** consists of recording those descriptors which are highly heritable, can be seen easily by eye and are expressed in all environments, whereas **evaluation** consists of recording those characters which are susceptible to environmental differences. Therefore, as a population will only have its characteristics recorded at one site, it can only be characterized; if evaluation is required, the germplasm will need to be taken into *ex situ* conservation. In practice, however, when a population is being characterized, a limited number of preliminary evaluation characters (such as any pest or disease tolerance) may be recorded.

11.7.4 Management information

This category of information relates to the particular management plan for the target taxa within the particular genetic reserve. The process of developing a management plan is discussed in detail in Chapter 9. The data that will be recorded as part of the management regime will describe the allowable human intervention in the reserve, such as the actual timing, frequency and duration of mowing, grazing, selective harvesting, genetic resource exploitation, etc.

11.7.5 Monitoring information

This category of information is related to the monitoring policy for the particular genetic reserve. The process of monitoring target taxon population levels in a genetic reserve is discussed in detail in Chapter 9. Even within a small reserve it would be impossible to record and monitor every plant, so the conservationist is forced to take samples of data that, if effectively selected, will reflect the overall picture for the target taxon within the reserve as a whole. The conservationist samples at both the taxonomic and population levels and must select key indicator taxa (including the target taxon) and sites within the reserve to monitor their population characteristics.

Estimations of population characteristics are usually taken from quadrats or transects within the reserve and the data recorded reflect levels of density, frequency and cover. Each will be expressed as numeric values whether absolute values, percentages, demographic structures, distribution pattern, biomass or some form of scale (Braun–Blanquet scale or Domin). The precise details of what monitoring data are recorded will depend on the target taxon, resources available and management plan. Having collected the data for a particular monitoring exercise the conservationist will want to compare the population characteristics of the survey with previous surveys to be able to draw conclusions about any significant changes that may have occurred in the target populations. So the process is cumulative, and is analogous to the routine monitoring of germination of accessions in a gene bank in *ex situ* conservation.

Just as with *ex situ* documentation, it is important to use a well-defined, tested and rigorously implemented set of descriptors to ensure consistency from one year to another so that the results of the sequential monitoring exercises are directly comparable. It would be pointless, when assessing population density, to use the Braun–Blanquet scale or Domin scale in alternate years. Where possible, as with all documentation it is important to use existing and widely accepted descriptors and descriptor states. When data are recorded, they have to be classified and interpreted; considerable time can be saved by consulting a predefined list of descriptors and descriptor states. The use of lists ensures uniformity and reduces error. People using the same lists will be able to exchange data readily and interpret the data with few, if any, problems.

The data that will be recorded as part of the monitoring regime will include taxa monitored, location of quadrats and/or transects, frequency of monitoring, target taxa statistics (number, density, cover and frequency), etc.

11.8 LINK TO *EX SITU* UTILIZATION

Routinely, within *ex situ* collections, samples of germplasm are requested and sent out to those who wish to utilize the germplasm. When the material is transported to those requesting the sample, it is desirable also to send the associated passport data, which will include nomenclatural, curatorial and descriptive data. The latter will describe the original site of collection and any characterization data specifically associated with that accession. This activity of providing passport data to those wishing to utilize the germplasm will be just as important for germplasm conserved *in situ* and so the data associated with a population should be stored in an appropriate manner to permit easy production of passport data summaries to accompany germplasm leaving the reserve to be evaluated and utilized.

12

Estimation of genetic diversity

H.J. Newbury and B.V. Ford-Lloyd

12.1 INTRODUCTION

Plant conservation means different things to different people. To some it means ensuring that a particular species is represented in the flora of a particular region. To the contributors to this book, this objective is not sufficient; the primary aim is the conservation of the range of genetic diversity present within a plant taxon, so that we are actually considering the maintenance of a gene pool and the conservation of alleles. This may appear a facile comment to informed readers, but seems a point worth making since if one is really only interested in species survival at particular sites then what follows in this chapter is quite irrelevant.

Both *ex situ* and *in situ* conservation programmes will generally focus on a target taxon such as a group of wild relatives of a crop species. If it is accepted, firstly, that the conservation of variation is the primary aim and, secondly, that only genetic variation can be conserved (or indeed inherited), then clearly methods of measuring genetic diversity have an important role within conservation programmes whether *ex situ* or *in situ*. Within *in situ* programmes, the ways in which genetic diversity measurement could be useful are as follows.

- To help to make a decision about whether an *in situ* reserve is actually needed. If measurements of variation show that all (or, more realistically, a large proportion) of the variation that can be found in the field already exists in *ex situ* collections, and these collections are deemed safe (e.g. seeds with long viability that have been effectively duplicated in established gene banks) then it may be that *in situ* conservation will be considered unnecessary.

Plant Genetic Conservation.
Edited by N. Maxted, B.V. Ford-Lloyd and J.G. Hawkes.
Published in 1997 by Chapman & Hall. ISBN 0 412 63400 7 (Hb) and 0 412 63730 8 (Pb).

- If *in situ* conservation is to be used, then appropriate sites have to be selected. Although political, economic and cultural factors will always play a role here, basic data on the distribution of variation within the target taxa within potential *in situ* reserves will be essential to facilitate the selection of an optimum site; molecular marker data can reveal the distribution of such variation.
- To monitor any change in the pattern of diversity within an *in situ* reserve, including studies of population dynamics, genetic erosion and viable population size.

Classical methods of estimating diversity among groups of plants have relied chiefly upon morphological characters. To facilitate this, for most important crop species there is a list of descriptors that can be scored for each accession. The use of morphological and other field measurements has, of course, been extremely valuable in the past but suffers from several difficulties. One widely recognized problem is that, for the measurement of genetic diversity, the use of morphological characters is an indirect method. Whilst one would expect the various measured characters to be under genetic control, it is clear that environmental conditions influence plant performance for many morphological traits. It is also clear that performances for some traits are more strongly affected by the environment than others. Because of these factors, it is quite possible to find differences in the patterns of variation revealed by analyses of morphological data collected in different environments using the same plant genotypes. Furthermore, the number of morphological characters that can reasonably be measured in field trials is necessarily limited, and may be quite low for some crop species. For example, in the case of beet they may be limited to two or three characters such as male sterility and multigenicity.

Nevertheless, morphological characters still play a central role in the analysis of variation in crop species and their relatives, largely because their collection does not require expensive technology. Some morphological characters are also of clear agronomic importance. In rice, which will be used as an example throughout this chapter, characters such as culm height, diameter and number, days to flowering and panicle size, which are all routinely collected during field evaluations (IRRI–IBPGR, 1980), have been selected as key features of the new plant type (Khush, 1993). Morphological and ecogeographic characterization has led to the recognition of three main races (Chang, 1976). The *indica* ecotype is predominant in tropical and subtropical regions, whilst the *japonica* and *javanica* ecotypes (often considered jointly as a single *japonica* group) occur in temperate localities. The identification of these classes has been important to breeders since crossability problems exist between *indica* and *japonica* types. However, if one accepts crossability as the critical difference between these two groups, then it has become clear that morphological data, though useful, are often inaccurate in distinguishing them. The effectiveness of various molecular marker techniques for this purpose will be compared later.

12.2 AN INTRODUCTION TO MOLECULAR MARKERS

Over the last two decades, diversity studies have made use of the concepts and techniques that have flowed from molecular biology. Initially, the methodologies of denaturing gel electrophoresis were applied and some groups used the banding patterns of seed proteins to compare material (Cooke, 1984). A more widespread method has been the separation of proteins in non-denaturing gels followed by staining for specific enzyme activities (Hamrick and Godt, 1990). Starch gels are often used for this purpose. The relatively poor resolution of such gels is not normally important since the stained banding patterns of the isozymes obtained are usually quite simple; an advantage of this matrix is that thick gels can be sliced and each slice stained for a different enzyme. The banding data frequently allow simple genetic analysis since the bands represent the protein products of specific alleles, but there is often more than one locus encoding proteins with a particular enzyme activity. The technique produces co-dominant markers; that is to say, if a plant is heterozygous at a locus, then the products of both alleles are visible as bands of differing mobilities within a gel. The technique is fast with only small amounts of biological material required. A major constraint is that there is a limited number of enzymes for which reliable staining protocols, resulting in the formation of a coloured product in a gel, can be formulated. Hence the overall numbers of markers that can be obtained, and the subsequent level of resolution of diversity assessments, are limited. However, the methodology is particularly effective for the classification of germplasm into groups. For example, in *Oryza sativa*, eight enzyme activities (encoded at 15 loci) were used to classify an extensive range of germplasm into six major groups (Glaszmann, 1987, 1988). Within this isozyme classification, the *indica* and *japonica* types are represented as groups I and VI, with the discrimination of four other types within *O. sativa* germplasm. This method for classifying rice germplasm has proved extremely useful during rice breeding programmes. Isozyme technology has also been used to study evolutionary relationships within *Oryza* and to identify several discrepancies in taxonomy based on morphological characters (Second, 1985).

In many laboratories, protein markers have now been superseded by markers resulting from polymorphisms in DNA. One advantage of using these is that in analysing variation within the genetic material itself one is avoiding potential difficulties with differential gene expression due to developmental state or environmental influences. A major factor influencing choice of scorable characters is the enormous number of independent DNA markers that are available for each genotype. DNA has a simple structure consisting of linear sequences of four types of base (nucleotide). It is the very large number of bases within a plant genome that makes it an enormously valuable database for the study of plant diversity. Genomes can range in size from the relatively small 1C values of 0.2 pg for *Arabidopsis* and 1.0 pg for rice to 17 pg for hexaploid wheat and over 40 pg for some lilies (Bennet and Smith, 1991), so that in practical terms the amount of information available is essentially unlimited. But how can differences between plant genomes be detected? Ideally, one would like to have available the entire nucleotide sequences of plant genomes for comparison, but as can be judged from the enormous effort being expended in the sequencing of the human genome, this is not currently practicable. Indeed, the sequencing of specific loci from different plant taxa for comparative purposes is still beyond the scope of many laboratories interested in biodiversity. Instead, we have to employ techniques that sample the genome in some way. In some cases, those parts of a genome that are complementary to a particular test sequence are sampled by labelling that sequence and using it in hybridization experiments. In other cases, the genome is sampled using short DNA primers to allow amplification of target sequences by PCR. In almost all cases, polymorphisms are visualized as banding differences either on gels or on autoradiographs.

12.3 GENERAL TECHNOLOGIES FOR DNA-BASED MARKERS

For all of the techniques that result in DNA-based markers, the first step must be DNA extraction. For many crop species, techniques are well developed and routine, but problems can sometimes occur when one is extracting from material that is rich in certain polysaccharides or tannins. There are many general references describing extraction methods (Draper and Scott, 1988; Boffey, 1991; Croy *et al.*, 1993; Weising *et al.*, 1995) but it is often necessary to refer to papers on the species of interest for detailed information. Different protocols may be followed depending upon whether the sample is going to be used for hybridization studies, which usually involve microgram quantities, or amplification using PCR, in which case only a few nanograms of DNA may be required.

Many DNA-based markers are produced using DNA hybridization technology in a search for RFLPs (restriction fragment length polymorphisms). These RFLPs occur where the distances between endonuclease restriction sites at specific loci differ between genotypes. The endonuclease restriction enzymes used typically recognize groups of four or six base pairs at which they cut double-stranded DNA. Dozens of restriction enzymes are commercially available and can be used independently to search for variation at specific loci. Cutting total DNA with any of these enzymes results in a complex mixture of fragments. In order to examine a subset of these, Southern analysis (Southern, 1975; Scott, 1988; Rivin, 1986; Sambrook *et al.*, 1989) is employed. The digested DNA is subjected to agarose gel electrophoresis and a specific locus can be examined by allowing complementary fragments to hybridize to a DNA probe. For technical reasons the separated DNA fragments are transferred from the gel on to a filter and this is incubated with a labelled probe sequence. The label employed is usually radioactive, and this may be an important consideration in the selection of a technique for measuring diversity. Hybridization to specific genomic fragments is recognized by autoradiography. Differences between two genotypes may result from point mutations creating or destroying restriction sites, or because of reorganizations of blocks of DNA, such as deletions or insertions, between restriction sites.

Alternatively, DNA-based markers can be obtained using PCR (polymerase chain reaction) technology (Erlich, 1989). Like many molecular biology methods, this procedure was made possible following the discovery of a new class of enzymes; here, the critical enzyme can copy DNA at elevated temperatures and the first employed was *Taq* polymerase (purified from the hot-spring bacterium *Thermus aquaticus*). In order to be copied, single-stranded DNA must be available as a template but there must also be an existing stretch of complementary (copied) DNA, hybridized to the template to form a section of double-stranded DNA, from which copying is initiated. In the PCR, specific regions of the genomic DNA are copied following the addition of short single-stranded sequences which hybridize to them. These sequences are called primers, and use of two primers which will hybridize to the ends of a specific gene will direct the copying of that gene alone. The PCR involves the repeated cycling of temperatures which result in copying, and then copying of copies, and so on, producing an exponential increase in the amount of specific copied sequence in the reaction tube. Since the amount of plant DNA placed in the tube at the outset is very small, by far the major DNA component at the end of the reaction cycles is the copied sequence. The reaction product(s) can be separated by gel electrophoresis and since they can often be detected by direct staining the use of radioactivity and autoradiography is usually avoided. As will become apparent later, the technique can be employed to produce markers from a range of different DNA types by selecting appropriate primers (Weising *et al.*, 1995; Sambrook *et al.*, 1989). The advantages and disadvantages of using different classes of markers will become apparent in the following sections, but Table 12.1 makes comparisons between markers with respect to a list of important technical features.

Table 12.1 Comparison between molecular marker techniques (after Scott and Williams, 1994; Westneat and Webster, 1994)

Technical criterion	Isozymes	Multi-locus minisatellites	Single-locus probes	PCR-amplified microsatellite	RAPD
Development time	2–4 weeks	1 month	1–3 months	2–6 months	1–2 weeks
Amount of material needed for analysis	About 100 mg of plant tissue	ng – µg DNA	pg – ng DNA	pg – ng DNA	pg – ng DNA
Processing time	2 days	2–4 weeks	2–4 weeks	<1 week	<1 week
Ease of use	+++	+	++	++	+++
Identification of allelic states	Yes	No	Yes	Yes	No
Rate of detection of genetic variation	Low	High	Moderate	High	High
Sensitivity of technique to slight changes in protocol	Low	Low	Low	Moderate	High

12.4 TYPES OF SEQUENCE WITHIN THE PLANT GENOME

Very early in the investigation of plant genome structure it became apparent that nuclear DNA comprises two general classes of base sequence. The first is usually referred to as unique sequence, although there may actually be more than one copy per haploid genome. Unique sequence DNA includes the vast majority of the genome that is expressed (i.e. the estimated 40 000–100 000 genes within the nuclei of each plant). The major exception is the ribosomal RNA genes which are usually present in sets of hundreds of randomly repeated copies. The second general class is repeated sequences, which often make up 80% of the total DNA within a plant genome.

Evidence that repeated sequence DNA was present within plants initially came from experiments in which DNA was made single-stranded (denatured) and then allowed to rehybridize to itself (renature) by adjusting the temperature. The initial rates of renaturation could only be explained by the existence of very large numbers of highly repetitive sequences. For example, using pea DNA, only 15% of the genome behaves as unique sequence DNA (Murray *et al.*, 1978). Experiments in which plant DNA was subjected to caesium chloride gradient fractionation revealed so-called satellite bands which were later shown to consist of tandem repeats of the same sequence that were separated since they happened to have a different density to that of the rest of the nuclear genome. Such data, along with studies involving restriction enzyme digestion, have allowed assessments of the number and lengths of repeated sequences in many plant species (Flavell, 1982; Latterich and Croy, 1993). Unfortunately, the nomenclature in this area has become very complex, with alternative terms being used for the same structures introducing the possibility of some ambiguity (Tautz, 1993).

12.4.1 Hybridization with single-locus probes

The first consideration in single-locus hybridization programmes is the method by which a set of useful probes can be accumulated. Since such probes tend to be species-specific, it is normally necessary to carry out preliminary work producing probes before diversity studies can commence. This requires the creation of a DNA library with each cloned insert representing a potentially useful probe. However, clones taken from a total DNA library need pre-screening since a large proportion will contain highly repeated DNA and will not detect single loci following hybridization. Alternatively, single-locus probes can be produced by creating cDNA libraries which, since they are synthesized using mRNA populations, represent the expressed regions of a genome. A useful exception to the species-specificity of probes is ribosomal RNA encoding

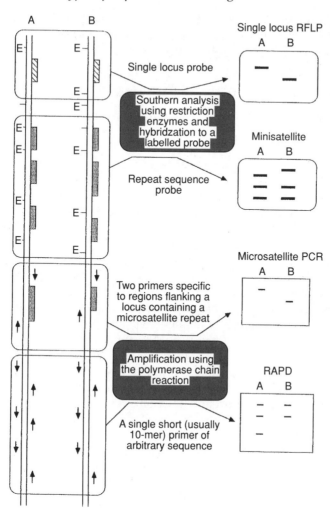

Figure 12.1 Diagram representing the processes by which marker bands are produced using four molecular techniques.

(rDNA) regions of the genome, which are relatively highly conserved between plant taxa and cloned sequences of which are available from many plant species.

The use of single-locus hybridization technology and the resultant co-dominant markers (Figure 12.1) have been the backbone of major mapping programmes for a range of plant species over the last decade (Helentjaris *et al.*, 1986; McCouch *et al.*, 1988; Tanksley *et al.*, 1989). The same methods have been applied in investigations of plant diversity. For example, genetic diversity and differentiation of *indica* and *japonica* rice have been detected using single-locus RFLP (Wang and Tanksley, 1989; Zhang *et al.*, 1992; Zheng *et al.*, 1994), and other studies have used RFLP data on a range of *Oryza* species (Wang *et al.*, 1992).

12.4.2 Hybridization with repeated sequences

Much of the repeated sequence within a genome is so-called satellite DNA and consists of a very high number of repetitions (sometimes more than 100 000 copies) of a basic motif which itself may be between two and several thousand base pairs (bp). However, some repeated DNA is organized somewhat differently; a class of repeats showing a lower degree of repetition, consisting of shorter motifs (usually 10–60 bp), and which probably occur at many thousands of loci, have been called minisatellites. Work on plant minisatellite markers was informed by pioneering studies on the human genome. Jeffreys' group studied tandem repeats of sequences ranging from 15 to 75 bp and found them to be hyper-variable in length between human individuals (Jeffreys *et al.*, 1985). The polymorphisms were found to lie in the number of repeated units and this led the sequences to acquire the alternative name of VNTRs (variable number of tandem repeats; Nakamura *et al.*, 1987). Having no clear function, repetitive DNA appears not to be under strong selective pressure so that changes in the lengths of these sequences are tolerated. By carefully selecting a probe derived from the core of a particular repeat and using conventional Southern analysis (section 12.3), it was possible to detect restriction fragments representing a large but manageable number of human loci. Astonishingly, the patterns of the minisatellite-bearing restriction fragments on autoradiographs allowed clear discrimination between different human individuals on the basis of their DNA fingerprint; the technology has subsequently been used to provide forensic evidence in legal cases (Jeffreys and Pena, 1993). Both the multi-locus probing possible here and the single-locus probing described above assess variation by detecting RFLPs.

The initial studies on humans required a great deal of preliminary work to identify appropriate probes which would reveal minisatellite fingerprints. However, by the time that the general technology was applied to plants (Figure 12.1) it was possible to avoid most of this effort. For example, it was fortuitously observed that probing with an internal repeat from the protein III gene of the bacteriophage M13 revealed minisatellite sequences in a wide variety of organisms, including plants (Rogstad *et al.*, 1988; Ryskov *et al.*, 1988). It has subsequently been shown that tandem repeats of even shorter sequences such as (GT)n or (GAC)n are common in eucaryotes (Hamada *et al.*, 1984; Tautz and Renz, 1984). These have been termed microsatellites (Litt and Luty, 1989) or simple sequence repeats (SSRs). Polymorphic patterns can be obtained by probing with repeats of much shorter motifs. Since these repeats are so simple, it is possible to synthesize relatively short sequences in the laboratory and use these for probing. This so-called oligonucleotide probing, using sequences such as $(GACA)_4$ or $(GATA)_4$, has been applied to humans (Vergnaud, 1989), a range of other animals (Mariat and Vergnaud, 1992) and to plants (Weising *et al.*, 1989).

Multi-locus fingerprint patterns have been obtained for rice genotypes by probing with repeated sequence probes, and the majority of probes detected a high level of polymorphism resulting in DNA fingerprints (Dallas, 1988; Zhao and Kochert, 1992, 1993; Dallas *et al.*, 1993; Ramakishima *et al.*, 1994). Such fingerprints can be used for genetic mapping and phylogenetic studies and are especially useful for the identification of specific rice genotypes. For instance, *indica* and *javanica* rices could be distinguished using a particular minisatellite probe (Dallas *et al.*, 1993).

12.4.3 Amplifying repeated sequences

Just as oligonucleotides representing specific microsatellite repeats and the core sequences of minisatellite repeats can be used for probing, they can also be used as single primers to direct amplification using the polymerase chain reaction. The use of this method has been reported to provide multi-locus markers (Matsuyama *et al.*, 1993; Neuhaus *et al.*, 1993). However, a more widespread application of repeat-sequence PCR has used single locus markers. If short unique sequences flanking repeated-sequence loci are established, then it is possible to define primers complementary to the flanking regions and use PCR technology to amplify the repeat. This approach yields markers at specific loci and, because of the hypervariability of the length of the repeats, is an efficient method for the detection of polymorphisms (Figure 12.1). Microsatellite sequences are more useful in such a protocol than minisatellites; many minisatellites are too long to allow amplification using current technology and they are not spread as evenly over the genome as microsatellites.

The use of PCR-amplified microsatellite markers, which are co-dominant, has been much favoured by workers producing genetic maps. High-resolution maps have been produced using this technology for a range of animals (Dietrich *et al.*, 1992; Weissenbach *et al.*, 1992) and plants (Morgante and Olivieri, 1993; Senior and Heun, 1993; Cregan *et al.*, 1994). The method requires sequence information for DNA flanking the repeat itself, some of which may be available in DNA databases for heavily studied species. However, it is inevitably necessary to obtain such information for further loci and this involves the significant effort of producing genomic libraries enriched in microsatellites (Rassmann *et al.*, 1991; Orstrander *et al.*, 1992), selecting potentially useful clones, and sequencing in order to define primers.

Yang *et al.* (1994) used PCR technology to identify microsatellite polymorphism across land races and cultivars of rice. Microsatellite loci exhibit a large number of alleles so that variation was more readily detected at these than at single-copy loci examined by RFLP (Zhang *et al.*, 1994). Wu and Tanksley (1993) have reported the identification of microsatellite alleles that are specific to either *indica* or *japonica* rices.

12.4.4 Amplifying arbitrary sequences

The methods of producing markers using the PCR described above have used pairs of selected primers in order to allow detection of variation at particular loci. However, it is also possible to use single primers of arbitrary sequence to direct amplification and obtain banding patterns. A generic term for this approach is MAAP (multiple arbitrary amplicon profiling; Caetano-Anolles *et al.*, 1992), though there are several names for variants of the technique (Caetano-Anolles *et al.*, 1993) such as RAPD (Williams *et al.*, 1990; Newbury and Ford-Lloyd, 1993), DAF (Caetano-Anolles *et al.*, 1991) and AP-PCR (Welsh and McClelland, 1990).

The various techniques all use short primers (usually 8–10 bases). Because these are of arbitrary sequence and commercial kits of primers are available, there is no need for the preparation of specific probes or primers common to other technologies. Many fragments are normally amplified using each single primer (Figure 12.1), and the RAPD technique in particular has proved a fast method for detecting polymorphisms and supplying data for diversity estimations. RAPD does suffer from a sensitivity to changes in PCR conditions; alteration of several of the characteristics of the PCR can result in changes to some of the amplified fragments. However, it is clear that reproducible results are obtained if care is taken to employ exactly the same conditions for each run (Munthali *et al.*, 1992). RAPD markers are dominant; that is to say, amplification either occurs at a locus or it does not, leading to scores of band presence/absence. This means that homozygotes and heterozygotes cannot be distinguished but this has not prevented the use of RAPD markers for genetic mapping purposes, including mapping in rice (Causse *et al.*, 1994).

The major use of RAPD has been for measurements of diversity. For rice, several authors have shown the method to be very efficient for the assessment of variation, and for the classification of material as *indica* or *japonica* (Fukuoka *et al.*, 1992; Ko *et al.*, 1994; Xu and Nguyen, 1994; Yamamoto *et al.*, 1994; Mackill, 1995; Virk *et al.*, 1995a). An example of the banding patterns obtained using RAPD with rice germplasm is shown in Figure 12.2. The technique has also been used to help to identify duplicate accessions in a rice germplasm collection (Virk *et al.*, 1995b).

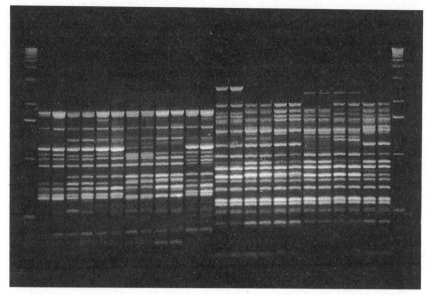

1 2 3 4 5 6 7 8 9 10 11 12 13 14 15 16 17 18 19 20 21 22 23 24 25 26

Figure 12.2 Amplification products following PCR directed by 10-base primers using six different accessions of rice. Lanes 1 and 26 = marker DNA; lanes 2–13 = duplicate amplifications using DNA from six accessions and primer OPK-03; lanes 14–25 = duplicate amplifications using DNA from the same six accessions and primer OPK-04. The amplification products were separated by electrophoresis in a 1.4% agarose gel and stained using ethidium bromide.

12.4.5 Other DNA-based marker technologies

There is not sufficient space to describe all of the molecular marker systems currently available. Those reviewed above are well established but, largely by ingenious combination of parts of different technical procedures, it has been possible to develop a range of further DNA-based marker types. For example, the use of primers comprising microsatellite repeated sequence with a few bases acting as a so-called anchor at one end has resulted in the development of dominant 'anchored microsatellite' markers (Zietkiewicz *et al.*, 1994). For AFLP (amplified fragment length polymorphism), selective amplification of restriction fragments from within a total digest of genomic DNA is used to yield typically 50–100 dominant marker bands per polyacrylamide gel track (Vos *et al.*, 1995). This method has not yet been fully exploited for the analysis of plant diversity.

Both of the above techniques use the PCR with primers which do not target specific loci and hence yield anonymous markers. However, for CAPS (cleaved amplified polymorphic sequences), pairs of primers that direct the amplification of specific loci are used and polymorphism between genotypes is revealed by subsequent test digestions of the amplified loci with restriction enzymes (Rafalski and Tingey, 1993). It seems likely that for the next few years there will be further proliferation of marker types, mostly based upon the PCR. As these new markers are applied, it seems particularly important to ensure that the patterns of diversity that they reveal are in general agreement with those previously reported using established techniques; a classification of germplasm that appears random with respect to accepted taxonomic, genetic or ecological relationships is of little use to anyone.

12.5 THE ANALYSIS OF MOLECULAR MARKER DATA

The data derived from different fingerprinting techniques can be analysed in various ways, depending on how the bands have originated and what information is required from them. If comparisons are to be made between just a small number of individuals, then visual comparisons are possible. Realistically, when studying diversity, large numbers of individuals need to be studied and therefore analysis will involve numbering or labelling the total number of bands produced using one or more probes/primers which occur in different positions on the detection gels, scoring whether each one is present or absent for each individual plant analysed, and the production of a simple two-dimensional data matrix of plants versus bands.

Studies of genetic variation within or between genetic reserves, and monitoring of changes within reserves for any given taxon, can be undertaken using data from both single-locus and multi-locus probes with a range of different analyses. More specific genetic analysis of population dynamics should really be confined to single-locus probes and where co-dominant markers can be scored. There are problems which arise specifically when multi-locus probes are scored, including the inability to distinguish between homo- and heterozygotes, difficulties in calculating allele frequencies at specific loci in anything other than inbreeding material, bands being scored independently even though they might represent alleles at the same locus, and (what in reality only appears to be a minor worry) that co-migrating bands may not represent homologous loci. The generalization is therefore that single-locus probes are needed for studies in population genetics where knowledge of population size, heterozygosity, levels of inbreeding, gene diversity and population structure is demanded.

In contrast, multi-locus probes/primers produce data where the genotype of the individual is only approximated, and often analysed by calculating similarity or distance values for pairs of individuals, where the results from this process are then put into a meaningful structure by means of one or other sorting or classificatory strategies. Such sets of procedures are often referred to as phenetic analysis and contrast with those called phylogenetic, where the emphasis is placed upon representing evolutionary structure. For *in situ* conservation, we feel that of most importance is the analysis of variation that exists at the present time within and amongst populations: phylogenetic studies would appear to be of little or no use for analysing such patterns of diversity.

(a) Population genetic analyses

Perhaps one of the most important statistics from the *in situ* conservation viewpoint which can be roughly estimated using molecular markers (Lynch, 1991) is that of effective population size, N_e (Chapter 6). Measurements of genetic diversity and population subdivision (Lynch, 1991; Weising *et al.*, 1995) will be of importance, as will be estimates of heterozygosity, as the proportion of homozygous loci within populations will provide information concerning the breeding system of the taxon being conserved. The gene diversity statistics of Nei (1973, 1987) and Nei *et al.* (1979) have been used in combination with molecular markers to particular effect by Dawson *et al.* (1993) and Francisco-Ortega *et al.* (1992) to study the partitioning of variation in natural populations. Similarly Russell *et al.* (1993) have used the Shannon–Weiner index of diversity, particularly appropriate for markers that do not reveal allelic information, to study variation in cocoa populations.

(b) Phenetic classificatory and ordination methods

If we have a data set of scores for markers for each one of a number of plants sampled from within one population or from several populations, then the relationships which exist between all pairs of those individual plants could be expressed in multidimensional space defined by the marker characters, and where each dimension is defined by one character. The position of plants in relation to each other can then be used to gain an insight into the degree of relatedness of plants occurring within one or other populations, and hence to obtain an idea of overall variation. This sort of picture can be used in the genetic resources context, and utilizing molecular marker data to prepare infraspecific classifications (Virk *et al.*, 1995a), to separate geographical or ecogeographic patterns of variation (Francisco-Ortega *et al.*, 1993) and to study patterns of genetic diversity generally (Weising *et al.*, 1995).

The picture can be revealed in various ways. Cluster analysis (Everitt, 1993) can be used by firstly calculating the similarity, Euclidean distance or genetic distance between pairs of plants, followed by the sorting of the coefficient values using one or other algorithms such as single linkage analysis, or UPGMA (unweighted pair-group arithmetic averaging). These methods are agglomerative in that groups of related plants are gradually built up in a hierarchical fashion. The end product is often a dendrogram, which depicts the relationships between all the individuals in two dimensions. Similar results can be obtained using the combination of ordination and polythetic divisive clustering afforded by TWINSPAN (two-way indicator species analysis; Hill, 1979). This allows the identification of key DNA fragments which are responsible for each division of the dendrogram. Furthermore, by arranging the original raw data in a reordered genotype-by-DNA fragment matrix, it facilitates the visual reinterpretation of the initial results and provides a 'bandmap' of shared DNA fragments. Francisco-Ortega *et al.* (1993) have used this to examine patterns of variation in a fodder legume in the Canary Islands.

Various other ordination techniques such as principal component analysis, principal co-ordinate analysis and correspondence analysis (Sokal and Rohlf, 1981) have been used to visualize diversity existing in germplasm across a range of characters (or bands) (Francisco-Ortega *et al.*, 1993; Crossa *et al.*, 1995). Many of these multivariate techniques have been used alone or in combination (Rohlf, 1992) to study various aspects of diversity within crop germplasm. Measurement of the genetic organization of land races of *Brassica* has been undertaken by Silva-Dias *et al.* (1994). They measured levels of genetic variation within, and genetic distance between, land races collected from a range of geographical regions using molecular markers, cluster and principal component analyses.

12.6 CONCLUSIONS

It should be clear that there is a large variety of molecular marker types which can be employed for the analysis of variation in plants. The value of the information that these markers provide cannot be questioned, but there are many factors which influence the choice of the marker type to be employed. Different markers have different genetic qualities (e.g. they are dominant or co-dominant, represent anonymous or characterized loci, and expressed or non-expressed sequences, etc.). Their acquisition requires different levels of effort, expertise, equipment and sometimes delicate fine chemicals. The choice of marker technology, and to some extent of an appropriate numerical analysis package, may be complex. However, as long as care is taken to apply techniques appropriate to the specific questions being asked, molecular markers offer significant opportunities during the analyses of plant variation associated with the siting, monitoring and management of *in situ* germplasm collections.

13

Conserving the genetic resources of trees *in situ*

P. Kanowski and D. Boshier

13.1 INTRODUCTION

The history of humankind is one of modification of the forested environment (Glacken, 1967; Ledig, 1992), through processes which Ledig (1992) has categorized as deforestation, exploitation, fragmentation, demographic and habitat alterations, environmental deterioration, translocation and domestication. Some (e.g. Perlin, 1989; Harrison, 1992) have postulated that these modifications represent a defining characteristic of civilizations; in any case, it is apparent that the political economies of resource use define the contexts and prospects for *in situ* conservation of forest genetic resources, just as they do for forests and trees more generally (e.g. Ledig, 1986; Byron and Waugh, 1988; Westoby, 1989; Romm, 1991; Colchester and Lohmann, 1993).

The increasing scale and rate of human impacts on forests have been reviewed elsewhere (e.g. Laarman and Sedjo, 1992; Rowe *et al.*, 1992; FAO, 1993a). Concern over the consequences for tropical forests, repositories of much of the world's biological diversity (Sayer and Whitmore, 1991; Wilson, 1992), and the temperate and boreal forests (Dudley, 1993; Norton, 1994a) has prompted many institutional responses (e.g. Johnson, 1993; CGIAR, 1994; Palmberg-Lerche, 1994a; Grayson, 1995) which promote, amongst other measures, *in situ* conservation of forest genetic resources. Effective conservation strategies demand clear definition of objectives, adequate knowledge and appropriate conservation methods (Eriksson *et al.*, 1993). However, definition of conservation objectives varies with culture and with scientific paradigm (Norton, 1994b; Reid, 1994)

Plant Genetic Conservation.
Edited by N. Maxted, B.V. Ford-Lloyd and J.G. Hawkes.
Published in 1997 by Chapman & Hall. ISBN 0 412 63400 7 (Hb) and 0 412 63730 8 (Pb).

along a continuum which Namkoong (1991) has characterized as ranging from 'engineering reductionism' to 'ethical holisticism'; emphases therefore vary from preservation of existing diversity to conservation of evolutionary potential (Eriksson *et al.*, 1993). Our perspective is essentially a pragmatic one: in the face of continuing pressures on forest resources, how can natural resource managers best meet the objective enunciated by the World Commission on Environment and Development (1987) of maintaining options for future generations whilst satisfying the needs of the present? This chapter summarizes current knowledge of forest genetic resources and considers implications for their *in situ* conservation in both genetic and strategic terms.

13.2 FORESTS

Forests are variously defined (e.g. Sharma *et al.*, 1992) to encompass a broad continuum of trees in the landscape, ranging from the classical image of a closed-canopy tree community, through more open woodlands and farming and pastoral systems which maintain trees *in situ*, to woody shrublands. Defined in these terms, forests occupy between 30 and 40% of the world's land surface (NRC, 1991; Sharma *et al.*, 1992). Few, if any, forest ecosystems are unmodified by human activities, which have impacted variously – and not necessarily adversely – on forest extent, structure and diversity (Ledig, 1992; McNeely, 1994). Although interpretation is handicapped by the absence of baseline data (Ledig, 1992), it is apparent that human intervention has sustained and enhanced the forest gene pool – for example, in the apêtê forest patches of the Brazilian cerrado (Posey, 1985), the *Leucaena* agroforestry systems of south-central Mexico (Hughes *et al.*, 1995), or the semi-deciduous humid forest islands amongst savanna in Guinea (Leach and Fairhead, 1994) – as well as, more stereotypically diminished it (e.g. Thirgood, 1981; Perlin, 1989; Rowe *et al.*, 1992).

Given the centrality of population genetic information to the design of appropriate conservation strategies (Bawa and Ashton, 1991), and that of the tree flora in defining the structural habitat and environment within which other species exist in forest ecosystems (Loveless, 1992; Wilson, 1992), this chapter focuses on the population-level genetic structure and dynamics of the tree component of forests. However, as Gentry (1992) has discussed, optimal conservation strategies for non-woody forest flora may differ from those for trees, reinforcing the more general point that effective *in situ* conservation demands that our forest population genetics perspective be embedded within the broader goal of conservation of ecosystem function and process (Franklin, 1988; Riggs, 1990; Bawa and Ashton, 1991; NRC, 1991). In this context, species addition may be as problematic as species extinction (Janzen, 1983; Hughes, 1995).

13.3 POPULATION GENETICS OF TREE TAXA

The distributions of individual tree taxa range from pan- or interconti-
nental to highly localized, with distribution patterns varying from essentially continuous, at least at the landscape scale (e.g. *Gliricidia sepium*:
Hughes, 1987; *Picea glauca*: Zobel and Talbert, 1984), to highly disjunct
(*e.g. Parkinsonia aculeata*: Burley *et al.*, 1986; *Pinus radiata*: Moran and Bell,
1987). Eriksson *et al.* (1993) defined five organizational patterns to encompass the range of natural population structures evident in trees, ranging
from single large random-mating populations to groups of disjunct and
predominantly isolated populations.

Given the seminal importance of mating systems and patterns of gene
flow in determining levels, patterns and dynamics of genetic diversity
(NRC, 1991), recent research using isozyme and molecular markers has
focused on these topics. The magnitude of the task in relation to the
limited resources available has dictated that researchers concentrate their
efforts on a small group of taxa taken to be representative of others sharing similar attributes (Moran and Hopper, 1987). Consequently, information remains both limited and biased towards coniferous taxa of the
northern temperate forests (Muona, 1989; NRC, 1991; Hamrick, 1992),
although recent tropical studies (reviewed by Murawski, 1995) have
begun to redress the balance. Nevertheless, some consistency is apparent
for the majority of tree taxa, which are characterized by effective mechanisms for outbreeding and gene flow, individual longevity and fecundity,
and little or no history of domestication (Brown and Moran, 1981; Yeh,
1989).

Tree taxa typically exhibit relatively high levels of genetic diversity,
with the average level of heterozygosity and proportion of polymorphic
loci both high in relation to those of non-woody plants (Hamrick *et al.*,
1992). Levels of genetic diversity vary with mating system, with higher
levels in the majority of predominantly outcrossing taxa. In contrast,
inbred species show lower levels of diversity, but greater interpopula-
tional variation. Most alleles (typically 70–80%) are common across most
populations (Hamrick, 1992; Loveless, 1992; Moran, 1992; Müller-Starck
et al., 1992) – even in rare taxa (Bawa and Ashton, 1991), those which
occur as small and isolated populations (Moran and Hopper, 1987), those
which are relatively fragmented (Hamrick, 1992) and those which have
been subject to substantial human modification (Harris *et al.* in press).
From the very limited evidence available, other forest plants with similar
reproductive systems and life histories (e.g. the rattan species *Calamus
manan* and *C. subinermis*: Bon *et al.*, 1995) appear, as might be expected, to
exhibit similar levels and patterns of genetic diversity to those of the tree
flora. Although genetic differentiation between populations is relatively
low, it is often of major significance for adaptation or production, as

provenance trials of many tree taxa have demonstrated (e.g. Zobel and Talbert, 1984; Eldridge *et al.*, 1993).

Mating systems, and hence outcrossing rates, vary with reproductive biology – the nature of incompatibility mechanisms, flowering synchrony and abundance, and pollination processes – and with the spatial structure of stands. Few tree taxa studied so far appear to have truly mixed mating systems (Murawski *et al.*, 1990); most exhibit high levels of outcrossing, associated with small but significant levels of inbreeding (Loveless, 1992; Moran, 1992), as a consequence of mating between relatives associated with local population structure (O'Malley and Bawa, 1987; Hamrick and Murawski, 1990; Boshier *et al.*, 1995). High levels of heterozygosity are therefore almost ubiquitous at the individual tree, population and taxon levels, a few notable exceptions notwithstanding (*e.g. Pinus resinosa*: Mosseler, 1992). While most taxa tolerate some inbreeding, most have effective mechanisms to preclude or disadvantage matings between relatives, which – particularly selfing – generally have adverse consequences for fitness (e.g. Park and Fowler, 1982; Griffin, 1991).

Gene flow between tree populations is typically extensive (Muona, 1989; Adams, 1992; Ellstrand, 1992a) but idiosyncratic, varying among taxa, populations, individuals and seasons (Ellstrand, 1992b). Results reported for some neotropical tree populations (Hamrick and Murawski, 1990; Murawski *et al.*, 1990; Hamrick, 1992; Hamrick *et al.*, 1993) are among the most detailed available, revealing pollen flow to be high at moderate distances (0.5–1 km) and significant over greater distances (2–3 km). While successful pollination occurred over considerable distances, a high proportion of fertilization was effected by nearest neighbours, and from a restricted suite of pollen parents, with substantial variation in the genetic composition of pollen received by individual trees. Correspondingly, allelic frequencies were relatively homogeneous among populations within a few kilometres of each other, but more heterogeneous further afield. Genetic structuring within populations is typical (Ledig, 1992; Moran, 1992; Yang and Yeh, 1992), particularly for those taxa with wind-dispersed seed. Near neighbours usually have more alleles in common than do more distantly separated individuals, even when they occupy relatively homogeneous habitats (Linhart, 1989; Hamrick *et al.*, 1993; Boshier *et al.*, 1995a). Spatial and temporal substructuring of populations, associated with corresponding variation in flowering, with the nature of incompatibility mechanisms and with stand composition and density, result in temporal variation in outcrossing rates for individual trees (Murawski *et al.*, 1990; Murawski and Hamrick, 1991; Boshier *et al.*, 1995b; Murawski, 1995).

A caveat to the synthesis of published results is that most investigations of genetic diversity and dynamics in trees have been based either on seed collected for provenance trials or on within-stand studies. As the operational definition of a population and thus pattern of sampling vary greatly between studies, synthesis of results is difficult (Loveless, 1992). Widely distributed species may be characterized by a higher ordering of population structure, so that patterns of gene flow not evident on a small scale may become apparent at a larger scale. Indeed, data for tropical woody species suggest that while gene flow at the local level limits differentiation among nearby populations, geographically distant populations do diverge genetically (Lavin *et al.*, 1991; Chalmers *et al.*, 1992; Loveless, 1992; Moran, 1992; Simons and Dunsdon, 1992; Chase *et al.*, 1995). Consequently, tree taxa may exhibit a complex and dynamic metapopulation structure (Lande and Barrowclough, 1987; Bawa and Ashton, 1991), consistent with the contention that all species persist as metapopulations when assessed at an appropriate spatial scale (Harvey *et al.*, 1991).

13.4 IMPLICATIONS FOR CONSERVATION OF FOREST GENETIC RESOURCES *IN SITU*

Much recent work has sought to apply our improving understanding of forest population structure and dynamics to the conservation of forest genetic resources (e.g. Ledig *et al.*, 1990; Bawa and Krugman, 1991; Adams *et al.*, 1992; Ledig, 1992; Kemp, 1993; Wilcox, in press). However, lack of information characterizing both species distribution and frequency and meta- and subpopulation structure is a major caveat to generalizations (Moran and Hopper, 1987; Bawa and Ashton, 1991). Estimates of effective population sizes required for long-term conservation of genetic diversity vary from around 500 to 2500 (e.g. Brown and Moran, 1981; Namkoong and Kang, 1990; Krusche and Geburek, 1991). In tree populations, dioecy, uneven sex ratios and long, overlapping generations reduce effective population size relative to census number (Lande and Barrowclough, 1987) and correspondingly increase minimal area requirements (Soulé, 1986); for example, Hamrick and Murawski (1990) suggested that the effective breeding unit for common neotropical tree species would be in the order of 25–50 ha.

As with other organisms, there are notable examples of tree species extinction consequent on extensive deforestation or extreme levels of exploitation, particularly amongst oceanic island flora (Cronk, 1993). Other than such species loss, the major current impacts on forest genetic diversity are probably those due to ecosystem fragmentation and associated demographic stochasity in remnant populations, which may lead to local extinctions (Lande, 1988). Thus, fragmentation may represent an erosion of diversity almost as serious as that due to species loss, with the possible destruction of allelic complexes of locally adapted populations through altered patterns and levels of gene flow (Ellstrand, 1992b; Ledig,

1992). At the extreme, such processes can lead to loss of rare species through interspecific hybridization and subsequent loss of fitness of the hybridizing individuals (Ellstrand, 1992b; e.g. *Leucaena* taxa: Hughes *et al.*, 1995; *Morus rubra*: Ambrose, 1995; *Populus nigra*: Arbez, 1994), though the converse may also be true as, for example, is the case with *Trochetiopsis erythroxylon* and *T. melanoxylon* (Rowe and Cronk, 1995). Similarly, the effects of fragmentation are complex and probably most often locally specific: whilst it may reduce populations below critical size and gene flow to a level below that necessary to prevent genetic drift, it may also increase or change patterns of gene flow between remnant populations, and hence maintain or increase within-species diversity (Hamrick, 1992; Bawa, 1994; Young, 1995). The effects of fragmentation on remnant stands and trees, and their consequent conservation value, are the subject of debate (e.g. Janzen, 1986; House and Moritz, 1991; Saunders *et al.*, 1991; Heywood and Stuart, 1992; Young, 1995). At the pessimistic extreme are views such as that enunciated by Janzen (1986) of the genetic conservation value of remnant trees in agro-ecosystems, which he characterized as 'the living dead'. More optimistically, the extensive and effective gene flow maintained between isolated trees of many taxa, even those in severely fragmented landscapes, suggests that remnant forest patches and trees can be effective and important in conserving genetic diversity (Hamrick, 1992). Preliminary studies (Young *et al.*, 1993; Prober and Brown, 1994), and Young's (1995) review, of the effects of fragmentation on forest tree populations suggest that the major impact of fragmentation may be the reduction of genetic diversity in remnant populations through the generation of genetic bottlenecks; however, subsequent effects on genetic processes are not yet well understood. Given the processes of fragmentation evident in most forested landscapes, we concur with Young (1995) that further research to clarify its genetic consequences is an urgent priority.

At the interspecific level, advances in phylogenetic analyses offer some guidance to the development of strategies to prioritize conservation of species for maintenance of total allelic diversity (e.g. Vane-Wright *et al.*, 1991; Williams and Humphries, 1994). At the intraspecific level, generalized conclusions have been drawn for widespread species with high rates of outcrossing and high levels of gene flow, for which the majority of allelic richness resides within populations. For such species, genetic diversity would be well conserved by a limited number of large populations distributed across the species range (Moran and Hopper, 1987; Schoen and Brown, 1991; Hamrick, 1992; Prober and Brown, 1994). Appropriate strategies are less easy to generalize where distributions are more localized (Moran and Hopper, 1987), or where genetic structure is the result of a more complex interaction of variously genetically divergent subpopulations. In such cases, only conservation of populations throughout the range will ensure adequate conservation of the genome (Allard, 1970); this might best be achieved through maintenance of a met-apopulation structure comprising a series of semi-independent popula-tions (Eriksson *et al.*, 1993), thereby mitigating the likely reduction in fitness of individual populations induced by isolation. The terms of this discussion (e.g. Janzen, 1983; Boecklen and Bell, 1987; Baker, 1989; Eriksson *et al.*, 1993) illustrate the tensions inherent between strategies that emphasize conservation of diversity at different levels, i.e. from the landscape scale to within taxa. For example, emphasis on perpetuation of natural fluctuations in landscape structure may cause fluctuations in the representation of species dependent on landscape structure, and frag-mentation may itself alter the mosaic regime of the ecosystem (Baker, 1989; Namkoong, 1994a).

13.5 STRATEGIES FOR *IN SITU* CONSERVATION OF FOREST GENETIC RESOURCES

The case for coordinated conservation strategies, in which technical options are integrated within a broader framework of action, has been well made elsewhere (e.g. FAO, 1991; Flint, 1991; NRC, 1991; Eriksson *et al.*, 1993; Kemp, 1993). Traditionally, technical options for forest genetic resource conservation have been classified in terms of activities *in situ* and *ex situ* (e.g. Flint, 1991; NRC, 1991); seen in this context, *in situ* con-servation is a means to an end, rather than necessarily an end in itself (Eriksson *et al.*, 1993). Correspondingly, we suggest that classical *in situ* conservation – that which seeks to maintain populations in the 'natural' state – has been often overemphasized at the expense of what has been characterised as *circa situ* (*sensu* Cooper *et al.*, 1992) conservation, *viz* that realized within production systems (and also characterized as 'farmer-based' or on-farm: e.g. Brush, 1991a; *in hortus*: Witmeyer, 1994; or sub-sumed within *in situ*: e.g. Lamola and Bertram, 1994). Given our deepening appreciation of how people have impacted historically on forests and forest genetic resources, and the pressures on forests and trees in much of the world – particularly in the gene-rich tropics – we see *circa situ* strategies as the only tenable approximation, in many circum-stances, to classical *in situ* conservation of forest genetic resources. The case of the economically important legume genus *Leucaena*, now essen-tially unrepresented within its native range other than within highly dis-turbed forest or farming systems, is a well-documented example (Hughes *et al.*, 1995; Figure 13.1).

The critical importance of effective *in situ* and *circa situ* conservation of forest genetic resources is demonstrated by the trivial number of tree taxa (perhaps 100: NRC, 1991) subject to effective *ex situ* conservation, by the scientific, technical and resource limitations which constrain *ex situ*

Figure 13.1 *Leucaena collinsii* var. *zacapana* at Chiquimula, Guatemala: a taxon under threat *in situ*, but conserved within its native range through traditional use on farms. Photograph: C.E. Hughes.

programmes (Bawa and Ashton, 1991; NRC, 1991; Eriksson *et al.*, 1993), and by the inherent deficiencies of *ex situ* populations as conservation gene pools (e.g. Brown *et al.*, 1989; Marshall, 1990; Yang and Yeh, 1992). In particular, the limitations of *ex situ* conservation based on seed storage and periodic regeneration (Brown *et al.*, 1989; Marshall, 1990) appear to be recognized by the forest genetics community more in principle (FAO, 1993b) than in practice.

In broad terms, *in situ* conservation options may be classified as those that focus on fully protected forests, in which human intervention is deliberately minimized, and those that seek to achieve genetic conservation objectives in more managed populations, including those we classify as *circa situ*. Each is considered below.

13.6 FULLY PROTECTED FORESTS

The ultimate expression of *in situ* conservation of forest genetic resources is the maintenance of a comprehensive network of reserves dedicated to genetic conservation objectives. The Finnish network of gene reserve forests (Koski, 1991), in which the genetic resources of major commercial species are conserved *in situ* through the representation at each site of 'functional pollination units' of varying age classes, provides one example; the USA's Research Natural Area network (Ledig, 1988) is another, at least in principle. Although state-managed or controlled forest reserves are almost ubiquitous globally, few have been established according to population genetic principles; reserve design and management are more often determined by political, social and economic constraints (Boecklen, 1986). Further, their location – typically on slopes, sites of lower fertility, and in stands of lesser economic value (Ledig, 1988) – bias their composition and limit their value for genetic resource conservation. Similarly, effective priority setting is seriously impeded by poor knowledge of the distribution and rarity of individual taxa (Gentry, 1992). Consequently, Gentry's (1992) suggestion of identifying priority areas through focusing on the occurrence of patterns of association rather than on that of individual taxa will often be the only feasible option for identifying priority areas for conservation; as he points out, the tree flora tend to be relatively well known, and are thus well suited to fulfilling this indicative role. Recent developments in software (e.g. Hawthorne, 1995) and its application – for example, the 'genetic heat' index of Hawthorne and Abu Juam (1995) – demonstrate how appropriate information systems can facilitate management for more effective conservation.

In contrast to the paucity of comprehensive networks, the establishment of reserves to conserve the genetic resources of particular populations of a taxon (e.g. the superior Yucul provenance of *Pinus tecunumanii*: Vilchez and Ravensbeck, 1992) is relatively common. Because of the large areas required to maintain sufficient effective population sizes of most tree taxa (Ledig, 1988; Loveless, 1992), and the impracticality or undesirability of managing exclusively for conservation goals in many circumstances, the prospects for *in situ* conservation on an adequate scale are very limited (NRC, 1991; Kemp, 1993; Palmberg-Lerche, 1994b). Existing pristine areas, assuming that they are sufficiently large and can be managed to maintain populations, and that they are integrated within the landscape matrix so as to allow gene flow, will play a critical role as 'sources' for migration to managed forests and trees in farming systems (Bawa and Ashton, 1991). It is through these variously managed forests, and in diverse farming systems, that most *in situ* conservation of the genetic resources of trees will be realized. As Namkoong (1994b) has argued, this does not diminish the importance of those forests that can be fully protected, but it does recognize the limits of their role, and emphasizes the need for formulation and enactment of international and national coordinating strategies to maximize their value to genetic conservation goals (NRC, 1991; Kemp, 1993).

13.7 MANAGED FORESTS AND FARMING SYSTEMS

Forests not fully protected from exploitation are harvested for a variety of non-timber and timber products (e.g. de Beer and McDermott, 1989; Poore, 1989), with interventions ranging from those of indigenous peoples to those associated with large-scale industrial wood production. These practices are variously sustainable in ecological, economic and social terms (e.g. Repetto and Gillis, 1988; Poore, 1989; Dudley, 1993), but our knowledge of the effects of exploitation on forest gene pools is poor, as the effects of harvesting and fragmentation on genetic diversity and mating system have seldom been quantified (Ledig, 1992). Those few studies reported for industrial wood production in temperate forests (e.g. Neale, 1985; Neale and Adams, 1985) suggest negligible genetic impact from conservative and ecologically appropriate timber harvesting regimes; in contrast, the only similar study to date in tropical forest (Murawski *et al.*, 1994) found increased levels of inbreeding in the residual stand, associated with the reduction in population size following selective logging. Clearly, the major ecological impacts associated with inappropriate harvesting, such as that which has typified exploitation of much of the tropical forest (Poore, 1989), are likely to have profoundly adverse effects for genetic conservation. The results reported by Murawski *et al.* (1994) demonstrate the critical importance of knowledge of reproductive biology in determining management regimes. Where these regimes are conceived and implemented with such knowledge, and with environmental and ecological sensitivity, the characteristics of most tree taxa suggest that, although local genetic structure may be altered by selection and by changes in demography and mating system, the effects of exploitation on genetic structure and diversity of tree populations should be relatively ephemeral (Ledig, 1992; Savolainen and Kärkkäinen, 1992).

Given the general inadequacy of resources available to support conservation *per se*, incorporation of genetic conservation criteria into forest and farm management practices will offer, in many cases, the best prospects for achieving conservation goals. Experience in planning forestry operations suggests that, whilst some income may be foregone in the short term by managing for conservation, these opportunity costs are relatively trivial (e.g. Cassells, 1992; Pinard *et al.*, 1995), especially in relation to longer-term benefits. Forest managers are now advocating and developing management strategies that accord priority to conserving genetic diversity within production systems (e.g. Riggs, 1990; Kuusipalo and Kangas, 1994; Namkoong, 1994b). Preliminary results of management based explicitly on such principles are encouraging (e.g. Finegan, 1992; Kageyama and Reis, 1993), although knowledge of the effect on associated gene pools is limited. Recent examples include the establishment and management of 'Genetic Resource Areas' in Peninsular Malaysia (Lim and Chin, 1995), in which ecological guidelines for the management of lowland dipterocarp forest are modified to retain a higher stocking of economically important species, without prejudicing the financial viability of harvesting operations; similarly, management of the Yucul, Nicaragua, provenance of *Pinus tecunumanii* recognizes gene conservation objectives by subdividing the forest into a mosaic of stands reserved from harvesting and those managed according to conventional sustained yield principles. In the latter case, management is effected by the local communities and supported by the sale of seed, timber, fuel wood and other forest products (Vilchez and Ravensbeck, 1992). Experience suggests that successful realization of genetic conservation objectives depends on designing regimes which achieve a locally appropriate balance between conservation and income-generation goals (Noss, 1991; Cunningham, 1994; Kremen *et al.*, 1994).

The realization of conservation objectives through strategies emphasizing the role of remnant forest patches and trees in farming systems has been fostered by a growing awareness of their importance in sustaining not only gene pools (e.g. Hamrick, 1992; Hellin and Hughes, 1993), but also environmental services and rural livelihoods (e.g. Arnold, 1991; Gilmour and Fisher, 1991; Shepherd *et al.*, 1991). In the most extreme cases, where forests have become highly fragmented and localized, effective genetic conservation is likely to require both proactive ecological restoration (Janzen, 1988) and 'conservation through use' by local communities (Hellin and Hughes, 1993; Hughes *et al.*, 1995). The latter, *circa situ*, strategy acknowledges the role which generations of communities and farmers have had in maintaining the genetic resources of trees important in farming systems (e.g. *Faidherbia albida*: Harris *et al.* in press; *Leucaena salvadorensis*: Hellin and Hughes, 1993), although it is only recently that genetic conservation strategies have sought to capitalize on these practices (e.g. del Amo, 1992; Hellin and Hughes, 1993). We clearly have much to learn if we are to realize the potential of *circa situ* strategies, and appreciate their limitations: for example, to what extent farming practices can be varied to address conservation objectives as well as sustain livelihood, and what are the consequences are of gene flow between managed and remnant natural populations? The level of intraspecific variation actively conserved may be related to the range of farming practices into which any one species is incorporated; this can be quite diverse, as Raintree and Taylor's (1992) research revealed of farmers' preferences for planting in different farm locations. Clarification of communities' and farmers' practices with tree germplasm is a prerequisite to advance such strategies more generally, and is currently the subject of preliminary investigation (Cromwell *et al.*, 1996). The extent of complementarity between *circa situ* and more traditional approaches to genetic conservation, and the best means of coordinating different strategies, should then be more apparent.

13.8 IMPROVING FOREST GENETIC RESOURCE CONSERVATION

Although knowledge of population structure and dynamics has increased rapidly with the advent of genetic marker technologies, resource limitations dictate that we shall have to continue to extrapolate appropriate strategies for the majority of taxa from the results of studies of relatively few (Moran and Hopper, 1987; NRC, 1991). It is therefore imperative that resources are used to support studies that will inform more generally, about issues of highest priority. Better definition of meta-population structure and dynamics, and clarification of the effects of fragmentation and its longer-term consequences, are two priorities about which consensus is emerging (e.g. Murawski, 1995; Young, 1995). Effective use of such information will require a more developed understanding of the relationships between societies and the maintenance of forest genetic resources, and the consequences for *in situ* and *circa situ* conservation. This is especially so where genetic resources are under greatest threat. As with plant genetic resources more generally, and as recognized by the Convention on Biological Diversity (e.g. Glowka *et al.*, 1994), more substantive recognition of the rights of those whose efforts have helped to maintain and create forest genetic diversity may be a prerequisite to sustaining it.

More effective interagency cooperation will also foster more effective conservation of forest genetic resources. The low status historically accorded to forest genetic resources in relation to that of crop plants has placed much of the onus for strategic development, coordination and action on individual institutions with an international mandate or on national agencies, which are often poorly resourced (NRC, 1991). At the international level, FAO's long-established programme in forest genetic resources (Palmberg-Lerche, 1994a) continues to be severely handicapped by lack of resources. Whilst the recent incorporation of forestry institutions (CIFOR and ICRAF) into the CGIAR system, and the expansion of IPGRI's mandate to encompass trees (CGIAR, 1994), are welcome and offer great prospective benefits in marshalling resources and coordinating strategies, these outcomes remain elusive. The challenge for those multilateral agencies with resources to support *in situ* conservation will be to empower and sustain local communities and agencies in support of

THE NATIONAL LOTTERY

BUY INSTANTS FOOTBALL
GAME GLORY GOALS TODAY

178-02381667-22941

A. 06 09 13 16 40 43

B. 02 10 14 29 30 42

SAT27 JUN 98
FOR 01 SAT DRAW

010766

£ 2.00

RET NO. 109611
178-02381667-22941

FILL BOX TO VOID

MINICOM Line for the hard-of-hearing is available on 0645 100045.

HOW TO CLAIM: Please refer to the How to Play leaflets for full details of how to claim. You can collect a prize of £75 or less from any National Lottery On-line Games Retailer. Some Retailers are authorised to pay prizes of up to £500. National Lottery Post Offices will pay out prizes of up to £10,000. Prizes over £10,000 must be claimed in person from a National Lottery Regional Centre. When claiming more than £500 proof of identity is required. If you win more than £500 or claim by post you will need to complete a claim form, available from all National Lottery On-line Games Retailers or by telephoning the National Lottery Line. If you claim by post, send your ticket and a completed claim form (if claiming over £500) **at your own risk** to the Accounts Dept, The National Lottery, P. O. Box 287, Watford WD1 8TT. You must claim your prize within 180 days of the applicable game draw or result. Safe custody of your ticket is your responsibility. You are likely to lose your entitlement to a prize if you lose your winning ticket, or if it is stolen, even if you have put your name and address on the back. **If you believe you have won a prize of more than £10,000, please telephone the National Lottery Line.**

SM 415416095

The game for which this ticket is issued is subject to the Rules and Procedures for that game, which set out the contractual rights and obligations of the player and the game promoter (and operator if different). They are available at National Lottery On-line Games Retailers, and copies can be obtained by telephoning the National Lottery Line. The promoter/operator is entitled to treat this ticket as invalid if the data hereon does not correspond with the entries in Camelot's central computer. The National Lottery is run by Camelot Group plc under licence granted by the Director General of the National Lottery. Over the licence period around 28% of the value of National Lottery ticket sales will go to the Good Causes designated by Parliament. The principal office of Camelot Group plc is The National Lottery, Tolpits Lane, Watford WD1 8RN.

Name _____

Address _____

Postcode _____

SM 415416096

stratfors

THE NATIONAL LOTTERY

The name of the game for which this ticket is issued appears overleaf. Please refer to the How to Play leaflets available at all National Lottery On-line Games Retailers for details of prize structures and how to find out whether you have won a prize. The results of National Lottery On-line Games are available through recognised media channels, from all National Lottery On-line Games Retailers or by telephoning the **National Lottery Line 0645 100000.** A separate MINICOM Line for the hard-of-hearing is available on 0645 100045.

HOW TO CLAIM: Please refer to the How to Play leaflets for full details of how to claim. You can collect a prize of £75 or less from any National Lottery On-line Games Retailer. Some Retailers are authorised to pay prizes of up to £500. National Lottery Post Offices will pay out prizes of up to £10,000. Prizes over £10,000 must be claimed in person from a National Lottery Regional Centre. When claiming more than £500 proof of identity is required. If you win more than £500 or claim by post, you will need to complete a claim form, available from all National Lottery On-line Games Retailers or by telephoning the National Lottery Line. If you claim by post, send your ticket and a completed claim form (if claiming over £500) **at your own risk** to the Accounts Dept, The National Lottery, P. O. Box 287, Watford WD1 8TT. You must claim your prize within 180 days of the applicable game draw or result. Safe custody of your ticket is your responsibility.

genetic conservation objectives, through truly collaborative programmes to mutual benefit.

DEDICATION AND ACKNOWLEDGEMENTS

This chapter is dedicated to the late George Gibson, friend and colleague, who personified commitment to and action for the conservation of forest genetic resources. We thank Colin Hughes for his many contributions to this work, and Tony Brown, Jeff Burley, Alan Pottinger, Janet Stewart, Mary Stockdale and Andrew Young for their comments. David Boshier is funded by a grant from the UK Overseas Development Adminstration, Forestry Research Programme (R.5729).

14

Integrating plant and insect conservation

V. Keesing and S.D. Wratten

14.1 INSECTS, RESERVES AND RESTORATION: DIMINISHING 'NATURAL' HABITATS

Since the sixteenth century the human population has increased exponentially (Cambell, 1983) and as a result landscapes have changed in dramatic ways. Depending on the fractal scale (Mandelbrot, 1982; Williamson and Lawton, 1991) of observation, the heterogeneity and diversity of landscapes has declined. It may be argued that a plain of grasses in an agricultural system may be no less species-rich than the previous community which was probably a late successional forest, but with regard to the animal component of the community, the plant architectural diversity (and hence the number of possible 'niches' for invertebrates) may have been greatly reduced.

In Europe over recent centuries it has been the mature serial stages, the forests, that have been most modified. Now only 8% of the land in England holds native forest, and of this a quarter is ancient semi-natural forest (Warren and Keys, 1991). These figures may be compared with 4% native forest cover in the Netherlands and 7% in Ireland. France, Germany, Italy, Spain and Eastern European countries have fared better, holding around 20–40% of their land area in forest, without a large component of exotic species, but these forests are still heavily modified by historical and current human use.

Around the world some 75 000–100 000 species of insect are dependent on forest (Warren and Keys, 1991), so with large reductions in forest area over a relatively short time, local or global extinctions are likely. In

Plant Genetic Conservation.
Edited by N. Maxted, B.V. Ford-Lloyd and J.G. Hawkes.
Published in 1997 by Chapman & Hall. ISBN 0 412 63400 7 (Hb) and 0 412 63730 8 (Pb).

the seventeenth century, native pine forests in Ireland are reported to have been completely cleared, resulting in the total extinction of the 'Caledonian' element of the pine-feeding sap-sucking fauna (Speight, 1985, in Warren and Keys, 1991). In tropical forests workers like Erwin (1982, 1988), and Stork (1988, 1991) have alluded to the prominence, taxonomically, of invertebrates, and that the rate of extinction of forest-dwelling invertebrates is likely to be higher than previously thought, given the gross underestimation of invertebrate species diversity and extinction rate in ecosystems (Hafernik, 1992; Smith *et al.*, 1993).

Modification of the age structure of tree populations in the remaining forests has also been an important factor in restricting specialized elements of the insect fauna and the flora; for example, *Phellinus robustus*, a very rare hard bracket fungus that occupies only very ancient oak trees (Anon., 1995). There are species of insect that rely solely on very old timber and on old-tree architecture (Warren and Keys, 1991).

It is not only forest systems that have been and are being modified, or reduced in area. Other biomes include wetlands (Williams, 1990), tall grasslands (Haggar and Peel, 1993) and heathlands (Moore, 1962; Webb, 1989). Habitats associated with the very early and latest successional stages have been most drastically altered, making specialist fauna of open spaces and specific fauna of mature tree species very prone to habitat loss. Within the landscape modified by humans, but nevertheless containing elements which have been relatively 'stable' for many human generations, reside the 'residuals' of the old forests, in the form of wood coppices, hedgerows, orchards and gardens. However, even these are threatened by changing land use as hedges are removed to make larger fields, and exotics are planted in place of indigenous species (Wratten and Van Emden, 1995). The loss of connectivity within the landscape and the fragmentation of tall vegetation biotypes (Fry, 1995) as well as increased pollution and increased recreational use of remaining reserves impinge on most communities.

Effective protection focused on vegetation often means inadvertent protection of vertebrates and invertebrates (Moore, 1991). Eventually, though driven by 'flagship' species (usually plants), functioning communities can be conserved.

The problem is that there are so many systems to protect, and the main questions become how many types of landscape, and of what size, need to be protected to halt extinctions and to ensure viable populations of as many species as possible. This area is one that has received much research (e.g. Strong *et al.*, 1984; Hart and Horwitz, 1991; Harris and Silva-Lopez, 1992; Dunn *et al.*, 1993) but it has not produced clear conclusions.

14.2 THE INTERACTIONS OF INSECTS AND PLANTS

Insects dominate the terrestrial animal world taxonomically and in bio-mass, and their impact in ecosystems is enormous. There are estimated to be between 1 million and 50 million species of insects (Erwin, 1982, 1988; Stork, 1988; Gaston, 1991b, 1992), with a currently accepted working fig-ure of around 8 million species (Hammond, 1992). They are divided into 29 orders (give or take a few: Gullan and Cranston, 1994), of which eight of the terrestrial ones (Lepidoptera, Thysanoptera, Hemiptera, Orthoptera, Phasmatodea, Isoptera, Thysanoptera and Coleoptera) are largely phytophagous (plant-eating) (Southwood, 1973; Strong *et al.*, 1984), while another nine (Blattodea, Psocoptera, Dermaptera, Grylloblattodea, Embioptera, Zoraptera, Archaeognatha, Thysanura and Collembola) are strongly represented by species that have close relation-ships with plant material, either dead or living, as their source of nutri-tion; two more (Diptera and Hymenoptera) have important roles in pollination, and are occasionally phytophagous.

Although insects remove 20% of the foliage in the world annually (Samways, 1994), birds and plants are more 'popular' than insects and so reserves and restoration projects are usually initiated using botanical or ornithological criteria (Moore, 1991). For instance, the Royal Society for the Protection of Birds in the UK has around 1 million members. Yet most reserves and restoration projects would flounder without insects provid-ing nutrient recycling via leaf-litter, the degradation of wood, carrion and dung, pollination and seed dispersal, the maintenance of structure and composition of plant species through phytophagy, and their role as a food source for vertebrates.

14.2.1 Community aspects of insect conservation

When the diversities of different components of an ecosystem are com-pared, they are often correlated (Murdoch *et al.*, 1972). In particular, animal and, especially, insect diversity is often correlated with plant diversity (Murdoch *et al.*, 1972; Strong *et al.*, 1984).

It is unclear whether plant species number or the diversity of plant structures (plant architecture) best correlate with insect diversity, as the two are often inseparable. What is clear is that a greater variety of taxa and structure of plants leads to a greater diversity of plant eaters (Southwood *et al.*, 1982). Further, large areas of established plants tend to have more 'stable' populations of insects (Figure 14.1a). There appears to be a minimum 'fragment' size that supports a temporarily stable insect community (Figure 14.1b), though this is a subject for debate (Schoener and Schoener, 1981; Simberloff and Abele, 1982; Quinn and Harrison, 1987). Similarly, the patch size of particular host species, as with patch

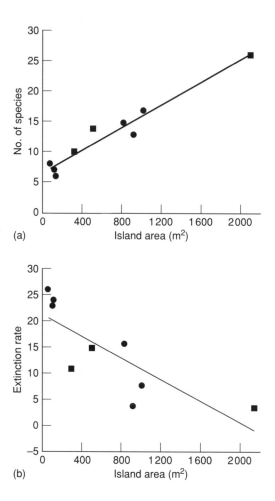

(a)

(b)

Figure 14.1 (a) Example of a species–area relationship for arthropods on islands of *Spartina alterniflora* (adapted from Rey, 1981). (b) Relationship between insect extinction rate on *Spartina alterniflora* and the area covered by the plant.

size of the total plant community, affects the number of herbivore species and there is usually a linear relationship (Figure 14.2; Strong *et al.*, 1984).

The majority of phytophagous insects are monophagous (Strong *et al.*, 1984) which reinforces the idea that plant species diversity leads to insect species diversity. The consequent variation in such things as plant architecture and chemical diversity creates a resource that is exploited by a range of phytophagous and predator species, providing resources on which insects can feed, mate, oviposit, shelter, bask, hunt, etc. As well as

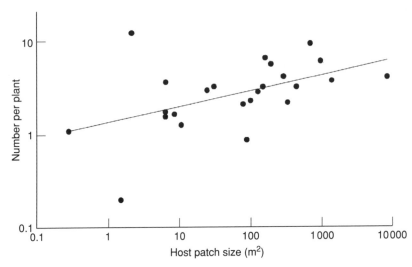

Figure 14.2 Total number of herbivores per plant on different patch sizes of rose-bay willowherb (*Chamerion angustifolium*). (From Strong *et al.*, 1984, with permission.)

dictating structure and the diversity of living tissue resources, plant species richness also affects the variety and abundance of dead plant material resources (wood, dead leaves, bark, etc.). This allows a diversity of insects adapted to this part of the ecosystem, such as Collembola, Diptera, Coleoptera, Annelida, etc. The plant component of any community strongly influences, or may even dictate, the invertebrate component (Figure 14.3).

14.2.2 Succession

The successional stages of plant communities are important as these correlate with plant species number. In Figure 14.4 the insect diversity (Williams' α measure, mean $\pm$ SE) positively correlates throughout early succession with an increase in plant species richness, but does not decline rapidly as does the plant species richness at later successional stages. The plant spatial plot supplies some explanation as to why insect diversity is maintained; it is because, as succession advances and the plant species change, the forms of those plants offer more additional resources, i.e. complex ecological factors such as litter accumulation and architectural complexity (Brown, 1991), micro-climate, spatial stratification of habitable zones (Lawton, 1983) and micro-habitats.

It is these additional factors that result in the maintenance of insect diversity in later successional landscapes even though the plant species richness often decreases (Figure 14.4; Southwood *et al.*, 1979). This is

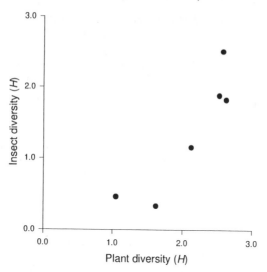

Figure 14.3 Number of species of Homoptera as a function of plant species diversity in an old field in Michigan. (Adapted from Murdoch *et al.*, 1972.)

because the insects make use of the wider range of resources of later successional species, i.e. trees are more architecturally complex and facilitate a more structurally complex habitat. Examples of late successional niches that develop and are filled are: stick insects (Phasmatodea) that exploit

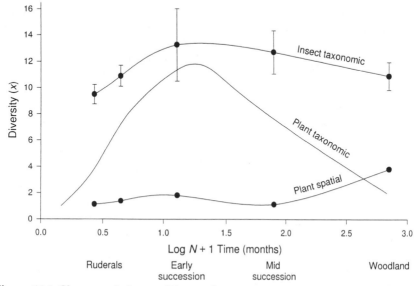

Figure 14.4 Changes of plant and insect diversity, and plant structure, through succession. (From Strong *et al.*, 1984, with permission.)

twigs; Pscoptera (booklice) and bark beetles (Coleoptera: Scolytidae) that live in and under bark; a myriad of fruit dwellers (e.g. fig wasps); ephiphyte specialists; wood borers (Coleoptera: Cerambycidae); and leafrollers (Lepidoptera: Torticidae). These trends apply to both 'natural' and agro-ecosystems (Andow, 1991; Wratten and Van Emden, 1995) where polycultures support more insect species than do monocultural systems (Altieri, 1994).

While some of the insect community does not directly affect plants, some may influence plant performance and competitiveness. Damage by herbivores is a dominant selection force on plants, affecting their survival and representing an evolutionary pressure (Gullen and Cranston, 1994). Conversely, food quality ('secondary' compounds, physical defences, etc.) has a dominant influence on the abundance and species composition of insects and the damage they cause (Erlich and Raven, 1954; Barker, Wratten and Edwards, 1995). There are strong co-evolutionary ties between these two sets of organisms.

Plants modify light, temperature, humidity, pH, etc., and supply a variety of resources that structure and regulate the abundance and diversity of insects. Conversely, the insect fauna influences plant communities and mediates processes such as nutrient cycling (De Angelis, 1992).

14.3 OLD (ESTABLISHED) PLANT COMMUNITIES AS NATURE RESERVES

These landscape features have sometimes persisted by chance, or by the actions of private landowners or the state; they may often approximate to pre-human conditions. From a New Zealand perspective, it is evident that the European 'cultural landscapes' are very different from those in New Zealand, where the term 'colonial landscapes' may be more appropriate. In Europe, plant communities comprise naturalized and indigenous species, but entirely exotic landscapes are relatively uncommon. In New Zealand there are much clearer distinctions between exotic and indigenous landscapes; for example, indigenous podocarp, broadleaf and fern forests compared with pastoral, horticultural, arable and forest landscapes which (including field boundaries) consist almost entirely of introduced plants. This polarization of the landscape in New Zealand has led to virtually all conservation research and practice taking place in 'native' areas, usually with a concentration on endangered species, often birds (e.g. the Kaka parrot, *Nestor meridionalis*: O'Donell and Rasch, 1991). Recently, however, the ecological 'bleakness' of the cultivated areas has begun to be recognized, and extensive surveys of what little native vegetation does remain have been carried out (Meurk, 1995). Also, restoration of original communities on farmland has begun, and is producing its own paradigms and protocols.

In Europe, and in the farmed hill country of New Zealand, old established plant communities tend to be small and spread throughout the landscape. In New Zealand, although there are many such fragments, there are still vast tracts of land, such as the Fiordland National Park (1.2 million ha), that are 'native' reserves at a World Heritage scale. The survival of insect assemblages in the fragmented habitats depends not only on the 'quality' of the site but also on corridors (Fry, 1995) connecting with other patches, allowing fluxes of individuals to and from patches (Burel and Baudry, 1995). This makes the fragmented landscape capable of supporting small, disjunct but connected populations of insects (metapopulations: Levins, 1970b; Harrison, 1994). In New Zealand the indigenous fauna is more confined to large reserves, leaving other large areas of the landscape devoid of most indigenous plants and of their associated insect communities. The Canterbury Plain in the central South Island is a good example (Wratten and Hutcheson, 1995): what remains is remnant patches (e.g. on the volcanic Banks' Peninsula) or as regenerated bush from 'unkempt' land – examples of this occur around the borders of the Canterbury Plain, and result from the 1930s depression and from the 1940s (World War II). These small pockets, often without original community components, tend to be isolated by large areas of exotic species, mainly comprising pasture and exotic shelter-belt species, the latter possibly acting as barriers rather than corridors for animal movement (Harwood *et al.*, 1992; Mauremootoo *et al.*, 1994).

The exotic landscape which comprises most of lowland New Zealand was established without most of its associated arthropod community. For example, 15–20 spider species have arrived from a pool of 151 species normally found in British arable landscapes (C. Vink, personal communication). In contrast, agro-ecosystems in Europe have an arthropod–plant relationship which is often well developed and includes strong coevolved associations.

14.3.1 Rare insects in reserves

In any country, remnant plant communities often harbour rare endemic insects that sometimes occur in only a few remaining patches of plants, which may be found only on islands. For example, on the Chatham Islands 850 km east of the South Island of New Zealand, Dieffenbach's speargrass weevil (*Hadramphus spinipennis*: Coleoptera: Curculionidae), occurs and is dependent on speargrass (*Aciphylla dieffenbachii*: Apiaceae). Once this weevil was more widespread, on several islands of the group, but now it is restricted to one. It is believed that cats and rodents are the cause of its decline, but its host plant is also regarded as vulnerable (Schöps, personal communication; Emberson *et al.*, 1993). The distribution of the speargrass is patchy and large patches have disappeared suddenly, sometimes to reappear some years later. This patchiness results in the weevil having to disperse widely as existing plant patches die out. This dispersal success will be influenced as patch size declines and as the distance between patches increases, so 'connectivity' within the landscape and permeability of landscape features to dispersing invertebrates obviously assume importance. Recent work on the benefits of landscape connectivity for invertebrates includes that of Baudry and Merriam (1988). In contrast, Mauremootoo *et al.* (1994), Frampton *et al.* (1995) and Fry (1995) have shown that landscape features may act as impediments to invertebrate dispersal.

14.4 FARM LANDSCAPES

Few New Zealand endemic species have adapted to pastoral, intensively managed exotic landscapes apart from the grass grub (Scarabaeidae: *Costelytra zealandica*), Porina (Lepidoptera: *Wiseana* spp.) and the carabid *Megadromus* spp. Perhaps because of their serial habitat preference and their polyphagous nature, *Megadromus* spp. can be found in agricultural land under debris, wood (fence posts), hay bales and wherever there is thick and semi-permanent ground cover. The grass grub and Porina moths are native grassland dwellers and have exploited, through successful host shifts, the introduced grassland species.

Protection of existing plant associations via repair and continuing maintenance will ensure that not only plant diversity will be preserved but also, if a community approach is used, insect assemblages will be included. This is a major justification for an *in situ* approach to plant genome conservation. Perhaps its most urgent and appropriate application is in farmland, where in Europe, and especially in the ecologically depauperate communities of New Zealand farmland, there are great opportunities for selective restoration (Norton, 1995; Wratten and Hutcheson, 1995).

Farmers control most of the land (Moore, 1991), but within this landscape, in Europe, nature reserves exist surrounded by a farmland matrix. In contrast, in New Zealand, Australia and perhaps in other relatively new farming landscapes, agro-ecosystems are more rarely punctuated by pockets of semi-native ecosystems (Figure 14.5). Therefore, insect conservation in these landscapes requires more proactive measures to return 'diversity' to the landscape.

Restoration is attracting increasing research interest in the 1990s (e.g. the arrival of the journal *Restoration Ecology*). An effort is being made using existing and established ecological principles as both tools and guides. The need to understand the processes, to avoid ecological restoration being little more than 'trees in a paddock', is important for successful community restoration. In Britain, more than 17 000 farms have had the benefit of farm conservation advisers covering how best to manage and enhance 'natural' features and where to get the financial aid to do this (Moore, 1991). The result has been the addition of hedgerows, enhancement of woods, increased connectivity and increased landscape diversity, all of which benefit insect populations. In New Zealand the initial steps have not yet been taken; it is not the changing of farming practices but the creation of refugia in a largely alien farmland that is now being considered.

14.4.1 New Zealand farming landscapes

Internationally, most reserves are forest ecosystems and it is these which attract most public attention. In New Zealand, some 5 million ha have been set aside as reserves, while another 1.4 million ha are managed indigenous forest. That leaves 21 million ha of the total 27 million ha, i.e. around 75%, that is occupied or strongly modified (e.g. farmed) by humans. Around half of this is in established pasture, of which 4 million ha are lowland, highly productive, intensively managed lands (Stephens, 1976).

In New Zealand, reserves occupy only around 25% of the land surface and are in large areas, on non-productive land, i.e. mountain ranges, coastal swamps. Of the remaining land, 50–60% is farmland, comprised mainly of exotic species. However, there are areas that have a better representation of indigenous species in the landscape, e.g. the Rangitikei ecological region (Lake and Whaley, 1995), though refugia here occur only as isolated patches in steep hill country in which farming is generally difficult.

14.4.2 Insects in depauperate farmland in New Zealand

The Canterbury Plain in the South Island of New Zealand is an area of 750 000 ha and has few remaining fragments of its original forest ecosystem (Meurk, 1996). They have been replaced with pastoral, arable, horticultural and non-native forest habitats. Ecologically, the Plain is degraded with the virtual absence of native flora and fauna. For example, Collembola (springtails), good indicators of disturbance (Greenslade, 1991), in southern New Zealand agro-ecosystems have around nine species in lowland pastures (M. Davidson, personal communication), of which eight are exotic. This compares with around 230 species in indigenous habitats (Wise, 1977), which illustrates the imbalance in the insect biodiversity of 'natural' and farmland areas.

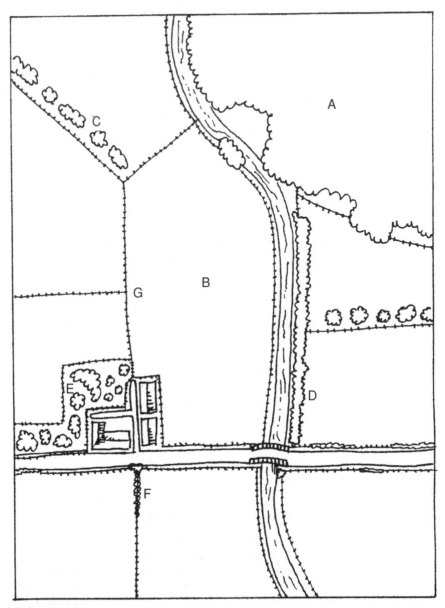

(a) New Zealand

Figure 14.5 Contrast between British and New Zealand farm landscapes. Note the heterogeneity of habitats and the connectance features in the British landscape compared with the uniformity of the New Zealand farm landscape. (a) New Zealand farm; (b) British farm. **A** = indigenous plant reserve – these tend to be large tracts of land not integrated with farmland; **B** = pasture, exotic grasses; **C** = typical shelter belt (e.g. poplar, *Cupressus macrocarpa*, *Pinus radiata*); **D** = ripar-

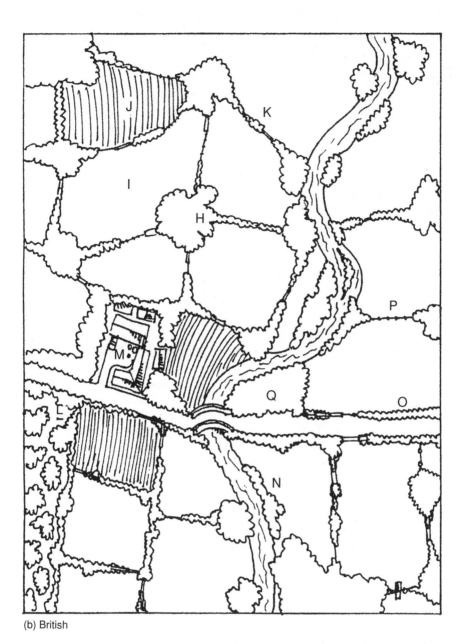

(b) British

ian vegetation (willow, grasses, some indigenous species); **E** = farmhouse garden; **F** = small areas of patchy gorse; **G** = wire fences – common field boundaries; **H** = small wood-lot, a highly used but sustained feature; **I** = pasture; **J** = ploughed field; **K** = hedge fence; **L** = orchard; **M** = farmhouse garden; **N** = riparian vegetation; **O** = roadside vegetation, hedges, trees, etc; **P** = wire fences or stone walls; **Q** = woodland.

14.5 RESTORATION: INCREASING INSECT DIVERSITY IN DEPAUPERATE LANDSCAPES

Restoration ecology aims to re-establish or improve the ecological status of damaged or lost plant and animal communities native to the area of interest (Jordan *et al.*, 1987). The emphasis is on the re-establishment of a functioning community, though the main aim is often a particular plant, insect, or bird species. The unique challenge of ecological restoration is to use ecological understanding by reconstructing, from the bottom up, entire communities.

The challenge of community restoration is to understand and exploit the principles of ecological succession at all serial stages by complementing and accelerating the processes of colonization and regeneration. The idea is to produce a functioning ecosystem with maximized diversity that persists, and becomes self-sustaining.

To date, in New Zealand at least, many 'restoration' projects have been initiated by private land-owners, land-care groups and regional and city councils. They frequently fail to apply ecological theory to the restoration process (Hobbs and Norton, 1996). This has often resulted in an inappropriate choice of plant species for the soil type, climate or landscape involved. The genetic origin, growth habit and successional stage of the species selected are often ignored, as is the 'final' successional community. The effect of the chosen plant species on the surrounding environment, the development of a mycorrhiza fauna, the role in supporting an insect and bird fauna and the competitive interactions between other species planted are not considered. Often the planting projects are too small and/or too isolated to regenerate and hold viable populations, especially of mobile animal species, including those with specialized needs, vulnerability to disturbance or large home ranges or territories.

14.6 THE RESTORATION PROCESS

The ideal process illustrated in Figure 14.6 extends further than most current attempts in that it recognizes early the target community being restored. It considers the addition of animal components and resources which are unlikely, or slow, to arrive naturally but which are considered appropriate for that community. A related aim is the provision of 'connectance', in the form of linear features linking more substantial blocks of restored habitat (Figure 14.6) and providing movement corridors (Burel and Baudry, 1995: Fry, 1995).

In terms of invertebrate conservation, the aim of these protocols is to create habitats for previously rare or absent invertebrates by producing a plant community appropriate to the area and by introducing, or encouraging the arrival of, an invertebrate community. As an example, work in

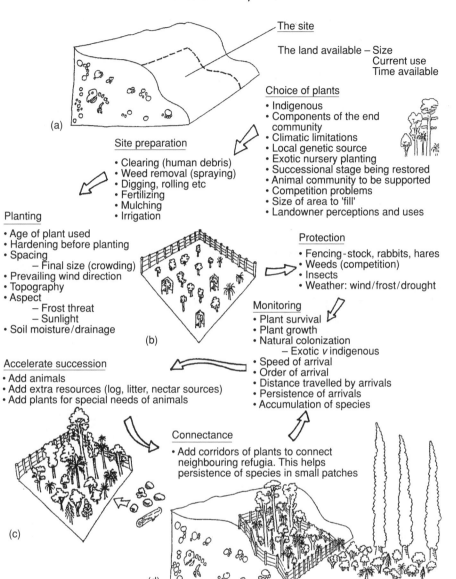

The site

The land available – Size
Current use
Time available

Choice of plants

- Indigenous
- Components of the end community
- Climatic limitations
- Local genetic source
- Exotic nursery planting
- Successional stage being restored
- Animal community to be supported
- Competition problems
- Size of area to 'fill'
- Landowner perceptions and uses

Site preparation

- Clearing (human debris)
- Weed removal (spraying)
- Digging, rolling etc
- Fertilizing
- Mulching
- Irrigation

Planting

- Age of plant used
- Hardening before planting
- Spacing
  – Final size (crowding)
- Prevailing wind direction
- Topography
- Aspect
  – Frost threat
  – Sunlight
- Soil moisture/drainage

Protection

- Fencing - stock, rabbits, hares
- Weeds (competition)
- Insects
- Weather: wind/frost/drought

Monitoring

- Plant survival
- Plant growth
- Natural colonization
  – Exotic *v* indigenous
- Speed of arrival
- Order of arrival
- Distance travelled by arrivals
- Persistence of arrivals
- Accumulation of species

Accelerate succession

- Add animals
- Add extra resources (log, litter, nectar sources)
- Add plants for special needs of animals

Connectance

- Add corridors of plants to connect neighbouring refugia. This helps persistence of species in small patches

Figure 14.6 The restoration process.

progress at Lincoln University, New Zealand, is attempting to restore native woody plant communities and their associated invertebrates, on exposed, ecologically degraded farmland. The experimental protocol for the first stage is as follows.

14.6.1 Experimental protocols in use for farmland restoration in New Zealand

Three areas are being used: one comprises open paddock, highly modified, with a varied history of farming disturbances, exposed to hot, dry
northerly winds and cold southerly winds; the second is pasture, with a
less diverse disturbance history and next to a creek; the third is under
'wild' plum trees in a willow/poplar grove. Two successional community types have been planted, one comprising early successional species
such as *Phormium tenax*, *Carex flagellifera*, *Olearia paniculata*, *O. odorata*,
Coprosma propinqua, *Hebe cupressoides*, *Cordyline australis*, *Kunzea ericoides*
and *Costadonia richardii*. The other has late-successional species:
Dacrycarpus dacrydioides, *Podocarpus totara*, *Melicytus ramiflorus*,
Pittosporum eugenioides, *Kunzea ericoides*, *Hoheria angustifolia*, *Plagianthus
regius*, *Aristotelia serrata*, *Coprosma robusta*, *Griselinia littoralis* and
Pseudopanax arboreus. The late successional species trial has further, multiple comparisons, the first comparing the planting's success in the open
paddock with the same species planted under a 'nursery crop' – the
plum/willow/poplar grove. Another tests the influence of mulch type
(bark or straw), while a third investigates the distribution of the plants
(clustered or spaced). A fourth examines the role of fertilizer, while a fifth
manipulates watering, all in a multifactorial design.

A trial in an open paddock has been established using just one species,
Hebe salicifolia, in which the effect of mulch type (straw, bark, or none),
with or without fertilizer (chicken waste), is compared.

At present plant survival, soil moisture, soil surface temperature and
invertebrate colonization are being monitored. So far five of the 12 late-
successional plant species have died (all broad-leaved species), and a
further three species are stressed. All species in the 'plum nursery' have
survived and are healthy, while mortality has been detected in some of
only one species of the early-successional plants. As yet, the mulch and
planting-distance effects are not measurable, but soil moisture is twice
that of the open paddock when bark or straw mulch are used.
Invertebrate colonization was expected to be slow, the first arrivals being
winged or 'ballooning' species of the 'tourist' type (Diptera, spiders),
with perhaps a few polyphagous Lepidoptera, while ground-dwelling
invertebrates were thought likely to require introduction. To date, one
species of Lepidoptera, several spiders and, surprisingly, carabid beetles
have been found, though the latter are found only in the plantings with
straw as mulch. The addition of facsimiles of fallen logs, adding structural resources, as habitat for ground beetles is the next process, aimed at
accelerating succession.

14.7 INSECTS AS THREATS TO RESTORATION

Recently it has been pointed out that seldom are the potential negative effects of herbivorous insects and pathogens considered in restoration projects (Louda, 1994). Rather, the emphasis is often on single rare plant species, or plant assemblages, ignoring the potentially negative effects of involving a community approach. Insects are crucial components of any plant assemblage, being potentially damaging, reducing plant growth and fecundity (Crawley, 1982; Verkaar, 1988) or affecting plant competitiveness (Bently and Whittaker, 1979). They can also be beneficial in that partial defoliation can promote plant growth and productivity (Owen, 1980, 1981; Van der Meijden, 1990). The balance between the beneficial and damaging impact of insects is still debated (McNaughton, 1979a,b, 1983; Crawley, 1982, 1985, 1989; Belsky, 1985; Verkaar, 1988; Karban and Myers, 1989), but clearly without pollination, largely by insects, restoration through regeneration would not be possible.

Louda (1994) illustrates some case studies where insect herbivory has affected reproduction and establishment of native plants (for example Pitcher's thistle, *Cirsium pitcheri*). Such evidence for the negative impact of insects on vegetation is focused on single endangered species, not on restored communities. If an endangered plant species is suffering from insect infestations it may be that the community in which the plant is placed is incorrect, i.e. the appropriate plant species, their herbivores and associated predators and parasites are not present. Often insect damage in native and restored communities can appear to be extensive, such as that caused by the scale insect (*Inglista* sp.) feeding on southern beech (*Nothofagus fusca*) in New Zealand (Milligan, 1974), but the tree recovers when atmospheric moisture increases. In fact, as with many insects, *Inglista* becomes damaging only when environmental conditions challenge the plant – in this case drought, which stresses the plant and suppresses fungi that usually parasitize the scale. In the initial stages of a restoration programme, environmental conditions can be harsh, increasing the risk of herbivore damage on stressed plants, but with appropriate water, shade, mulches and shelter, the danger of long-term damaging herbivory can be alleviated. In New Zealand the problem of insect 'pest' outbreaks in restoration areas is unlikely to assume importance, as the surrounding land is usually agricultural, accommodating few insect species capable of using native plants.

ACKNOWLEDGEMENTS

We would like to thank Dr J. Hutcheson, Dr C. Muerk, Mr S. Raey and Ms M. Davidson for helpful and usefull discussion, and Mr C. Vink for illustrations.

Part Three

Case Studies

15

The Ammiad experiment

Y. Anikster, M. Feldman and A. Horovitz

15.1 INTRODUCTION

Many of the crop plants that helped to found Western civilization have their origin in an arc of land that connects the valleys of the Euphrates and Tigris with the Jordan and which has been termed the Fertile Crescent by Breasted (1938). The progenitors or close relatives of these crop plants survive to this day in wild populations in those areas. Increasing urbanization and modern farming practices, herbicides, etc., are a threat to the richness of these gene pools or their very survival.

A cardinal species in the assembly of wild relatives of agricultural crops is *Triticum turgidum* var. *dicoccoides* (Koern. in Schweinf.) Bowden or wild emmer. Wild populations of this plant, a progenitor of durum and common wheat, thrive in northern Iraq, south-eastern Turkey, Syria, Lebanon, Israel and Jordan. In Israel, they radiate south and west from a centre of diversity in the catchment area of the Upper Jordan Valley.

In 1984, a study of the natural dynamics of wild emmer populations was launched in Israel (Anikster and Noy-Meir, 1991). It was to serve as a precursor to conservation of selected wild cereal populations in their native ecosystems. Biological aspects, demography, ecological affinities and genetic and phenotypic variation of the species to be conserved were to be examined. The aims were to augment our factual knowledge and to help us to arrive at recommendations for a management that would protect or even increase the amount of variation generated. After a survey of numerous localities, a pilot site for the study was chosen near the settlement of Ammiad in eastern Galilee (Figure 15.1). The site was selected because of its central location in the area of wheat distribution. Also, its

Plant Genetic Conservation.
Edited by N. Maxted, B.V. Ford-Lloyd and J.G. Hawkes.
Published in 1997 by Chapman & Hall. ISBN 0 412 63400 7 (Hb) and 0 412 63730 8 (Pb).

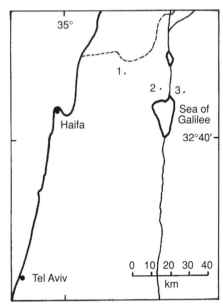

Figure 15.1 Locations of wheat study sites in northern Israel. 1 = Har Meron Nature Reserve; 2 = Ammiad study site; 3 = Yahudiya Nature Reserve.

diverse slope exposures and uneven rock micro-relief were features promising genetic and phenotype variation.

15.2 DESIGN AND METHODS

On the site as a whole, physical, edaphic, climatic, floristic, wheat demographic and phytopathological features were recorded. A more detailed study was made along six linear transects (Figure 15.2) that spanned an area of about 6 ha. Permanent sampling points were set up at roughly 5 m intervals along the transects. Each point consisted of a peg and a circular area with a 50–55 cm radius around it. From 1984 to 1990 and again in 1993, a single wheat spike was collected at each sampling site. Wild emmer is an almost totally obligate selfer, and presumably true-breeding lines were raised from each collection. In all, eight annual collections were made: the first four from up to 249 sampling points along Transects A, B, C, D (Figure 15.2) and the last four from up to 405 points along all six transects. None of the annual collections, with the exception of the first one, contained a full set of samples; some of the micro-populations at collection points underwent cycles of extinction and recolonization. So far, some 2700 accessions have been examined. S1 generations, as counted from the wild-growing parent plants, were raised and further propagated in container nurseries.

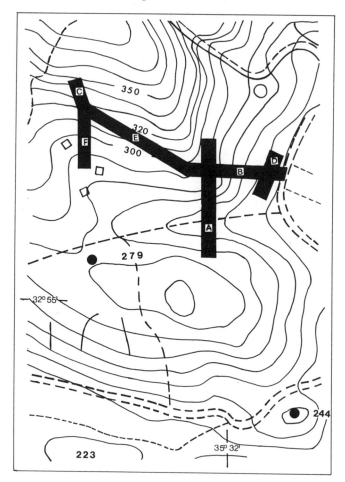

Figure 15.2 Ammiad area, showing Transects A, B, C, D, E, F.

The S_1 or more plentiful S_2 generations of each accession were examined by participating researchers from different Israeli institutions. Allozymes extracted from seedling leaves and seed storage proteins were used as genetic markers and indicators of variation patterns. Morphological, phenological and yield variation were studied in container nurseries and different open-field plantings. Seedlings and adult plants were subjected to artificial infection in a study of phytopathological responses of the different wheat accessions.

Part of the original spikelets and surplus seed from later nursery generations were stored at 12°C and 25–35% RH.

15.3 ECOLOGICAL CHARACTERISTICS

The soil at Ammiad is a special type of phosphate-rich terra rossa that drains rapidly and does not swell on wetting. The depth of the soil cover above the rocks is very variable. The climate is sub-Mediterranean, with recorded temperatures ranging from −2 to +42°C. Rain is limited to the cooler half of the year and is most abundant between December and February. Annual rainfall totals and their monthly distribution vary greatly between years (Table 15.1). Rainfall and fires are considered important ecological variables, causing temporal changes in plant stands.

The site is covered by Mediterranean grassland with almost no shrubs and trees. It is floristically rich; some 150 species were recorded, many of them annuals. A lack of very highly dominant species among the perennial vegetation is probably the result of long-term cattle, sheep and goat grazing. Today, the area serves as a cattle ranch.

A unique feature of our study was the detailed screening of each sampling point (Noy-Meir *et al.*, 1991b). Soil depth, mineral nitrogen in the soil, amount of rock cover and the texture, height and angle of surrounding rocks were measured, as well as the effect of micro-relief on topsoil moisture content. Soil moisture was found to be highest at sampling points close to rocks, especially tall, smooth rocks. The floristic affinities of the associated vegetation were determined at each sampling point by the method of Noy-Meir (1973), and three floristic components were identified. On the basis of rock micro-relief and floristic affinity, the sampling point areas could be grouped into four main habitats and 11 specific habitats, described in Table 15.2. The Karst habitats with highest available soil moisture at both the beginning and the end of the growing

Table 15.1 Annual and monthly distribution of rainfall (mm) at the Ammiad study site over 10 years

Year	Oct.	Nov.	Dec.	Jan.	Feb.	March	Apr.	May	Total
1983/4	0	97	28	129	46	118	87	0	505
1984/5	50	74	103	80	233	20	63	0	623
1985/6	26	17	89	99	133	13	12	26	415
1986/7	47	241	141	156	68	124	15	0	792
1987/8	20	13	215	102	151	88	9	0	598
1988/9	23	56	145	46	46	99	0	0	416
1989/90	34	103	64	91	108	52	40	0	492
1990/1	13	17	35	168	38	97	42	4	414
1991/2	3	147	330	214	321	30	0	11	1047
1992/3	0	116	252	137	87	49	0	8	649
Monthly average	21.6	88.1	140.2	122.2	123.1	69.0	26.8	4.9	595.1

season, a high content of available nitrogen and the lowest grazing pressure (because of rocks interfering with the accessibility to grazing livestock) emerged as the most favourable to wheat growth. The Valley habitats showed the lowest soil moisture content and suffered the highest grazing pressure but had a higher available nitrogen content than other units.

The *a priori* ecological classification was generally confirmed: groups of plant accessions originating from the same ecological unit shared or lacked certain traits or exhibited them in specific frequencies or distribution patterns.

15.4 DEMOGRAPHY

The number of wheat seedlings and the number of spike-bearing plants in the 1 m² sampling-point areas were monitored throughout the study period and varied from zero to over 100 (Noy-Meir *et al.*, 1991a). Mean densities of mature plants varied between habitats and were generally lowest in the Valley. In the North habitats, lower radiation was assumed to delay early desiccation in spring, while in the Ridge and Karst habitats rockiness lowered accessibility to grazing animals. From 1985 on, parts of transects were fenced in to exclude cattle. The impact of this change was a gradual rise in stand density in micro-populations in the fenced as compared with the accessible areas. The lowering effect of cattle grazing on stand density was also noted by the present investigators in the very luxuriant wild wheat populations in the Yahudiya Nature Reserve (Figure 15.1) in the Golan Heights. However, in other wheat studies (Noy-Meir, 1990), moderate grazing was found to be advantageous, in as far as it reduces perennial competitors.

Wild emmer appears to have a two-year soil seed bank. In experiments under nursery conditions, in which the upper (larger) and lower seeds of some 340 spikelets were sown separately, 85% of the former and only 19% of the latter germinated in the first year. Some 50% of the lower seeds germinated in the second year and a very small number of plants from lower seeds also appeared in the third year. When intact spikelets were planted, the percentage of simultaneous germination from the two seeds was even higher. These findings were more or less corroborated by a field study in the Yahudiya Reserve. Here, micro-populations were decimated or wiped out within a year, after all plants had been removed from them together with the spikelets from which they had sprouted. The plants were removed before seed dispersal and so were all spikes in a belt 3.5 m wide surrounding each test plot. Thus, many individuals are exposed to the varying environments of two consecutive growing seasons.

Table 15.2 Ecological subdivision of the Ammiad site

Major habitats	Specific habitats	Slope	Exposure	Floristic ordination by leading species	Rock cover (%)	Rock height (mm)	Soil moisture after early rain*	Grazing pressure	Surface-soil nitrogen (ppm)**
Valley (V)	Main valley centre (Vc) Main valley margins (Vm) Narrow valley (Vn)	None or moderate	None or variable	*Aegilops* spp. (**Fa**)	0–30	0–50	Low	High	44
North (N)	Upper north-facing moderate slope (N1) Middle north-facing steep slope (N2)	Moderate Steep	North North	**Fa+Fb**	20–60	20–60	Intermediate	Moderate	26
Ridge (R)	Ridge, east-facing slope (**Re**) Ridge, shoulder of plateau (**Rp1**) Ridge, top of plateau (**Rp2**) Ridge, south-facing slope (**Rs**)	Moderate to steep	East None None South	*Brachypodium distachum* (**Fb**)	40–80	20–50	Intermediate	Moderate	32
Karst (K)	Upper karst (**K1**) Lower karst (**K2**)	Steep	South	*Cephalaria joppensis* (**Fc**)	40–80	30–80	High, varies with distance from rocks	Low	40

* Surface-soil moisture content measured in October–December, 1985–1987.
** Mean total mineral nitrogen measured in September, 1986.

15.5 BIOCHEMICAL MARKERS AS INDICATORS OF GENETIC SPATIAL AND TEMPORAL VARIATION

Two groups of proteins were studied electrophoretically: allozymes (Nevo *et al.*, 1991), occurring in all tissues and encoded by alleles distributed over the entire wheat genome, and high-molecular-weight glutenin subunits that occur in only a single tissue (the endosperm) and are encoded in a restricted region of the genome, in only two of 28 chromosomal arms (Felsenburg *et al.*, 1991).

15.5.1 Allozyme variation

The electrophoretic patterns of 42 loci were tested in wheat collections from five consecutive years. Thirty-one of the loci were found to be polymorphic. At four of these loci, polymorphism was only detected in the fifth year.

The distribution of alleles in the different Ammiad habitats was found to be non-random, and the differences among habitats were greater than those between collection years. The greatest within-habitat differences were found in the Karst with its variable soil moisture regime. Gene diversities computed for the major habitats, i.e. the probability of gametes in a habitat unit carrying dissimilar alleles, ranged from $H_e = 0.152$ in the Karst to 0.076 in the Valley. The Karst population was separated from the three other major habitat populations by a mean genetic distance D (Nei, 1972) of 0.088.

Altogether, 1060 electrophoretic profiles based on 31 polymorphic loci were analysed, and 400 of these differed from each other. Certain alleles and some entire genotypes showed marked affinities to specific habitats. Some of the genotypes may have been sorting-out products of rare outcrosses. This is indicated by 14 instances of heterozygosity at a single locus.

15.5.2 The glutenin study

Variation in high-molecular-weight (HMW) glutenin subunits of endosperm storage proteins was studied in collections from five consecutive years (1984–1988) and again in 1993. In all, 1126 accessions were analysed by SDS PAGE. The glutenin subunit genotype is encoded at four loci, two in genome A and two in genome B. In the present study, four alleles were recorded for the Glu-A1–1 locus, two for Glu-A1–2, six for Glu-B1–1 and seven for Glu-B1–2. All alleles recorded in the 1984–1988 period reappeared in the 1993 survey (Table 15.3). In the latter survey, an additional allele *i* was found at the Glu-B1–1 locus in a single accession from the N2 habitat.

Table 15.3 Frequencies of HMW glutenin genotypes in 11 habitats at Ammiad during a 10-year study (**bold print** indicates the most frequent genotypes in a habitat)

(a) Frequency of genotypes in 1984–1988

Habitat	No. accs.	No. genotypes	Genotypes a l a k	a a l n	k l l n	o a l a	o a l m	o a u m	o a u s	t a a k	t c p	t a j m	t a l a	t a l g	t a l m	t a l n	t a l u	t a u k	t a u m	t a u s
N1	68	8	–	–	–	0.029	0.015	–	–	–	–	**0.368**	0.059	0.015	–	–	0.015	0.235	0.265	–
N2	77	8	–	–	0.078	–	–	–	0.039	–	–	0.091	–	**0.429**	0.052	–	–	0.169	0.104	0.039
Vn	62	11	–	–	0.097	–	0.048	–	0.032	–	0.016	0.048	0.016	0.161	0.129	0.010	–	–	**0.403**	0.016
Rs	102	8	–	0.049	0.108	–	–	–	–	–	0.010	0.118	0.017	–	0.029	–	–	–	**0.667**	0.010
Rp2	115	7	–	0.009	0.052	–	–	–	–	–	–	0.070	0.027	–	0.026	0.017	–	–	**0.809**	–
Rp1	73	10	–	0.014	0.055	–	0.055	–	**0.356**	0.014	–	0.178	0.027	–	0.110	–	–	–	0.164	0.027
Re	145	7	–	–	–	–	–	0.041	0.262	–	–	0.083	0.138	–	0.035	–	–	–	**0.407**	0.035
Vc	180	9	0.006	–	0.044	–	0.001	**0.533**	0.067	–	–	0.006	0.033	–	0.033	–	–	–	0.267	–
Vm	147	5	0.007	–	–	–	–	**0.660**	0.252	–	–	0.007	–	–	–	–	–	–	0.075	–
K1	43	4	–	0.093	–	–	–	–	–	–	–	–	–	–	–	0.023	–	–	**0.861**	0.023
K2	96	6	0.219	**0.635**	0.021	–	–	–	–	–	–	–	–	–	0.010	0.010	–	–	0.104	–
Whole population	1108	18	0.021	0.065	0.039	0.002	0.007	0.183	0.107	0.001	0.002	0.074	0.032	0.040	0.034	0.005	0.001	0.026	0.351	0.012

(b) Frequency of genotypes in 1993

Habitat	No. accs.	No. geno-types	Genotypes															
			a a a	a a l k	a a l n	k l l n	o a j m	o a l m	o a l n	o a u m	o a u s	t a i g	t a j m	t a l a	t a l g	t a l m	t a u k	t a u m
N1	14	3	–	–	–	–	–	–	–	–	–	–	**0.357**	–	–	–	0.143	**0.500**
N2	17	6	–	–	–	0.059	–	–	–	–	0.059	0.059	0.059	–	**0.412**	–	–	**0.353**
Vn	13	5	–	–	–	–	0.077	–	–	–	–	–	0.154	0.077	0.154	–	–	**0.538**
Rs	21	4	–	–	0.047	0.095	–	–	–	–	–	–	0.095	–	–	–	–	**0.762**
Rp2	25	6	–	–	–	0.040	–	–	–	0.040	–	–	0.080	–	0.040	0.040	–	**0.760**
Rp1	15	5	–	–	–	0.133	–	0.133	–	–	**0.333**	–	–	0.067	–	–	–	**0.333**
Re	31	6	–	–	–	–	–	–	–	0.129	0.226	–	0.032	0.161	–	0.032	–	**0.419**
Vc	31	6	0.032	–	–	0.032	–	–	–	**0.774**	0.097	–	–	0.032	–	–	–	0.032
Vm	24	4	–	–	–	–	–	–	0.041	**0.583**	0.208	–	–	–	–	–	–	0.166
K1	7	2	–	–	**0.714**	–	–	–	–	–	–	–	–	–	–	–	–	0.285
K2	18	3	–	**0.222**	**0.722**	–	–	–	–	–	–	–	–	–	–	–	–	0.055
Whole population	216	16	0.005	0.019	0.088	0.032	0.005	0.009	0.005	0.199	0.097	0.005	0.060	0.037	0.046	0.009	0.009	0.375

In the 1984–1988 survey (Table 15.3a), a total of 18 genotypes was found, representing some 30% of 4-locus genotypes known from other studies (Levy and Feldman, 1988). Each of the Ammiad habitats was characterized by specific genotype frequencies, with a leading, most common genotype. Within a habitat, the genotypes had a patchy distribution, and the patches themselves were dispersed non-randomly along the transects. An example is shown in Figure 15.3. Yearly fluctuations were small.

In 1993 (Table 15.3b), 16 genotypes were recorded, four of which had not been sampled in the earlier survey. Each of the 11 populations was again characterized by its most common genotype from the earlier survey. The table shows that the frequencies of the four 'new' genotypes, as well as those of six genotypes recorded in the 1984–1988 survey but not in the 1993 survey, were below 0.05 each. Thus, two kinds of genotype can be distinguished: stable, representative ones, recurring every year at high frequencies, and rare ones that escape sampling or are in fact unstable.

15.6 PHENOTYPIC VARIATION

All accessions were subjected to a detailed study of morphological, phenological and productivity traits (Anikster *et al.*, 1991). The phenotypic attributes can give at least a partial indication of the desirability of conserving germplasm. While given phenotypes can have a direct present-day breeding potential, phenotypic diversity as such is desirable in a natural conservation site.

Nursery-grown progenies of the original collections were examined for some 20 traits that more or less conform to wheat descriptors used in gene banks (IBPGR, 1985c). Phenotypic variation was compared between and within the 11 specific habitats defined by the ecological survey. As with the biochemical markers, significant habitat-specific variation patterns could be traced. Figure 15.4 shows that, in all but one habitat, most sampled plant lines have an anthocyanin-pigmented first leaf sheath. In contrast, the morph lacking the pigment predominates in K2, the Lower Karst habitat.

Quantitative traits that differed significantly between groups of accessions originating from different habitats were leaf width, length of the spike and number of spikelets in it, and number of days from a set date to emergence of the first awn, to anthesis and to maturation of the first spike. When identical accessions were grown in different environments, these properties were modulated but the differences between groups of accessions derived from the different ecological units were maintained. Environmental effects did not blur the presumably genetic differences between groups from different habitats (Eshel *et al.*, 1989).

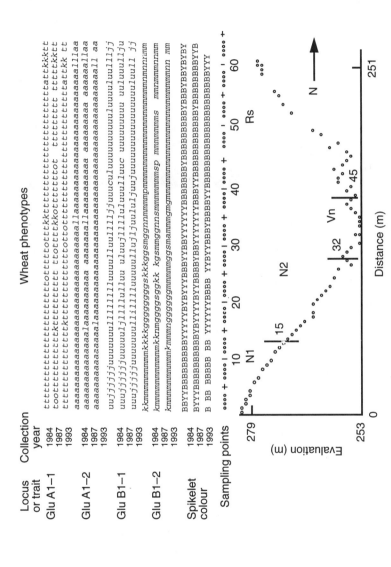

Figure 15.3 Sequences of wheat phenotypes collected in different years from sampling points 1 to 66 along a transect that traverses the four habitats N1, N2, Vn and Rs. Lower-case letters = markers for alleles at four loci that encode HMW glutenin subunits; upper-case letters = spikelet colour morphs. B = black-containing; Y = yellowish, lacking black. The lower part of the figure shows elevations of the sampling points and the distances between them.

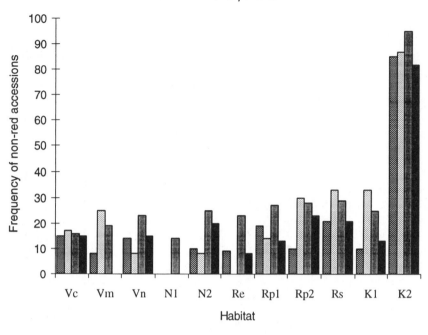

Collection years: ▨ 84+85 ▨ 86 ▨ 87 ■ 89

Figure 15.4 Frequencies of wheat accessions from different habitats, in which the sheath of the first leaf lacks red pigment. Accessions from the same sampling points were studied during five years.

In productivity traits, such as above-ground biomass, number of tillers and spikes and weight of spikes, habitat specificity was almost completely blotted out by environmental variation. Notwithstanding large standard variations within each planting, probably indicating sensitivity to lack of uniformity within the nursery, differences between nurseries were significant. Also the interaction between productivity traits, such as allocation of biomass to tiller number, differed significantly in the different environments.

Within a habitat, alternative morphs are likely to inhabit the same sampling point. Thus, in different collection years, different morphs are likely to represent the mixed micro-population. However, where the same morph was sampled at several adjoining collection points, as well as in different years, a significant non-random clustering in patches could be assumed. In Figure 15.3 the predominant morph B, with a black-tinted spikelet, is interspersed by patches in which it either co-exists with the black-lacking Y morph or in which the Y morph is in sole occupancy. In the figure, the patchiness recorded for spikelet colour coincides partly with that shown for glutenin alleles.

15.7 FUNGAL DISEASES

Responses to different pathogen races were monitored as parameters that can indicate genetic and phenotypic variation, and for possible detection of resistance genes. Numerous fungal disease agents were present on the site, yet disease incidence was low and did not impair the fertility of the host plants. The foliage diseases recorded were races of the rust fungi *Puccinia recondita, P. graminis tritici* and *P. striiformis*, powdery mildew *Erisyphe graminis*, an *Aschochyta* species and the glume blotch *Septoria nodorum*. Below ground, *Gaeumannomyces graminis*, the take-all disease that devastates cultivated wheats, was recorded for the first time in a naturally growing wild wheat population. No loose smut of wheat, which is common on cultivated wheats in the area, was encountered during the 10-year survey.

The most thoroughly studied disease was wheat leaf rust *P. recondita*. This fungus invades populations from outside sources anew each year. Some 30 different races were monitored during a three-year survey at Ammiad, their frequency varying from year to year. The same races have also been recorded on cultivated wheat in Europe. There was no variation in overall host-plant response to this disease. All plants of 1050 test accessions examined were susceptible, even to the least virulent rust races.

Artificial inoculation with a race isolated from locally cultivated wheat also revealed predominant susceptibility to the take-all disease, *vis-à-vis* a virtual absence of disease symptoms in the wild location.

There was more variation in host response to some of the other diseases, especially to powdery mildew. This disease, which persists during summer dormancy of host plants in the form of cleistothecial resting bodies, had a restricted distribution and occurred with high severity on single rock-shaded sites in one of the study years. Upon controlled inoculation, all of 225 accessions representing all transects were susceptible to a local mildew isolate but 10% showed resistance to a geographically alien isolate (Dinoor *et al.*, 1991). Moreover, when 24 randomly picked Ammiad accessions were inoculated with a wide geographical range of European powdery mildew races in Germany, high degrees of resistance were found, and a third of the accessions was immune to one or another of the pathogen races.

The overall low disease incidence in wild-growing, genetically susceptible hosts appears to be due to the protection afforded by ecological buffering in nature. Among suspected contributing biotic factors that minimize the impact of disease are the heterogeneity of host genotypes in the wild stand, their slower growth rate and lesser biomass in the wild than under cultivation, and the presence of unknown microbes that are antagonistic to the pathogen. The paucity or absence of genetic resistance factors to pathogen races that abound in the natural population is probably connected with this absence of grave disease.

It should be noted that the Ammiad situation does not hold for all local wild emmer populations. A typical habitat in the centre of the wild wheat distribution is Tabor oak park forest, where populations extend into the tree-shaded patches. In the Yahudiya Nature Reserve in the Golan, we regularly encountered plants heavily infested with leaf rust, stripe rust or powdery mildew in the shaded spots. Plants in these niches are quite likely to harbour genetic resistance factors.

15.8 DISTRIBUTION PATTERNS OF POLYMORPHISM AND IMPLICATIONS FOR CONSERVATION

The electrophoretic analysis of proteins in the different wheat accessions confirmed our expectations as to genetic variation in the centrally located Ammiad population. The mean number of alleles per locus, the number of polymorphic loci sampled within the habitat groups treated as specific populations and gene diversity within these populations were comparable to estimates obtained in other populations in the wild emmer core area (Nevo and Payne, 1987; Levy and Feldman, 1988; Nevo and Beiles, 1989). With respect to the markers used, the overall allele pool sampled at the site represents a considerable fraction of the country-wide variation known. If the sizes of the habitat units into which this allele-rich site was divided meet the criteria of minimum viable populations (Soulé, 1987), there is a good chance for eventual retrieval of these alleles under dynamic conservation. In each of the small units of Upper North and Middle North, with areas of about 300 m^2 and annual stands of about 3000 plants, eight glutenin genotypes were sampled in the course of the study period (Tables 15.3a,b), and the characteristic components of each unit were also retrieved after 10 years. Since the different populations are in close proximity to each other, the entire site satisfies prerequisites for *in situ* conservation.

These findings are in contrast to data collected in small scattered wheat patches in the Har Meron Reserve (Figure 15.1) in a typical Mediterranean habitat on the outskirts of the sub-Mediterranean local wild emmer distribution. Here individual patches are monomorphic in structural traits and glutenin profiles and are separated from each other by considerable genetic distances. In these outpost units, the different genotypes are best salvaged by *ex situ* storage. We have witnessed extinction of patches, so that conserving micro-populations for self-perpetuation of a single rare combination may be hazardous.

15.9 ADAPTIVE STRATEGIES

The environment in which wild emmer succeeds has a marginal Mediterranean climate. In the core distribution area, rainfall totals and distribution vary from year to year, as shown in Table 15.1, and so do temperatures at the time of grain ripening. Environmental uniformity is further impinged upon by topography, fires and grazing regimes. It appears that this autogamous annual adapts itself to variable natural environments in different years and in different micro-niches within a habitat by a dual strategy of genetic polymorphism in some traits and phenotypic plasticity in others.

The study of allozyme variation at Ammiad points to an abundance of genotypes, maintained by autogamy. In parallel, the study of quantitative traits and of response to pathogens points to great phenotypic plasticity. Significant differences in plant performance in different growing regimes were noted in estimates of quantitative traits. Plasticity must also be involved where genetically susceptible host plants show no symptoms of fungal disease in the natural environment but succumb to the same disease under different growing conditions in the nursery.

In addition to its role in natural dynamics, phenotypic plasticity is of interest in conservation sites of crop-plant relatives because of its intrinsic value to breeders. Modern cosmopolitan crop cultivars, bred for use in diverse geographical regions, have to be flexible in phenology or allocation of biomass to different yield components. Our knowledge of the genetic control of plasticity is speculative. At the molecular level, Smith (1990) ascribes it to the differential expression of genes grouped in multigene families that encode similar proteins. In the allopolyploid wild emmer, variation between homologous multi-gene families of the A and B genomes may amplify possibilities of gene expression. Comparisons of plasticity in diploid as opposed to polyploid wheat should verify whether this assumption holds.

In conclusion, our decade-long study points to genetic stability in centrally located wild emmer populations. However, our study period has been lacking in environmental catastrophes. Moreover, in old and flexible populations, the detection of dynamic shifts in structure may require very much longer periods of monitoring.

16

In situ conservation of genetic diversity in Turkey

A. Ertug Firat and A. Tan

16.1 INTRODUCTION

Turkey is one of the most significant places in the world for plant genetic resources: it is located on two of Vavilov's Centres of Origin (Near East and Mediterranean) and three phyto-geographical regions (Euro-Siberian, Mediterranean, Irano-Turanian). Therefore it is not surprising that it is the centre of diversity or micro-gene centre of many crop species, as well as being the site of domestication for many temperate agricultural crops such as wheat, chickpea, vetch, faba bean and barley. It has a unique combination of diverse geographical and climatic conditions that have given rise to a large number of rare or endemic species.

The potential of Turkey's plant diversity was determined and recognized by well-known plant scientists such as Vavilov (1926), Zhukovsky (1950), Harlan (1951), Harlan and Zohary (1966) and Zohary (1969), who worked in the area, noted the rich diversity of cultivated plants and their wild and weedy relatives and designated it as containing the centres of diversity. Through human agency, some non-native cultivated plants were introduced; these became widespread and increased tremendously in their diversity. Thus Turkey was established as an important secondary centre of diversity for cultivated crops. Moreover, lying at the northern edge of the Fertile Crescent, it is one of the original locations of ancient agriculture. Paleobotanical findings of early Neolithic village sites show archaeological evidence for the early domestication of plants in Anatolia, and early sites at Çatal Hüyük, Mersin, Hacilar, Can Hasan

Plant Genetic Conservation.
Edited by N. Maxted, B.V. Ford-Lloyd and J.G. Hawkes.
Published in 1997 by Chapman & Hall. ISBN 0 412 63400 7 (Hb) and 0 412 63730 8 (Pb).

and Çayönü have been identified and dated to 7000–5250 BC (Tan, 1992).

This rich plant diversity provided the sources of important traits for world agriculture and it must be maintained and made available for exploitation. Owing to genetic erosion and even loss of plant genetic resources in Turkey, many countries that are aware of its wealth of floral diversity are collaborating with the National Plant Genetic Resources programme to further explore and conserve the country's natural plant genetic resources and guarantee the Turkish heritage.

Turkey was a pioneering country and has practised active conservation of its plant genetic resources for an extended period. Therefore, it has considerable experience with *ex situ* conservation for both vegetatively propagated plants in collection gardens (field gene banks) and generatively propagated plants in seed gene banks. However, *in situ* genetic conservation of wild crop relatives did not begin until 1994.

The recent application of *in situ* conservation projects, within the framework of the Global Environment Facility (GEF, 1993) project, aims to maintain the wild genetic resources of crops in their natural habitats on existing state-owned lands. This project is unique; it is the first to attempt to conserve both woody and non-woody crop relatives in an integrated multi-species and multi-site approach. This will be undertaken using ecogeographic surveys and inventories to provide a basis for the establishment of *in situ* genetic reserves.

Gene Management Zones (GMZs) or genetic reserves will be selected in pilot areas that are rich in target wild species related to crops. The highest priorities have been given to globally significant non-woody crop species which are in the primary gene pool of wheat, barley, chickpea and lentils as well as important woody species such as chestnut, plums and pistachio. The project will initiate and develop a mechanism to foster the ongoing National Plant Genetic Resources Research Project (NPGRRP) and aid the identification, designation and management of the areas specifically for *in situ* conservation of nationally and globally significant wild crop relatives that originated in Turkey. The project also aims to integrate *in situ* conservation with the existing national *ex situ* conservation programmes.

The project will use the complementary strengths of the Ministry of Agriculture and Rural Affairs (MARA), which has experience in genetic resources activities (especially in *ex situ* conservation), the Ministry of Forestry (MOF), which has experience in land management, and the Ministry of Environment (MOE), which has a strategic outlook on resource management. MARA and MOF are the implementing ministries.

The lead institute of MARA is the Aegean Agricultural Research Institute (AARI) at Izmir, which is National Coordinating Centre of NPGRRP. AARI will coordinate the activities of MARA for the *in situ*

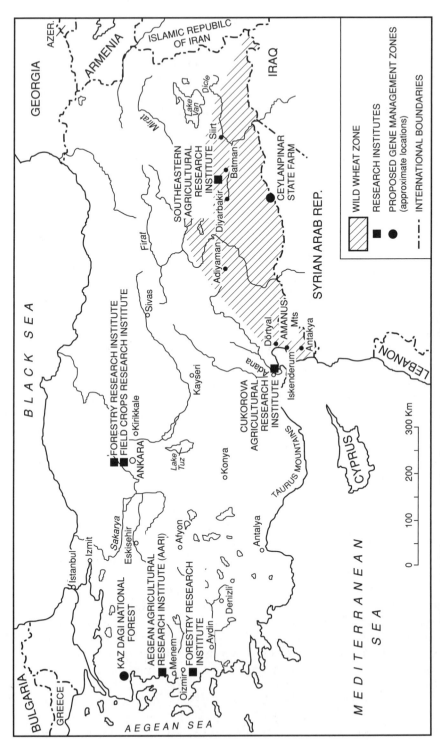

Figure 16.1 Pilot areas and implementing institutes of *in situ* conservation project in Turkey.

conservation project and will collaborate with the Field Crop Central Research Institute (FCCRI) at Ankara, the South-east Anatolian Research Institute (SEARI) at Diyarbakir, and the Cukurova Research Institute (CRI) at Adana.

The pilot areas (Figure 16.1) are selected and described as the Kaz Dagi area of the north-western Aegean region, Ceylanpinar of south-eastern Turkey, the Amanus Mountains (Gavur (Nur) Daglari) and the Central Taurus Mountains in Southern Anatolia (on the southern part of the Anatolian Diagonal).

The project has been designated around five components:

- site survey and inventories;
- designation of GMZs;
- data management;
- development of a National Plan for *in situ* conservation;
- provision of the institutional strengthening within and between MARA, MOF and MOE.

The project commenced in 1993 with the training of project staff from MARA, MOF and MOE. The survey activities commenced in the three designated areas in 1994.

16.2 PROJECT ACTIVITIES

The project will seek to identify and establish *in situ* genetic reserves in Turkey for the protection of genetic resources and wild relatives of major crop and forest tree species that originated in Turkey. It is proposed to provide sustainable *in situ* conservation of genetic resources of cereals, horticultural and ornamental crops, medicinal plants and forest trees. It is also proposed to develop and implement a national strategy for *in situ* conservation, as well as to test and develop a new integrated approach to the conservation of genetic diversity of wild crop species using *ex situ* and *in situ* techniques. Land races are not a major focus of this project because the *in situ* conservation of land races is very complex, involving biological, social and policy issues. However, during the implementation of the project it is hoped to increase the institutional capacity of Turkey in the *in situ* (on-farm) conservation of land races. Research is an essential part of the project; therefore the research on genetic diversity and conservation biology will be conducted at all stages of *in situ* conservation project.

16.3 SURVEY AND INVENTORIES

Surveys and inventories constitute the initial assessment of suitable reserve sites for the wild crop relatives of primarily wheats, chickpeas, lentils, barley, chestnut and plum as well as other herbaceous and woody species.

The GEF project, as a pilot project, cannot cover all of Turkey, nor even all of those areas where wild crop relatives occur; therefore it is limited to selected areas. Given the state of previous ecogeography and taxonomic knowledge for target species, site choice includes selected areas in Kaz Dagi National Forest to represent the Aegean region, the Ceylanpinar State Farm to represen south-eastern Turkey, and the southern part of the Anatolian Diagonal (Central Taurus and Amanus Mountains) to represent Southern Anatolia where the north-west point of the Fertile Crescent is located. Within these areas, ecosystem-based surveys for target species have been started to determine suitable habitats with regard to diversity, naturalness and management considerations for *in situ* conservation management. Following the surveys, a species inventory will be conducted at each site with regard to species abundance, distribution and management needs. To support a complementary approach between *in situ* and *ex situ* conservation, some selected representative samples will be collected for *ex situ* conservation. The survey output will consist of:

- physical description of area surveyed including size, topography, access, land ownership designation and ownership of adjacent land;
- ecological description of area surveyed: detailed description of plant community types currently present at the site and the occurrence of rare and endangered plant species according to the Red Data Book;
- sampling for population biology, quantitative description and genetic diversity of target plant species associated with the inventoried areas;
- additional descriptions including geology, soils, climate information, fire history, logging, grazing or human impact, uses, historical information, etc.

Guidelines for the sampling of woody and non-woody species for the determination of genetic diversity have been prepared. The objective of sampling is to provide a quantitative description of targeted species in the selected sites. The sampling methods would be appropriate to the number, size and distribution of wild crop relatives on the site as well as the life history of the species in question. The guidelines provide a good basis for determining the existing variation and for monitoring the changes in abundance and distribution of species within the GMZs.

A complementary approach requires that a few select representative germplasm samples in the determined sites, but not all populations of some species, will be collected for *ex situ* conservation at AARI Gene Bank. Each plant species will have a voucher specimen taken which will be preserved at AARI Herbarium.

16.4 GENE MANAGEMENT ZONES (GMZS)

This component builds on the survey and inventory phase, and will involve the establishment of reserves on the sites selected with detailed management regimes for the different species included. A series of reserves will be selected to represent the ecological ranges needed for targeting wild relatives in order to support sufficient environmental heterogeneity for both wild crop, woody and non-woody species.

The *in situ* conservation of wild crop relatives will be accomplished through the establishment of genetic reserves called Gene Management Zones (GMZs), where wild crop relatives are maintained in their natural habitats. These reserves will include sufficient environmental heterogeneity to provide the major ecological niches that influence the distribution of wild crop populations in the selected pilot areas.

Size and management strategies will vary depending on the target species to be conserved in the GMZs. The appropriate number and size of sites will be determined on the basis of genetic diversity analysis and the information gathered from survey and inventories activities. It is considered that the protection of islands within the larger reserves will provide more effective control and management. Based on current knowledge from the initial surveys of the genetic diversity and population biology of wild wheat, it is expected that there will be a network of small GMZs at each sites.

In selecting GMZs to conserve genetic resources of target non-woody species, such as wild wheat (Figure 16.2), priority will be given to identifying areas with a variety of factors affecting genetic diversity in that species, e.g. rainfall, rockiness and grazing. Within the south-east and south central regions, the management objectives for non-woody species (e.g. wild wheat, wild barley) are to maintain or increase the abundance and perhaps genetic diversity in each GMZ and to prevent perennial plant species from taking over the habitats of wild target species. These objectives can be accomplished through controlled grazing, mowing or fire management to discourage perennial species, especially perennial grasses, from displacing the annual wild target species. Management strategies for wild relatives of food legumes require the elimination of grazing during critical life history phases such as flowering and fruiting periods. The management of woody crop species has the same objectives as for herbaceous species. For woody species, planting of trees from seed source outside of GMZs will not be permitted, so as to prevent population of the gene pool from exotic sources. To encourage development of seedlings, associated species and a forest understorey, grazing will not be permitted in forest areas. To reduce the influx of foreign pollen for wind-pollinated chestnut species, orchards of cultivated varieties will not be permitted near GMZs.

Figure 16.2 Wild wheat (*Triticum dicocoides*) growing in the Ceylanpinar genetic reserve in south-eastern Turkey.

16.5 DATA MANAGEMENT

A database will be built for the complex array of information that will be acquired while surveying and managing the GMZs. The project will depend initially on the acquisition of existing data and will result in the generation of substantial quantities of new data from survey and inventory activities. As there are no models to follow for the incorporation of *in situ* conservation data, new techniques of data management will be required in addition to the existing National Project system. Detailed new quantitative and spatial data will be managed through use of the Geographic Information System (GIS); therefore the site data will be incorporated into a system that can interface with a GIS and this may also involve the use of remote sensing data.

Data management and databases for *in situ* conservation projects will be linked with and complementary to existing *ex situ* management and documentation systems.

16.6 NATIONAL PLAN FOR *IN SITU* CONSERVATION

Development of a National Plan for *in situ* conservation will necessitate a review of conservation data handling and will facilitate the coordination and cooperation of the GMZs and other conservation strategies in Turkey. The development of this plan is expected to help the Government of Turkey to expand *in situ* conservation activities into a network of GMZs, which will include other areas of the country and will incorporate additional important wild crop and forest species, especially endangered ones, beyond the targeted species in the current project. The plan will also be consistent with and complementary to other national, regional or global efforts to conserve wild plant genetic resources.

The plan will lead the way for the continuation and replication of the GEF pilot project which can provide a supportive platform for continued assistance from the international donor community.

16.7 INSTITUTIONAL STRENGTHENING

An essential ingredient of *in situ* conservation is a community of scientists and extension agents who are well trained and equipped to implement *in situ* conservation objectives. The existence of these workers is integral to conservation and without it the practical activities cannot take place. Institutional capacity and flexibility in planning and implementing *in situ* conservation projects will be built and sustained in three ways: through research, through training, and by linkage between the formal and non-formal (NGO) sectors.

Since this is a complex and innovative project, it requires considerable coordination and communication between different governmental levels and scientific disciplines as part of the institutional strengthening process. A cooperative linkage between project scientists and local communities is important for the success of this project. Local communities living near the GMZs need to understand and be supportive of the project. This requires public education and awareness through the development of a wide range of extension material in various media.

The project will generate a considerable amount of information on the diversity and current status of Turkey's wild crop genetic resources. Therefore an international seminar would be a good opportunity for Turkish scientists to share and exchange information with the broad international community on *in situ* conservation. Also it will contribute to the broader development of an international network on *in situ* conservation of wild crop relatives.

16.8 CONCLUSION

Turkey's GEF *in situ* conservation project targets the conservation of wild relatives of crop species and other selected plants with significant genetic variability. This type of activity is still in its infancy. This project is the first attempt to integrate the various components of *in situ* conservation for multi-crops (target species identification, establishment and management of conservation areas) and the monitoring of genetic diversity through protecting wild crop relatives in their natural habitat. By complementing the ongoing *ex situ* programme it will develop institutional and technical mechanisms for a comprehensive strategy for plant genetic resources conservation in Turkey. Success here could provide a model applicable in other parts of the world. Although there have been a few national *in situ* programmes, notably in Brazil, Mexico and Israel, these have targeted single species and have not focused on the ecosystem approach to management. *In situ* genetic conservation is different from the concept of protected areas. The significance of *in situ* genetic conservation is that its conservation target is genetic diversity in specific plants, and not necessarily the protection of entire ecosystems.

The integration of *in situ* and *ex situ* strategies for genetic resources conservation in this project is also innovative through the use of three different ministries within Turkey, each contributing to the siting, design and maintenance of *in situ* conservation areas. In this way the Turkish experience may also act as a model for other countries rich in the genetic diversity of wild crop relatives.

17

Genetic conservation: a role for rice farmers

M.R. Bellon, J.-L. Pham and M.T. Jackson

17.1 INTRODUCTION

The genetic resources of rice have been well utilized in efforts to solve today's food problems. Rice land races, collected over several decades, have become 'parents' of the high-yielding, pest-resistant and well-adapted varieties which resulted in unprecedented increases in rice yields. The cost of rice to millions of consumers is now approximately half what it was in 1960 because of these gains in productivity.

The diversity of the rice crop has evolved over thousands of years, as Asian and African peasant farmers – mostly women – selected different types to suit local cultivation practices and needs. This process of selection has led to numerous rice varieties adapted to a wide range of agro-ecological conditions, and with resistance to insect pests and diseases. The number of varieties of Asian rice, *Oryza sativa*, is impossible to estimate, although claims of more than 100 000 have been made (Chang, 1985, 1995). Asian rice varieties show an impressive range of variation in many characters, such as plant height, tillering ability, maturity and size of panicles, among others. Variation in grain characters such as size, shape and colour is most useful for distinguishing different varieties. Some wild species occur as weeds in and around rice fields, and even hybridize naturally with the cultivated forms. This complex association between cultivated and wild forms has enhanced the diversity of the rice crop in traditional agricultural systems, where farmers often grow mixtures of varieties to provide a buffer against the risk of complete loss

Plant Genetic Conservation.
Edited by N. Maxted, B.V. Ford-Lloyd and J.G. Hawkes.
Published in 1997 by Chapman & Hall. ISBN 0 412 63400 7 (Hb) and 0 412 63730 8 (Pb).

of the crop due to biotic and abiotic stresses.

Rice farmers in Asia continue to grow thousands of different varieties for specific traits, such as aroma or cooking quality, or because of a particular cultural aspect (Figure 17.1). However, there is widespread concern over the loss of the genetic diversity represented by these varieties, particularly as they are replaced more and more by a few genetically uniform, high-yielding varieties in many farming systems (Hawkes, 1983; Plucknett *et al.*, 1987; Brush, 1991a; Harlan, 1992; National Research Council, 1993). The need to conserve the diversity found in crop land races has been recognized as important for many decades. *Ex situ* conservation – the storage of seeds in gene banks – has been the principal strategy for the preservation of crop genetic resources, and this applies especially to rice. *Ex situ* conservation is static conservation that aims to retain as far as possible the structure of the original population (Guldager, 1975). Seed storage is a safe and efficient way of conserving rice genetic resources, and has the advantage of making the germplasm readily available for use by breeders and for study by other researchers (Ford-Lloyd and Jackson, 1986). Rice has so-called orthodox seeds that can be dried to a relatively low moisture content (± 6%), and stored at subzero temperatures. Under these conditions, the viability of rice seeds can be assured for many years – certainly decades, if not considerably longer. Therefore, this strategy has been favoured for the conservation of cultivated rices.

Lately, on-farm conservation (Altieri and Merrick, 1987; Oldfield and Alcorn, 1987; Brush, 1991a; IPGRI, 1993) has been advocated to complement *ex situ* conservation. For more than two decades, on-farm conservation of crop land races was considered as impractical and inappropriate (Arnold *et al.*, 1986). However, concern in developing countries about the concentration of genetic resources in gene banks in the industrialized countries and the fact that static conservation halts evolutionary processes have opened a debate concerning the value and objectives of on-farm conservation methods. Rural societies maintain agricultural biodiversity because it is essential to their survival. They select and breed new varieties for the same reason. There is no useful distinction for them between conservation and development. Indeed, conservation as such may not be a concept known to farmers. On-farm conservation of local varieties is an existing strategy for food security. It is a potential strategy for genetic conservation. By its very nature, on-farm conservation is dynamic because the varieties that farmers manage continue to evolve in response to natural and human selection. It is believed that in this way crop populations retain adaptive potential for the future.

In south and south-east Asia, the community of non-governmental organizations (NGOs) has been particularly active in support of farmer and community groups that have begun to conserve traditional rice varieties in community seed banks, or in dynamic on-farm management systems (Salazar, 1992). The Thai-based Technology for Rural and Ecological Enrichment (TREE) and the Philippine-based South East Asia Regional Institute for Community Education (SEARICE), Farmer–Scientist Participation for Development (MASIPAG), and the Sustainable Agriculture Coalition have been working with farmers and community groups to collect and manage traditional rice varieties as part of farming systems. However, their work has not been well documented, and we are unable to comment on the nature and scope of their activities.

Brush (1995) has suggested that, besides directly providing genes for crop improvement, on-farm conservation should be seen as satisfying four other needs:

- It preserves evolutionary processes that generate new germplasm under conditions of natural selection.
- It maintains important field laboratories for crop biology and biogeography.
- It provides a continuing source of germplasm for *ex situ* collections.
- It provides a means for wider participation in conservation, allowing for a more equitable role for nations with abundant crop germplasm resources.

In this context, therefore, on-farm conservation of crop genetic resources can be defined as the continued cultivation and management of a diverse

Figure 17.1 Surveying rice land race in Indonesia.

set of crop populations by farmers in the agroecosystems where a crop has evolved. This set may include the weedy and wild relatives of the crop that may be present together with it, and in many instances tolerated. It is based on the recognition that, historically, farmers have developed and nurtured crop genetic diversity, and that this process continues in spite of socio-economic and technological changes. It emphasizes the role of farmers for two reasons:

- Crops are not only the result of natural factors, such as mutation and natural selection, but also and particularly of human selection and management.
- In the last instance, farmers' decisions define whether these populations are maintained or disappear.

In spite of increasing interest in on-farm conservation, which is even addressed in the Convention on Biological Diversity, there is still limited knowledge about what this approach means, and even less understanding of its various social, economic, cultural and genetic aspects. There have been only a few studies aimed specifically at documenting and understanding the conservation and management of crop genetic resources among small farmers (e.g., Brush *et al.*, 1981, 1992; Dennis, 1987; Quiros *et al.*, 1990, 1992; Bellon, 1991; Zimmerer and Douches, 1991; Bellon and Brush, 1994; Brush, 1995;).

Many questions remain to be answered about the viability of on-farm conservation for genetic conservation. In what way do varieties change over time? Do farmers conserve varieties or do they conserve traits, such as aroma, plant architecture, or disease resistance, for example? What is the importance of seed exchange among farmers for enriching their germplasm? Why do some farmers continue to grow their local varieties and yet others have abandoned them in favour of improved varieties? What is the degree of outcrossing between varieties in farmers' fields, and so on?

In rice there are several studies on the adoption of modern varieties (e.g. Huke *et al.*, 1982; Herdt and Capule, 1983; David and Otsuka, 1994) and some studies, among them several ethnographies, that give detailed accounts of the management and use of traditional varieties (Conklin, 1957; Rerkasem and Rerkasem, 1984; Lambert, 1985; Richards, 1986; Lando and Mak, 1994a,b,c). There are few studies, however, aimed at describing and understanding the way farmers maintain and manage rice diversity, as well as the factors that influence these (Dennis, 1987). The lack of such studies on rice contrasts with its importance as a world crop, and with the impact that modern varieties have had on farmers.

This chapter discusses some important issues related to the on-farm conservation of rice, namely: the nature of on-farm conservation; the genetic and evolutionary implications of farmers' management of diversity; the role of institutions in on-farm conservation, and some of the research needs in this area.

17.2 THE NATURE OF ON-FARM CONSERVATION: FARMERS' MANAGEMENT OF DIVERSITY

There is increasing evidence that small-scale farmers throughout the world, and especially in areas of crop domestication and diversity, continue to maintain a diverse set of crop varieties (Boster, 1983; Hames, 1983; Bellon, 1991; Zimmerer and Douches, 1991; Brush, 1992, 1995; Bellon and Taylor, 1993). These varieties are crop populations that farmers identify and name as units (farmers' varieties). In rice, a few studies have shown this as well (Conklin, 1957; Rerkasem and Rerkasem, 1984; Lambert, 1985; Richards, 1986; Dennis, 1987; Lando and Mak, 1994a,b).

The number of varieties found in these studies and other information on the associated farming systems are presented in Table 17.1. However, we need to distinguish between the total number of varieties reported, many of which may not have been planted during the time of the research, and those that were actually planted. This distinction is important because the first category is an indicator of the cumulative number of varieties present in a location, based to a great extent on farmers' memories. The second may be more important because it actually refers to what was happening during the time of the research. To report these numbers without qualification may give the wrong impression about a given level of diversity. For example, Lambert (1985) reported 45 varieties, of which only 32 were actually planted.

In any case the important conclusion is that for both numbers there is variation across studies. The number of varieties reported is larger than the number of those planted. Most of the varieties planted are traditional, but in many cases modern varieties are already present. The number of modern varieties is low, in general, but they may cover a large area and be planted by most farmers. The data do not permit an assessment of the relative importance of modern versus traditional varieties in area or number of farmers.

The average number of varieties per farmer is much lower than the number of varieties present per village or cluster of villages. This suggests that there is a low overlap in the sets of varieties each farmer is planting. Even the maximum number of varieties per farmer falls short of the village or cluster total. Therefore, even if only studying villages, one needs to sample several farmers to capture village diversity. This suggests that, while the farmer is the basic unit of decision-making in terms of variety selection and maintenance, the village is the minimum unit of analysis for diversity. There is a need to explain the variation

Table 17.1 Indicators of farmers' management of diversity from studies of rice in Asia

| Study | Country | Level[1] | Unit[2] | Sample size (no. farmers) | Rice ecosystem[3] | No. requested[4] | Varieties | | | | | Farm size[5] (ha/household) | Ave yield[6] (t/ha palay) | Purpose production[7] | Other economic activities[7] | Ethnicity |
							No. planted[5]	Traditional	Modern	Average/farmer	Max/Farmer					
(a)	Cambodia	Cluster of villages	Kandal	17	RL	14	11	9	2	1.7	3	1.1	1.9	S/MGY	Off-farm labour, palm-sugar	Cambodian
(a)	Cambodia	Cluster of villages	Kompong Speu	14	RL	9	7	6	1	2.14	4	1.1	1.7	S/MGY	Palm-sugar	Cambodian
(a)	Cambodia	Cluster of villages	Takeo	14	RL	15	11	11	0	2.36	3	1.2	1.3	S/MGY	Palm-sugar, cattle	Cambodian
(b)	Cambodia	Cluster of villages	Prey Kabas	67	DWR+RL+IR	23[8]	12	8	4	NS	NS	1.8	2.1	S/M	Garden crops, pigs, off-farm labour	Cambodian
(b)	Cambodia	Cluster of villages	Piam Montia	36	DWR+RL+IR	18	18	18	0	1	3	3.2	1.1	S/M	Little off-farm labour	Cambodian
(c)	Malaysia	Village	Pesagi	28	Swamp	45	32	32	0	3 to 4[9]	6[9]	0.534	0.45[10]	OS	Rubber	Malay
(d)	Thailand	Village	Pa Laan	22	IR	18	9[11]	6[12]	3	1.4	NS	0.85	NS	NS	Fruit orchards, vegetables	Thai
(d)	Thailand	Village	Mae Salap	20	IR	7	5[11]	4[12]	1	1.6	NS	1.3	NS	NS	Off-farm labour	Thai
(d)	Thailand	Village	Tha Mon	20	IR+UP	19	13[11]	10[12]	3	1.8	NS	0.45	NS	NS	Tobacco	Thai

(d)	Thailand	Village	PaNgiw 27	IR+UP	IR+UP	30	24[11]	23[12]	1	2.9	NS	0.6	NS	NS	No cash crops	Karen
(d)	Thailand	Village	Gong Hae 22	IR+UP		17	11[11]	11[12]	0	1.3	NS	0.33	NS	NS	Tea	Thai
(d)	Thailand	Village	Buak Jan 19	UP		13	12[11]	12[12]	0	1.4	NS	0.7	NS	NS	Vegetables, UN program[13]	Hmong
(d)	Thailand	Village	San Pa Hiang 20	RL		9	3[11]	2[12]	1	1.3	NS	0.86	NS	NS	NS	Thai
(d)	Thailand	Village	Pa Daeng 19	RL		8	4[10]	2	2	2	NS	2.9	NS	NS	NS	Thai

(a) Lando and Mak, 1994a.
(b) Lando and Mak, 1994b.
(c) Lambert 1985.
(d) Dennis, 1987.

RL= rainfed lowland; UP= upland ; IR= irrigated; DWR= deep water ; TV= traditional varieties; MV= modern varieties; S= subsistence; OS= only subsistence; MGY= marketed only in good years; M= marketed; NS= not specified.

Notes

1. A **village** means that the sample of farmers used to generate the data was taken from one village. A **cluster of villages** means that the sample was taken from a cluster of villages within a region.
2. **Unit** is the name the authors gave to either the villages or clusters of villages used in their studies.
3. **Rice ecosystem** is the classification of the rice production systems sampled. It is reported as the authors did, e.g. Rainfed lowland, Swamp.
4. **Varieties** refer to farmers' varieties – crop populations named and identified as units by a group of farmers. The **number of varieties** used here refers to the number of variety names. Using this number can slightly under- or overestimate the actual number of populations present, i.e. two different populations having the same name, or one population having two different names. Nevertheless, usually within a village names give an accurate representation of the number of populations present (Dennis, 1987; Quiros et al., 1990). In a cluster of villages, this may be less precise. The '**number of varieties reported**' means all the varieties listed by farmers, whether currently or previously planted, or simply declared.
5. The information on **farm size** was not precise across studies. It was not clear whether it was actual farm size or only area planted to rice. In any case area planted to rice was equal to or smaller than the number reported.
6. The data on **yield** sometimes referred to average yield for a village reported by informants, or to cuttings in some sampled farmers' fields. This information just provides a rough idea of the level of productivity in these sites.
7. **Other economic activities** may actually be underestimated because reporting of them was not consistent across studies, and they were taken from different sections of the studies.
8. Including 15 varieties lost during Pol-Pot times.
9. Lambert (1985) provided a range of the average number of varieties only, but not specific information. In terms of the maximum number of varieties, he just reported that it was not uncommon to observe up to five or six varieties planted per household.
10. The yield reported by Lambert (1985) was for a poor year (1976) when he did his study. According to him the average normal yield is 0.89 t/ha
11. The data for the villages surveyed by Dennis refers to 1984. He presented data from 1979 to 1984, but here only the last year is presented.
12. Traditional varieties in the case of Dennis (1987) included what he defined as Locally Improved Varieties and Local Varieties.
13. This is a United Nations Opium Replacement Project.

among farmers and among villages. It is important also to point out that there are farmers who maintain a much larger number of varieties than the average. As Dennis (1987) has shown, there are contrarians (his term), i.e. farmers who maintain more varieties and who exhibit a contrasting behaviour with respect to the rest. They either adopt new varieties early on (not necessarily modern ones) or maintain varieties that have been discarded by the rest. They have a better than average knowledge of varieties and ability to explain decisions concerning them.

In terms of other factors that may explain the variation observed, these studies comprise all types of rice ecosystems and some combinations of them (e.g. irrigated and upland). They show that the average farm size, or at least the area planted to rice, is relatively small. The yields are also low, but not atypical of the levels observed for traditional varieties; however, one must be cautious about this comparison (see Table 17.1, footnote 6). In general, rice production is undertaken for subsistence purposes, although surpluses may be sold in good years. Nevertheless, subsistence should not be confused with market isolation. In almost all cases, these farmers engaged in market activities such as cash cropping or off-farm labour. Subsistence indicates that the rice produced is consumed by the farm family, but this does not preclude that the farmer may have to purchase some rice or that the farmer may sell it as well. It is not clear what the relationship is between diversity and increased rice production for the market while maintaining subsistence. There is variation in ethnicity of the studies reported, but these emphasize ethnic majorities. Only Dennis (1987) compared an ethnic majority with ethnic minorities.

Infraspecific crop diversity maintained by farmers is not just the set of varieties they plant, but also the management processes these varieties are subject to and the knowledge that guides these processes. In fact, the specific varieties in the set may change over time (Dennis, 1987). Hence, the diversity observed in farming systems is a process rather than a state. We can refer to this process as farmers' management of diversity, which can be characterized as one in which farmers cultivate a diverse set of more or less specialized crop populations. These populations are named and recognized as units by the farmers (farmers' varieties). They are usually segregated in space, time and/or use. The set of varieties is formed through a constant process of experimentation, evaluation and selection of existing and new varieties. There are two levels of selection: choosing the varieties to be maintained; and then for each one, choosing the seed stock that will be planted the next season. The selection process is dynamic and is influenced by the supply of populations from other farmers, villages, regions, or even countries. This supply may involve new populations, as well as existing ones that a particular farmer may have lost and wishes to replant. Four components of farmers' diversity management can be identified: seed flows, variety selection, variety adaptation, and seed selection and storage.

17.2.1 Seed flows

The exchange and transport of germplasm form a common historical pattern throughout the world that currently continues, particularly with the introduction of modern varieties. Several studies have documented the flow of seed of different varieties among small-scale farmers (Dennis, 1987; Cromwell, 1990; Sperling and Loevinsohn, 1993; Louette, 1994). These flows can happen within a village, a region, a country, or even between countries. They take place as farmers exchange or market seed among themselves, purchase seed from commercial or government outlets, receive seed as a gift, or collect it from other farmers while travelling. The increasing importance of migration as an economic activity for many farmers may foster these flows.

In rice, Dennis (1987) documented an active exchange of rice germplasm among farmers of northern Thailand across village, district and provincial lines. This means that a variety does not need to stay in the same village to persist successfully within a region. He distinguished three categories of varieties: local, which have been grown in an area for many years or have been bred or selected from varieties long used in the area; locally improved, which were developed from traditional ones by pure-line selection; and modern, which have been released since 1965, have high-yielding potential, and are generally short-stemmed and fertilizer-responsive. Here, the first two categories are lumped together and referred to as traditional.

Dennis found that variety flows were mostly from north to south. Most of the varieties adopted were traditional ones: from 40 instances of village-level adoption between 1976 and 1984, 29 were traditional varieties and only 11 were modern varieties. However, almost all of the discarded ones were traditional. The largest diversity and the lowest level of variety replacement were found among the Karen and Hmong ethnic minorities. Nevertheless, even among the Thai ethnic majority, the more isolated villages had lower replacement rates than those that were more integrated. Some farmers among the two ethnic minorities planted seed plots, while this was not the case among the ethnic majority. Surprisingly, the villages with fewer varieties and a greater percentage of varieties discarded were not located in the irrigated but in the rainfed system. In both cases, these were Thai farmers relatively close to the city.

Seed flows are important in understanding the diversity in a given location because they are the basis of incorporating new varieties and obtaining materials that have been lost but are desirable. It is not uncommon for a farmer to lose a desired variety by accident, or even purposely discard one, and then wish to recover it (Dennis, 1987; Sperling and Loevinsohn, 1993). Furthermore, these flows may have major genetic implications because they may be an important mechanism for the migration of genes, and may counter genetic drift in varieties planted over very small areas (Louette, 1994). In theory, a network of seed exchange coupled with a rigorous and consistent seed selection method, which produces high-quality seed, may allow farmers to abandon poorer lines whenever there is access to better ones, eventually creating a cumulative effect of generating and maintaining highly adapted and productive cultivars (Lambert, 1985).

The collection of land races, their use for the development of modern varieties and their introduction into the farming systems themselves have expanded the scope of these flows and the level of diversity. Modern varieties incorporate germplasm that originated from many different countries. It is common to observe modern and traditional varieties being grown by the same farmers. For example, Dennis (1987) used analysis of isozymes to sort the different rice varieties collected in northern Thailand into different genotypes. He found that, while most of the traditional varieties in his sample belonged to one genotype, the collected modern varieties established a new isozyme group. He concluded that, in his area of study, the introduction of modern varieties was more likely to broaden genetic diversity in the landscape than the introduction of a traditional variety brought from another area in northern Thailand.

17.2.2 Variety selection

The process of variety selection can be seen as a farmer's decision to maintain, incorporate or discard a variety to be planted in a particular growing season. The diversity of varieties present in a farmer's field is the outcome of this decision. If the number of varieties incorporated and maintained is larger than that of the ones discarded, then diversity increases, and vice versa. The varieties maintained or incorporated are either kept from the previous agricultural cycle or obtained through exchange or purchase.

Farmers continually evaluate each variety, and the process has two components. One is to find out how a variety performs with respect to each concern or selection criterion, such as its performance under drought or flood conditions. The second is to rank the performance of the varieties in terms of different stresses, such as drought resistance. Farmers constantly try to match their crop populations or varieties to these concerns, which in turn reflect the conditions in which they farm. In describing the management of traditional rice varieties in Pesagi, a Malay village, Lambert (1985) observed that farmers constantly experiment with rice cultivars. Even with well-known cultivars individual households test one variety against another, a process of matching varietal performance to small but significant differences in localized habitats.

The fact that farmers have multiple criteria for selecting what varieties and where, when and how to plant them has been well established (e.g. Brush *et al.*, 1981; Lambert, 1985; Bellon, 1991; Brush, 1992; Sperling *et al.*, 1993; Lando and Mak, 1994c), and those criteria reflect their concerns. Bellon (1991) grouped them in three major types of concerns:

- agro-ecological, which refers to the performance of a variety with respect to agro-ecological conditions, such as rainfall, temperature, soil quality and topography;
- technological, which pertains to the performance of a variety with respect to management and inputs – for example, the response to the amount of fertilizer applied, to delays in weeding and to association with other crops;
- use, which relates to the performance of a variety with respect to the purposes and uses of the output, such as taste, texture, yield, quality, production for subsistence or for the market, and production of straw or fodder.

Rice studies report different selection criteria by farmers (Table 17.2), but many are common to most of them. In terms of agro-ecological concerns, common ones include maturity and adaptation to different water level regimes such as drought and submergence. In terms of use concerns, yield and texture are very common. Texture is associated with different purposes such as subsistence or market production, or different uses such as direct consumption or elaboration of cosmetics or cakes. Certain categories manifested by the farmers were not well defined by them, such as 'good field adaptation'. This category may be a combination of factors specific to a habitat (e.g. Lando and Mak, 1994c). Reliability is only mentioned by one author (Lambert, 1985), although it may be very important for all subsistence farmers. In terms of technological concerns, the ability to 'fit' with other crops and to avoid labour bottlenecks, factors that are related to maturity, are reported as well. Only Lando and Mak (1994c) provide some quantitative data on the percentage of farmers who declared each of the concerns. An interesting finding for all varieties is that, while yield was cited as the most frequent reason to plant a variety, this trait was mentioned as frequently as field adaptation and maturity for early-maturing varieties, and as often as flood tolerance for late-maturing ones when data were desegregated by maturity. Unfortunately, in these studies the association of farmers' selection concerns with the varieties they planted is not systematically reported: for example, how each variety performed with respect to drought or floods. It is not clear whether different farmers have different selection concerns or how they even rank their concerns. For example,

Table 17.2 Farmers' selection concerns

Source	Agroecological	Use	Technological
Lando and Mak, 1994c	Field adaptation Maturity Drought tolerance Flood tolerance Lodging resistance	Yield Eating quality Price Volume expansion	Not reported
Lambert, 1985	Performance under different levels of water depth Drought tolerance Dependability: production on adverse conditions	Texture (glutinous, vitreous, viscous), related to use for subsistence or market Yield Price Colour of husk	Resistance to weeds, insects and disease
Rerkasem and Rerkasem, 1984	Drought tolerance Flood tolerance Maturity (earliness) Lodging resistance	Texture (glutinous subsistence, non-glutinous market) Quality Price Production of straw for mulch	Fit with multiple cropping patterns Fit with patterns of off-labour

poor farmers may have different concerns from rich ones, as may be the case of female farmers versus male farmers, or an ethnic majority versus minorities.

Farmers' selection concerns are not homogeneous, and may vary with the different agro-ecological, socio-economic and cultural conditions they face. Rich and poor farmers in a productive region probably have very different concerns, as may be those of two poor farmers in a marginal area. Even within a farming household, there may be differences between male and female concerns. In many rice farming systems, there may be a clear sexual division of labour (Lambert, 1985) that underlines the possibility not only of different concerns, but also of conflicting ones. This area merits further research, given the increasing recognition of the role of women in rice farming.

Since farmers' concerns are varied, and a good performance with respect to certain concerns often implies poorer performance with respect to others, several varieties are maintained. Diversity may sometimes be maintained as an option because farmers may not know the future benefit or availability of particular varieties, or because humans can value diversity for its own sake, with no ulterior purpose. However, in the case of diversity that is directly useful, it is important to underline that, in order to explain its development and maintenance, there should be trade-offs among the varieties. For example, Harlan (1992) points out that alleles for disease resistance generally have negative effects on yield in the absence of the disease, and sometimes even in its presence. Hence, there are costs associated with resistance. Therefore, it is important to know and understand not only the positive traits of a variety, but also its negative ones, as those relate to the trade-offs among different farmers' concerns. The combination of two types of traits defines the opportunities for complementation among varieties.

Variety selection is a process of continual experimentation and evaluation. Much of this information is transmitted from farmer to farmer. Experimentation and communication have two important roles in the management of diversity, since they are the basis of the development of farmers' crop knowledge and they allow farmers to know and evaluate new and unproved germplasm without jeopardizing their livelihood or scarce resources. The fact that many small-scale farmers have a well-developed knowledge of their crops and crop varieties has been well documented by human ecologists, anthropologists and ethnobiologists (Conklin, 1957; Berlin *et al.*, 1974; Brush *et al.*, 1981; Boster, 1983; Hames, 1983; Bellon, 1991). This knowledge includes ecological, agronomic and consumption characteristics of the crops and crop varieties they plant. In many instances, this knowledge is systematized in a regular system of nomenclature, organized in a taxonomic manner, i.e. folk taxonomies (Brush *et al.*, 1981). It may be used to make decisions regarding management, use, storage, culinary aspects and rituals (Boster, 1983; Hames, 1983; Bellon, 1991).

17.2.3 Variety adaptation

Whenever a farmer finds a variety that is superior for whatever reason, it will be cultivated under the conditions or for the purposes for which it is superior. This process contributes to the development of increasingly adapted crop populations. The stronger and more distinct the selective pressures, the more specialized populations are likely to be. It has been observed that traditional and modern varieties are usually segregated in different areas of the farm, subject to different management and aimed at different uses (Brush, 1991a). The fact that many rice farmers match different varieties to different field levels, that in turn reflect different regimes of water availability, is well documented (Lambert, 1985; Lando and Mak, 1994a,b). Certain varieties have been maintained only for very specialized uses such as making rice-starch cosmetics, medicinal preparations, or traditional snack foods and cakes (Lambert, 1985).

17.2.4 Seed selection and storage

Farmers not only choose which varieties to plant, or where and how to manage them, but also the seed that will be planted the next season. Variety selection and management are reinforced by a careful and rigorous selection of the seed that will be planted the next season. Seed selection procedures vary by crop and its reproductive system. In open pollinated crops such as maize, seed selection may be fundamental to maintain the integrity of a variety (at least from the point of view of the farmer), but this can be easily lost due to hybridization (Bellon and Brush, 1994; Louette, 1994). This may not be such a problem in the case of a self-pollinated crop such as rice. Nevertheless, rice farmers may decide to keep varieties separate to facilitate their identification and allocation to specific niches. Even if mixtures are planted, in general, they are not a random collection of varieties, but specific combinations. For example, in Uttar Pradesh, India, a popular variety in drought-prone areas, called *gora*, is a mixture of brown, black and straw-coloured genotypes that differ in drought resistance and grain quality (Vaughan and Chang, 1992).

The seed selection and storage methods reported in the studies analysed here are shown in Table 17.3. Although all of them recognize that seed selection is an important component of farming, the level of detail of their descriptions of it is variable. Some have detailed descriptions (e.g. Lando and Mak, 1994c), while for others it is minimal (e.g. Lambert, 1985). Nevertheless, different methods are reported. Particularly, they differ in the timing and place of seed selection. These two aspects are important because they define whether plant characteristics can be taken into account (special harvest of areas with a good standing crop), or only panicle traits (selection during threshing at home). In general, farmers are reported to maintain segregation of their varieties, and in some cases to go to great lengths to accomplish this (e.g. Lando and Mak, 1994c). Careful seed selection may not happen at the end of every season, but every three or four seasons, with a bulk selection of seeds in the intervening ones (Dennis, 1987).

Seed selection may also be important to identify a new population or variety that may arise due to hybridization or mutation. At harvest, a farmer may single out seed from one or more plants that are perceived as being entirely different from a known cultivar, in an attempt to develop a new strain (Lambert, 1985). Richards (1986) pointed out that harvesting rice with a knife, panicle by panicle, as done by farmers in Sierra Leone, West Africa, permits the careful roguing of off-types. Frequently, this material is kept for experiment, and in some cases leads to the selection of new varieties. Therefore, this process may be important to increase diversity in self-pollinated crops, where hybrids between varieties occur at low rates. Nevertheless, the introduction of modern varieties may modify the systems of seed selection because farmers may purchase seed

Table 17.3 Seed selection and storage of rice

Aspects of selection	Dennis, 1987	Lando and Mak, 1994a	Lambert, 1985
General comments	One method, in two cycles: (a) one careful selection every three to four years (b) simple bulk selection in the intervening years	Two methods: (1) after harvest at threshing (2) identify area with standing crop with desirable characteristics and harvest it	Recognized as important, but poorly described Underlie the importance of appropriate training of seed selectors
Time and place	(a) before general harvest in the field (b) after harvest at threshing	(1) after harvest at threshing in the house (2) before general harvest, in the field	Not specified
Plant parts used	(a) panicles and other plant characteristics observed in the field (b) seed from threshing mat	(1) panicles (2) panicles and plant characteristics. If varietal mixing occurs in the field, one person (usually a women) must separate seed stock panicle by panicle	Portion of best grain from mature crop
Frequency	(a) every three to four years (b) every year in the intervening ones	Every year in both methods	Not specified
Criteria	(a) large healthy panicles true to type (b) disease-free sheaves of grain	(1) and (2) full panicles with well–filled seeds (2) area in the field with standing crop, with desirable characteristics	Colour, size and shape of grains
Storage method	Not specified	Sun dried for 3–4 days, stored separately by variety; use home-made bags called *khbong*, or gunny bags. Non-glutinous varieties under house or separate granary. Glutinous varieties and seed stock stored separately	

instead. Dennis (1987) noted that the practice of on-farm seed selection was declining as improved seed supplies became more available from government agencies. It is also important to emphasize the role of women as seed selectors in rice. Their knowledge and expertise in this respect are increasingly being documented (Conklin, 1986).

It is clear that there is variation in the management of diversity among farmers. There are different numbers and types of varieties maintained, selection concerns, seed selection methods and different rates of seed flows. The variables that describe farmers' management of diversity can be seen as a set of dependent variables. On the one hand, they are affected by the environmental, socio-economic and cultural factors that influence the farmers' decision-making. These factors operate at different scales. Some have to do with farmers' characteristics such as socio-economic status, access to resources and knowledge; others with village level characteristics such as local institutions (e.g. patterns of labour exchange, land tenure, social obligations); and others are related to processes that occur at the regional or national levels, such as availability of infrastructure (irrigation, roads, telecommunications), degree of development of markets and government policies. On the other hand, farmers' diversity management has consequences for the genetic structure and diversity of the crop. Nevertheless, genetic structure of the crop is also influenced by environmental factors through natural selection. Furthermore, although farmers cannot observe or appreciate the genetic structure of the crop, they gain knowledge of morphological traits expressed (e.g. yield, plant stature, resistance to drought, insects). This knowledge is in turn used in their decision-making processes regarding their management of diversity (Figure 17.2).

Government policies, particularly those aimed at increasing the food supply for a growing urban population, also affect diversity. Governments have provided infrastructure, modern inputs and subsidies to farmers, favouring and sometimes forcing specialization. In many cases, they have imposed restrictions on what farmers can and cannot do, either through a legal process, or by imposing conditions on the access to desirable inputs (e.g. conditioning credit to the compulsory planting of modern varieties, in the belief that it will increase food production). Government policies vary from country to country, even by region, and from time to time. It is difficult to predict their impact on diversity, except that usually there has been a bias against diversity.

17.3 GENETIC AND EVOLUTIONARY IMPLICATIONS OF FARMERS' MANAGEMENT OF RICE DIVERSITY

Considerable progress has been made during the last 10 years in the study of the rice genome. There is a strong contrast between our knowl-

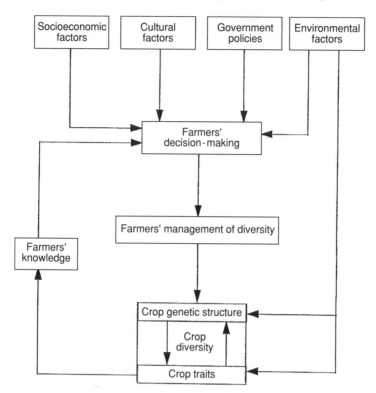

Figure 17.2 Conceptual model of the factors that influence farmers' management of diversity.

edge of diversity at the molecular level and our ignorance about the role of genetic diversity in farmers' fields. This is not really surprising. Studying the diversity in farmers' fields implies analysis of an evolution-ary process involving a triple interaction between genetic diversity *per se*, natural selection and farmers' management. From a specific point of view of conservation of genetic resources (and one that initially excludes the important question of a farmer's well-being), on-farm conservation of genetic resources aims to use this evolutionary process to promote the adaptation of these resources to environmental conditions. It is essential, therefore, to increase our knowledge of this process in farmers' fields in order to define the potential outputs from on-farm conservation of rice genetic resources.

Several important questions must be posed. Firstly, why can we assume that farmers' management of rice diversity is an evolutionary process? Secondly, what are the genetic components of this process? Thirdly, to what extent can on-farm conservation of rice genetic resources be expanded towards utilization? Lastly, what are or should be the relationships between on-farm conservation and *ex situ* conservation ?

17.3.1 Farmers' management of rice diversity as an evolutionary process

To be convinced that farmers' management is an evolutionary process that has produced diversity, there is no better way than to consider the overall diversity of *O. sativa*. It results from several thousands of years of farmers' management and natural selection. Useful lessons for the genetic and evolutionary implications of on-farm conservation can be drawn from studies on the structure of this diversity.

On the basis of isozyme studies, Second (1982) interpreted the diversity of *O. sativa* as resulting from a continuous recombination process between two independently domesticated subspecies, so-called *indica* and *japonica*. This hypothesis is supported by Glaszmann (1987), who classified Asian varieties of *O. sativa*. Highland rices from Madagascar also offer clear examples of reciprocal introgression between *indica* and *japonica* varieties (Ahmadi *et al.*, 1991). More recently, Second and Ghesquière (1995) have used RFLP markers to demonstrate that the genome in some *O. sativa* varieties consists of 'pieces' of genetic material originating from both subspecies.

Studies on the diversity of *O. sativa* varieties have also demonstrated the important input of wild species having the same AA genome. Marker studies have highlighted the contribution of *O. rufipogon* and *O. longistaminata* to the diversity of *O. sativa*, in Asia and Africa, respectively (Ghesquière, 1988). In contrast, the species complex of African rice, which consists of the cultivated species *O. glaberrima* and its wild progenitor *O. barthii*, is strongly isolated from other AA genome species by strong reproductive barriers.

17.3.2 Components of the evolutionary process in farmers' fields

Let us suppose here that rice varieties are not pure lines but polymorphic populations, as might be expected in land races. Like other crops, rice varieties have continued to evolve under two kinds of selection pressure: environmental and human. These act on genetic polymorphism by discarding the less adapted genotypes. It is the role of genetic recombination to create new material to be exposed to selection. Therefore, to be adaptable (i.e. to be able to respond to changes in selection pressures), a variety must be genetically polymorphic. Selection will act on gene frequencies in the variety. Another factor of change in gene frequencies is genetic drift, a random process of loss of genes, and one that depends in particular on population size. Within-variety polymorphism is generated through both rare mutation events and, to a much greater extent, gene flows from other varieties. However, gene flows will enhance polymorphism only if they come from genetically different varieties. In other words, research into on-farm conservation must deals both with the polymorphism within rice varieties and with the genetic diversity of cultivated varieties (including the modern ones) as the source of this within-variety polymorphism.

17.3.3 Within-variety polymorphism

Numerous germplasm collecting reports detail the occurrence of morphological polymorphism within traditional varieties. The strongest evidence obviously relates to grain and panicle traits, which are most easily observed in the field. However, few studies have been carried out to assess the genetic components of this polymorphism. Ghesquière and Miézan (1982) studied the genetic structure of West African traditional rice varieties as revealed by isozymes (40 loci). They studied 44 *O. sativa* varieties including 19 from Guinea and 25 from Côte d'Ivoire. The within-variety polymorphism accounted for 12% of the *O. sativa* variety diversity. It was higher for *indica* varieties than for *japonica* varieties (37% versus 16%), as also observed for the genetic diversity of these sets of varieties (0.090 versus 0.077). Six *O. glaberrima* varieties were also studied. High within-variety polymorphism was observed (32%), particularly in contrast to the low genetic diversity of this sample of varieties (0.0269). Miézan and Ghesquière (1985) obtained the same overall picture of the genetic structure by studying a subset of these varieties for several agro-morphological traits. Morishima (1989) studied the isozyme polymorphism of 15 land races from India, Nepal and Thailand. Six land races showed a gene diversity less than 5%. The gene diversity of the nine other land races ranged from 10 to 25%. In some cases (Oka, 1991), land race populations were found to be more heterogeneous than annual wild populations of *O. rufipogon*.

Gene flows are the main factor to increase diversity. They are achieved if cross-pollination occurs between different genotypes, and if more or less fertile progeny are produced. The two cultivated rice species are predominantly self-pollinated crops. At anther dehiscence, pollen grains preferentially deposit on stigmas of the same panicle because of proximity. Nevertheless, when flowering, both stamens and stigmas are exerted outside the spikelets. Pollen grains are dispersed by wind, and stigmas are receptive to pollen coming from other plants. Natural outcrossing is therefore possible in rice. The actual outcrossing rate of rice in farmers' fields and the migration distance of rice pollen are poorly known. Outcrossing rates could be increased in the case of products of hybridization between genetically distant *O. sativa* varieties. F1 hybrids within *O.*

sativa, especially intersubspecific hybrids between *indica* and *japonica*, often show partial pollen sterility, leading to lower seed fertility (Oka, 1988; Pham, 1991), and may result in increased cross-pollination, since female fertility of F1 hybrids is more rarely affected (Oka, 1988). Morishima (1989) reported the occurrence of *indica–japonica* hybrids in an upland rice population from Yunnan.

In West Africa, there are several reports of *O. sativa* and *O. glaberrima* growing together (Borgel and Second, 1978; Sano *et al.*, 1984), and presumed hybrid plants between the two cultivated species have been observed *in situ* (Borgel and Second, 1978; Pham and de Kochko, 1983). However, little evidence of gene flow has been found, except for the single study by Second (1982) who identified a plant of the weedy form of *O. barthii* that could have resulted from the introgression of *O. sativa* genes into *O. glaberrima*. The absence of gene flow can be explained by the strong reproductive barriers isolating the two species (Sano *et al.*, 1979, 1980; Pham and Bougerol, 1993), including phenological ones (Oka, 1988).

The management of fields and varieties by farmers is a potential cause of variation in the rate of gene flow between varieties. Coincidence in flowering dates (which depends on sowing date and earliness) and proximity (adjacent varieties are likely to exchange more genes than varieties in distant fields) are factors that influence gene flows. However, bearing in mind the expected low level of gene flow due to the mating system of rice, accurate studies would be time-consuming in terms of their potential impact. More attention should be given to particular situations, such as mixtures of varieties, accidental or controlled, that seem the most efficient way to promote gene flow between two varieties, as Dennis (1987) has reported for traditional rice varieties in northern Thailand.

17.3.4 Genetic diversity of rice varieties

The study of genetic diversity at the farm, village and regional level is a key aspect of research on farmers' practices and on-farm conservation. The aim should be to quantify the diversity in each of these units, and in particular, the impact of modern varieties must be evaluated. Many modern varieties have been bred using germplasm from traditional ones, and their impact on diversity may vary, depending on the region where they are released.

Any genetic approach to the study of on-farm conservation must have strong links to the social, economic and cultural aspects of farmers' diversity management. It is indeed important to study the relation between the level of diversity perceived by farmers and the actual level of genetic diversity. Dennis (1987) suggested that Karen farmers retained colourful varieties because colours are a useful indicator of genetic diversity. However, such morphological traits are governed by few genes and generally are not useful indicators of genetic diversity *per se*.

The rate of turnover of varieties is also an important factor. If a traditional farming system is a closed system, the adaptability of the set of varieties cultivated will depend on its level of genetic diversity. If new varieties are introduced, it can be enhanced. But if the turnover is too fast, the time of co-occurrence between varieties will be too short to allow significant gene exchange. Nothing new would be produced in the extreme case where, at every cultivation cycle, a set of genetically diverse varieties would be replaced by another set of genetically diverse varieties. In other words, we have to distinguish between the adaptation of the farming system and the adaptation of the germplasm.

17.3.5 The role of diversity

The adaptation to changes in different biotic and abiotic pressures, particularly pests and diseases, is perhaps the principal objective of dynamic conservation of genetic resources. Future changes in climate are predicted to affect agriculture world wide (Ford-Lloyd *et al.*, 1990; Jackson and Ford-Lloyd, 1990), but changes in pest pressures have actually been documented in rice culture in recent decades, and have widely affected the orientation of rice research programmes.

Several approaches have been proposed to manage the coevolution of pathogens with host plants to prolong resistance (Mundt, 1994). However, few data have been obtained to assess their actual efficiency. Some are based on dynamic management of resistance genes, that aim to avoid rapid selection for pathogen races virulent to the varieties. These approaches are analogous to what happens or could happen in farmers' fields. Among the examples are the following:

- **Gene rotation**. A set of cultivars, each with a single race-specific resistance gene, is cultivated in rotation. This approach was implemented in Indonesia for the control of rice tungro disease (Mundt, 1994). A reduction of tungro disease was observed, which is linked to gene rotation or to changes in practices induced by rotation. It would be interesting to see if such a rotation in varieties is *de facto* observed in traditional fields.
- **Gene combinations**. Two or more race-specific resistance genes are combined into a single host genotype. To know if such combinations can occur naturally in farmers' fields is an important issue. Situations where both traditional and modern varieties are cultivated would be of particular interest.
- **Cultivar mixtures**. Mixtures of cultivars with different resistance genes are another strategy employed against diseases; for example,

against barley powdery mildew in Germany (Wolfe, 1992). The cases of intentional mixtures of traditional varieties of rice reported in northern Thailand by Dennis (1987) are not related to disease management. However, more situations must be investigated.

• **Gene deployment**. The strategy is to distribute resistance genes among different fields or regions. The pattern of distribution of genetic diversity at the field, village and regional level should provide valuable information in relation to this strategy.

It is unlikely that each of these different strategies will be found separately in farmers' fields. Mixed variety situations are more likely, with variation in both space and time. Methodologies will have to be developed to assess the potential relation between genetic diversity in fields and conscious or unconscious management of pest pressures.

More generally, an issue of great interest is to know whether a farmer's management of diversity leads to the selection of specialist varieties, generalist varieties, or both. Strong local selection pressures would be likely to select specialist varieties, that fit particular farmers' needs or particular agronomic conditions. Generalist varieties could be related to what is perceived by farmers as the 'reliability' of some varieties. David (1992) used simulations to show that, in certain conditions, gene flows can maintain generalist abilities in populations submitted to strong directional selection. Therefore, the role of within-varietal polymorphism could be to maintain an overall generalist ability in rice varieties.

17.3.6 Demonstrating genetic changes in farmers' fields

On-farm conservation is dynamic. Nevertheless, it will be difficult to demonstrate genetic changes in varieties and changes in the amount of cultivated diversity over a short time frame. However, a 'historical' approach comparing the present with past situations could provide such evidence. Changes of within-variety diversity could be studied through: sampling cultivated varieties and comparing them with samples of the 'same' varieties cultivated 10–20 years ago, by using the gene bank collections; comparing varieties that were released as unique 10–20 years ago; and comparing different samples of widespread traditional varieties, such as *Azucena* in the Philippines.

Changes in the total amount of diversity could be approached by comparing at the village or regional level the diversity cultivated nowadays and the diversity cultivated in the past, evaluated by using collecting reports and gene bank collections.

17.4 LINKS BETWEEN ON-FARM CONSERVATION AND RICE GENE BANKS

It seems that on-farm conservation of traditional rice varieties is something that farmers choose to do individually. On-farm conservation is not the same sort of strategy as *ex situ* conservation, in terms of the way that public sector institutions can decide to establish a gene bank, for instance. We believe that institutions, including NGOs, cannot do 'on-farm conservation', but they can identify the opportunities – social, economic and cultural – under which the cultivation of land race varieties of rice may continue to thrive. They may also be able to facilitate farmer access to a broad range of rice genetic diversity, and establish the links between farmers and gene banks.

This view should not remain a theoretical one. Recent experiments on participatory breeding support the idea that farmers can be efficiently involved in processes previously managed by institutions only. In Rwanda, farmers who are bean experts have been identified and invited to the research station to assess cultivars and select those they prefer for their plots (Sperling *et al.*, 1993). Compared with cultivars selected by the breeders, those chosen by farmers were often higher yielding on-farm. Moreover, they were retained longer by farmers. Participatory breeding therefore seems to be a useful approach in promoting the adoption of new cultivars by farmers. This was also the conclusion reached by British plant breeder John Witcombe (personal communication) about an experiment of participatory breeding of rice varieties in India and Nepal. In Nepal, the farmer participation was extended to include breeding of segregating material that was supplied to farmers for on-farm selection.

17.4.1 Management by farmers of 'foreign' diversity

The strategy proposed here is to involve farmers in managing a sample of genetic diversity in addition to their own varieties. There are two reasons for this strategy. Firstly, if the farmers' management of diversity does produce changes with adaptive significance, why not artificially increase the amount of diversity exposed to this process? Secondly, genetic polymorphism is required to permit adaptation to evolving selection pressures. This condition is necessary but not sufficient alone. Indeed, being polymorphic does not necessarily mean being adaptable when the available polymorphism does not permit an appropriate response to selection. Even if the polymorphism may be sufficient to permit a slight adaptation under selection pressures, it may be insufficient for the variety to reach an optimal adaptation. This means that cultivated land races are not necessarily the ones best adapted to the local conditions where they are grown.

Three complementary ways may be proposed for this strategy: reintroduction of varieties, introduction of alien varieties, and introduction of composite populations. Besides its interest for the conservation of genetic resources, this approach would provide useful information on the consequences of farmers' management. This would complement the descriptive part of any research with an experimental, controlled approach.

The most promising approach is likely to be the management of composite populations. It is essential to bear in mind that traditional varieties have resulted from several thousand years of cultivation. Managing artificial populations should permit changes to be observed over a far shorter time scale. The idea of conserving bulk populations is not recent (Suneson, 1956; Simmonds, 1962). One important experiment has been conducted on barley in which a composite population has been cultivated since 1928. The main results relate to changes in disease resistance (Allard, 1988, 1990). More recently, an experiment on the dynamic management of composite populations of winter wheat has been carried out in France (Henry *et al.*, 1991), based on the so-called metapopulation concept. Significant changes were observed after only six years of multiplication. The maintenance of resistance genes to mildew and the appearance of novel gene combinations (Le Boulc'h *et al.*, 1994) were among the most important results. Some guidelines for an experiment of conservation of composite populations of rice have been proposed by Pham *et al.* (1994), based on the multiplication of these populations in a multilocation network in order to promote their multidirectional differentiation through different selection pressures.

The composition of populations should have at least two objectives:

- The variability of material included in the initial population must permit evolution under various conditions. There is no interest in seeing local populations disappearing after the first year! Consequently, it will be necessary to check if traditional entries can grow under intensive conditions with completely different disease pressures. Involving modern varieties in the initial population should give time for the development of resistance gene combinations.
- Farmers must be interested in cultivation of the populations. This means that farmers now growing only dwarf varieties are unlikely to accept populations with only traditional traits (tall plants, lodging susceptibility, low yield potential). Introducing dwarfing genes in the initial population could be a critical point to make the genetic material more attractive for farmers.

The choice of initial material will require consideration and the following ideas should be considered:

- The initial population could be made by mixing a great number of traditional varieties. Some of them will be eliminated very quickly in local populations. Others will contribute to the future generations. Hybrid sterility will favour outcrossing, particularly if *indica* and *japonica* varieties are mixed.
- Another way to compose the populations would be to limit the number of entries to popular and widely cultivated land races. Traditional varieties that are well known by breeders or good donors for particular traits (tolerance to drought or blast, for instance) could also be used. There is no need to use completely evaluated varieties. A pyramidal cross could then lead to the initial population, as in the experiment with wheat (Henry *et al.*, 1991). Utilization of genic male sterility will favour intercrossing and introduce dwarfness if necessary.

Management should be as simple as possible. After dispatching the initial population in the network, farmers should cultivate each local population every year. They would store the harvested seeds for sowing the next season. Two modes of management could be compared: normal management, and no conscious selection on panicles and seeds while harvesting, storing and sowing.

Use of molecular markers should permit the monitoring of the populations in terms of specific marker frequencies based on samples of appropriate size. Furthermore, modelling effects on population structure could be demonstrated through using molecular markers like RFLPs (Resurrecion *et al.*, 1994) and RAPDs (Virk *et al.*, 1995a). They provide many different markers spread over the genome, showing allelic polymorphism similar to isozymes and closely related to *indica–japonica* differentiation. More recently, the identification of micro-satellites in rice has provided many allelic differences coming from small repeat units (Wu and Tanksley, 1993) and these sequences can provide more successful markers for following genetic changes. Adaptive variation of biotic and abiotic factors in relation to environmental heterogeneity should also be monitored.

Finally, the composite population approach may be the most appropriate one for dynamic conservation of *O. glaberrima*, given the genetic isolation of this species from *O. sativa*, and the decrease in its cultivation area.

17.4.2 Wild and weedy species

The genetic contribution of wild species to the diversity of cultivated rice has been significant, but little is known about this over short time-frames. *O. longistaminata* is an outbreeder, and the outbreeding rate of *O. rufipogon* ranges from 7 to 56% (Oka, 1988). However, because of the autogamous mating system of *O. sativa*, and also because of the reproductive barriers between *O. sativa* and these two wild species, gene flow is probably low. The wild–cultivated species relationship in rice cannot be compared with the frequent exchanges expected between maize and teosinte (Wilkes, 1967), for instance, or between wild and cultivated pearl millet (Pernès, 1984). Does this mean that on-farm conservation of rice genetic resources should not consider the wild–cultivated species relationships? We suggest that this issue should be addressed by considering both on-farm conservation of cultivated rice and *in situ* conservation of wild rice populations.

17.4.3 Complementarity between *ex situ* and on-farm conservation

The complementarity of on-farm conservation and *ex situ* conservation clearly exists at the level of objectives. *Ex situ* conservation aims to capture and maintain the genetic diversity at a given instant, whereas on-farm conservation aims to promote the adaptation of this diversity by using an evolutionary process. This complementarity does not mean that these genetic conservation strategies should remain isolated from each other. On the contrary, it should be enhanced through reciprocal flows of genetic material. The proposed flows are summarized in Figure 17.3.

If adapted varieties are produced by on-farm conservation, the question is whether useful changes can be detected and used. On-farm conservation does not only deal with the conservation of allelic diversity; it also deals with the occurrence of adapted combinations of alleles. Epistatic relationships are expected in selfing species. The criteria for the evaluation of the products of on-farm conservation must be defined, and must take into account the heterogeneity of the material.

Release of modern improved varieties, bred thanks to genetic resources collected from farmers' fields, can be considered as a feedback from the institutional sector to farmers. The development of original populations or genotypes through on-farm conservation and their use in plant breeding programmes or their conservation in *ex situ* collections would provide an exciting example of reciprocity in the production of improved genetic material.

The current attention that on-farm conservation is attracting and the apparent rush to implement conservation projects seem to be inversely proportional to the research effort being expended. As emphasized in this chapter, on-farm conservation is a process managed by farmers themselves, and not one imposed by institutions. The establishment of on-farm conservation reserves has been proposed where farmers would be encouraged, through a range of incentives, to continue to cultivate their local varieties. What is urgently needed is information on the circumstances and opportunities that promote on-farm conservation, or at the very least that permit farmers to make objective decisions about the

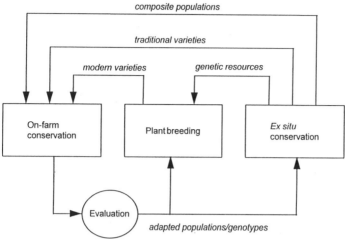

Figure 17.3 Possible exchanges of genetic material between on-farm conservation, plant breeding and *ex situ* conservation.

crop varieties (including locally adapted ones) that they choose to grow. On-farm conservation research must address these social, economic and cultural issues, and seek to determine the genetic consequences of different types of management by farmers. After all, we are seeking to preserve the adaptive potential of crop varieties in dynamic systems, while not neglecting the welfare of farmers. On-farm conservation must bring tangible benefits to the farmers who have nurtured this genetic heritage for generations.

18

Ethiopian *in situ* conservation

M. Worede

18.1 INTRODUCTION

The Ethiopian region is characterized by a wide range of agroclimatic conditions that account for the enormous diversity of its biological resources. Probably the most important of these is the country's various crop plants that farmers have adapted over centuries of selection and maintenance under various agro-ecosystems. Ethiopia is often referred to as a major Vavilovian gene centre, one of the eight centres in the world where crop plant diversity, notably barley (*Hordeum vulgare*) and wheat (*Triticum* spp.), is concentrated and where several important crop plants including sorghum (*Sorghum bicolor*), sesame (*Sesamum indicum*), coffee (*Coffea arabica*) and minor millets became domesticated (Hawkes, 1983).

The indigenous land races of the various crop plant species, their wild relatives and the wild and weedy species that form the basis of Ethiopia's plant genetic resources are all highly prized for their potential value as sources of important variations for crop improvement programmes. Among the most important traits that are believed to exist in these materials are earliness, disease and pest resistance, nutritional quality, resistance to drought and other stresses, and characteristics especially useful in low-input agriculture.

The existence of such diversity has special significance to maintenance and enhancement of productivity of agricultural crops in a country like Ethiopia, with its highly varied agroclimatic and diverse growing conditions. Such diversity provides security for the farmer against adverse growing conditions. It also allows farmers to exploit the full range of the country's highly varied micro-environments, differing in characteristics

Plant Genetic Conservation.
Edited by N. Maxted, B.V. Ford-Lloyd and J.G. Hawkes.
Published in 1997 by Chapman & Hall. ISBN 0 412 63400 7 (Hb) and 0 412 63730 8 (Pb).

such as soil, water, temperature, altitude, slope and fertility.

Diversity among species is especially significant to Ethiopia as it represents an important subsistence resource to farming communities in the country. A wide variety of plant and animal species provides material for food, fibre, medicine and other socio-economic uses. Such diversity is also crucial to sustain current production systems, improve human diets and maintain life support systems essential for the livelihood of local communities.

Maintenance of genetic diversity both within and among species is, therefore, crucial to sustainable agriculture. This is especially true for resource-poor farmers practising agriculture under low-input conditions on marginal lands. The conservation of such diversity is essential in securing the country's food and livelihood. Genetic resource activity already represents a major national effort that the Plant Genetic Resources Centre (PGRC/E) has undertaken systematically over the last 16 years (Worede, 1991). The existing options pose a serious challenge, requiring major inputs in terms of technical know-how and material.

There is also a unique opportunity to salvage and sustainably utilize land races through a complementary approach involving farmer-based activities and off-farm gene bank conservation.

In this chapter, an attempt is made to describe this particular aspect of genetic resource conservation, highlighting the important role that Ethiopian peasant farmers and scientists as well as various governmental and non-governmental organizations play in these regards. The importance of land races is further discussed, stressing the need for research to conserve, enhance and effectively utilize in-field diversity. PGRC/E's rationale in undertaking such an activity is also presented. The risks of genetic erosion and related problems are discussed, illustrating the importance of preserving Ethiopia's indigenous germplasm resources.

18.2 ROLE OF FARMERS IN SUSTAINING DIVERSITY

In Ethiopia, as in many other developing countries, farmers are instrumental in conserving germplasm as they control the bulk of the country's genetic resources. Peasant farmers always retain some seed stock of numerous crops for security unless disruptive circumstances prevent them from doing so. Mechanisms have also been developed for safe storage of seeds. Individual farmers often store seeds in clay pots and rock-hewn mortars or underground pits which are sealed, buried or stored in other secure places (Worede and Hailu, 1993). Length of storage may be for a period of 1–5 years as seed stock or for planting at some appropriate time, or longer (up to 7 years) for use in some anticipated social event like a daughter's wedding. In times of famine, farmers even bury their seed in some secured place within farm premises (communally or at the household level) before they migrate to other regions, returning to reclaim and plant the seed after the drought is over.

Ethiopian farmers have been instrumental in creating, maintaining and promoting crop genetic diversity through a series of other long-standing activities which include (Worede, 1992):

- intercropping and cropping with varietal mixtures which result in rapid diversification due to introgression from accidental crosses (e.g. *Brassicas*);
- promoting the intercrossing of cultivated crops with wild or weedy relatives, which results in new characteristics (e.g. *Guizotia abyssinica*);
- identifying and propagating new, mutant types which occur in their fields, or hybridization between wild and/or cultivated types, or cultivars obtained through exchange;
- diffusing both crop varieties and knowledge through local seed exchange networks;
- growing a diversity of local varieties of crops (e.g. *Coffea arabica*) preserved in small areas alongside new, improved/introduced varieties;
- making available their knowledge and skills in identifying, collecting/rescuing and utilizing plants which they have helped to develop and maintain for generations.

18.3 THREAT OF GENETIC EROSION

The broad range of genetic diversity existing in Ethiopia, particularly the primitive and wild gene pools, is presently subject to serious genetic erosion and irreversible losses. This threat, which involves the interaction of several factors, is progressing at an alarming rate. The most crucial ones include displacement of indigenous land races by new, genetically uniform crop cultivars, changes and development in agriculture or land use, destruction of habitats and ecosystems, and drought.

The drought that prevailed in the regions of Wello, parts of Shoa and northern Ethiopia in the last decade has directly or indirectly caused considerable genetic erosion, and at times has even resulted in massive destruction of both animals and plants. The famine that persisted in some parts of Ethiopia has forced farmers to eat their own seed in order to survive or to sell the seed as a food commodity. This often resulted in massive displacement of native seed stock (mostly sorghum, wheat and maize) by exotic seeds provided by relief agencies in the form of food grains.

The extent to which the displacement of native seeds by exotic/improved materials occurs has not yet been fully documented. This would also vary between regions and crops. In many cases, farmers still plant both types interchangeably or alongside each other, at times in mixtures, depending on their particular need, market demand, or other prevailing situations.

In general, native barley and durum wheat are probably among the most threatened by new improved/introduced varieties and/or by other crop species such as teff and bread wheat, which are expanding within the cereal growing highlands of Shewa, Arsi and Bale regions, because of greater market demand. Similarly, in the central highlands, including the northern Shewa and Gojam regions, introduced varieties of oats are expanding rapidly, often replacing a wide range of cereals, legumes and pulses grown in these areas.

With sorghum and millets, exotic varieties do not pose any immediate threat because expansion of such materials is at present somewhat restricted. In the case of sorghum, however, genetic erosion is progressive on account of extensive selection based on breeding of the native populations. A similar situation exists with the various pulses, legumes and oil crops grown in the country, where the bulk of the material used in breeding programmes is represented by indigenous land race populations.

18.4 NEED FOR RESEARCH TO CONSERVE AND ENHANCE IN-FIELD DIVERSITY

The maintenance of species and genetic diversity in fields is an effective strategy that Ethiopian farmers have long adapted to sustain a stable system of conservation for low-input agriculture. *In situ* (on-site) conservation of land races (relating in this context to maintaining traditional cultivars or land races in the surroundings where they have been adapted, or the farming systems under which they have acquired their distinctive characteristics) on peasant farms, which Don Duvick (personal communication) calls 'evolutionary conservation', would, therefore, provide a valuable option for conserving crop diversity (Worede, 1991). More importantly, it will help to sustain the evolutionary systems that are responsible for the generation of genetic variability. This is especially significant in regions of the country subject to drought and other stresses, because it is under such environmental extremes that variation useful for stress-resistance breeding is generated. In the case of diseases or pests, this would allow continuing host–parasite coevolution.

Also under these conditions, access to a wide range of local land races would probably provide the only reliable source of planting material. The ability of land races to survive under these stresses is conditioned by their inherent broad genetic base. This is often not the case with the more uniform, new or improved cultivars which, despite their high yield potential, are less stable and not as reliable as sources of seed under the adverse growing conditions generally present in many of the drought-prone regions of Ethiopia.

Land race evaluation and enhancement programmes will certainly be needed to stimulate a more extensive utilization of germplasm resources that are already adapted to these regions. Under such extreme environments, locally adapted land races would also provide suitable base materials for institutional crop improvement programmes. There is, therefore, an outstanding need to maintain land races growing under these conditions in their dynamic state, and this is probably best achieved through farm or community-based conservation programmes.

Work has recently begun in Ethiopia to develop farmer-based conservation activities from two major approaches: conservation and enhancement of native seed stock (land races); and maintenance of indigenous land race selections (elite materials) on selected farms (Worede, 1992). The salient features of these and other related activities are described in the following sections.

18.5 LAND RACE CONSERVATION AND ENHANCEMENT ON THE FARM

This aspect of genetic resource activity involving farmers, scientists and extension workers has been in progress since 1988 and is now being consolidated within a network of some 23 peasant farms (selected and organized through their respective farmers' cooperatives) on strategic sites in north-eastern Shewa and south-eastern Wello, in areas under recurrent drought and other stresses including disease and pest epidemics. The crops include sorghum, chickpea, field peas and locally adapted maize.

The conservation measure was designed, primarily, to maintain in-field crop diversity in areas where land races are still widely grown, by protecting major cultivars from disappearing as well as improving their genetic performance. Materials collected (or rescued) during the drought period from within the above-mentioned regions are included in the programme.

The land races are maintained on each peasant farm (Figure 18.1) following exclusively the traditional practices of selection, production (including weed management), storage and utilization. The particular site would vary each season based on the traditional cropping pattern which involves the various crops grown in rotation on the farm. The plot size and seed rates employed are those already established by the farmers over centuries of planting their land races. For each crop, this is determined by the farmer, depending on need, amount of seed and labour available, method of seeding and soil type. With sorghum, for example, this would vary from 0.5 to 4 ha, and a seed rate of 5–20 kg/ha. Farmers have also set the minimum size of land that they require to ensure a safe harvest (e.g. 0.25 ha for sorghum). Management and maintenance of the *in situ* plots in this way has been practised routinely to optimize the *in situ* conservation. The rationale for this is based on the known fact that this is how the farmers maintained diversity of their land races as they exist now, thus providing the basis for a sound and viable approach to conservation.

Studies are under way to document and build on existing knowledge and practice relating to land race production and management on these farms. Scientific inputs are needed in the areas of socio-economics, ethnobotany and population biology focusing on the population structure and dynamics of the various land races for a more rational planning and effective management of *in situ* conservation. Research along these lines is now being sought, on a multidisciplinary basis, involving farmers and giving due emphasis to women's knowledge and roles (Figure 18.1). Much of this work is in the process of being established, given the greater emphasis placed so far on the multiplication and distribution of élite land races and the limited resources available.

Farmers collaborating in the project also practise various forms of mass selection and multiply their land races (mainly sorghum and local maize) separately for production. Seeds of selected plants are bulked to form a slightly improved population, which is included in plantings for seed increase and continued selection. An appreciable amount of improvement in crop yield has been observed among the selected

Figure 18.1 A farmer-based land race sorghum *in situ* conservation plot in southern Wello (1993).

materials which are also produced basically following the traditional systems, surpassing both the original populations and the improved, high-input varieties (SoS/E, 1993). This also provides an opportunity for transferring genes that control characters of interest (e.g. disease/pest resistance, high lysine in sorghum, and drought tolerance) from existing selections or from external sources to enhance the élite populations.

Farmers are paid on a contractual basis for conserving and multiplying land race materials, and élite land races are distributed to local farmers in the region, currently involving some 500 peasant farmers. This is determined on the basis of the additional inputs (labour and various costs) incurred in such a task. Their activities are guided by PGRC/E and closely supervised or monitored jointly with the local field staff of the Unitarian Service Committee of Canada (USC/C), which is financially supporting the land race programme in Ethiopia.

18.6 MAINTAINING ÉLITE INDIGENOUS LAND RACE SELECTIONS ON PEASANT FARMS

This aspect of land race conservation represents a measure for restoring land races to regions where such materials were once widely grown but are now displaced by new, improved (high-input) varieties or by varieties of crops introduced as food grain through relief agencies. In this programme, which is currently limited to tetraploid wheat (*Triticum durum* Desf.), élite indigenous plant materials are maintained on peasant farms (Table 18.1), following traditional low-input farming practices.

These populations are subjected to modification by mass selection based on performance in yield tests under different conditions of environmental stress. Selected genetic lines (agro-morphotypes) were bulked for further multiplication and distribution to farmers (Tesemma, 1987). The programme is funded by USC Canada, and uses land races collected and developed by the durum wheat breeding team at the Debre Zeit Agricultural Research Centre, Alemaya University of Agriculture, in the mid 1980s.

Some land race selections of durum wheat (composites of two or more élite agro-morphotypes) developed in this way have been more productive in preliminary yield trials than released cultivars (Tesfaye Tesemma, personal communication). Also in a recent survey, the élite materials were shown to have outyielded (by as much as 10%) the improved, high-input durum wheat varieties which normally out-yielded the land races originally grown by the farmers (Ataro and Bayush, 1994). In some cases, as in Dire, the élite land races were the only wheats that farmers risked planting, and these survived the drought that occurred in 1991 in that area.

The farmers will continue to multiply and use the composites that are best suited to their conditions along with other entries provided by the

Table 18.1 AUA–PGRC/E durum wheat land race (elite agro-types) multiplication and testing sites (farms) during the 1992/93 crop season in Shewa region

District	Locality	No. of composites	Agro-morphotypes	Remarks
Ada	Dirre	20	60	Low rainfall zone
Ada	Godino	29	60	–
Ada	Cheffe Donsa	14	43	Water logging; cold/frost area
Ada	Ejere	15	55	–
Ambo	Awarro	–	48	Marginal soils

breeder. The PGRC/E also takes representative samples from these lines for long-term storage at the gene bank. This would allow the farmers to continue to evaluate critically their sources of planting material, which at present consist largely of relatively poorly adapted cultivars distributed to farms in the region. It also encourages them to make continued use of land races, thus ensuring effective utilization of superior germplasm and avoiding the threat of losing unexplored germplasm represented by the indigenous population. A number of selected farmers are also multiplying the elite seeds for distribution to the surrounding farms, so far involving some 80 farm families in the region, and are paid on a contractual basis.

Studies are also under way on a multidisciplinary basis for further integration of the conservation work with yield-sustaining production technologies, which would involve restoring and/or enhancement of the traditional cropping patterns and other activities that were disrupted with the expansion of the modern varieties. Other studies under consideration include socio-economic and cultural aspects as well as distribution and marketing of the élite land races, recording previous knowledge that farmers still retain about their land races.

Further breeding work is the primary responsibility of the breeder. Considerable progress could be achieved through selection from some of the better indigenous land races (Tesemma, 1987). Continued hybridization among various selections, possibly by employing chemical male gametocides, would probably generate unique genetic combinations that might not surface under a controlled manual hybridization system. Cultivars developed from locally adapted land races would serve as controls or standards in the national yield testing programmes. Through this mechanism, it should be possible to restrict the expansion of high-risk seeds and still provide farmers with élite populations that represent improved versions of adapted local types. This is especially significant for marginal areas or extreme environmental conditions where improved cultivars fail to meet adequately the requirements of farmers.

Although the focus is on élite materials, the original seed stock is included in such a strategy and is maintained *in situ*, employing the methods and practices described earlier. Measures are also being taken to cover other crops, particularly pulses (chickpea, faba bean and fenugreek) and (in the higher altitudes) barley, grown in rotation.

18.7 FUTURE PROJECTIONS AND PERSPECTIVES

The involvement of farmers in the conservation and utilization of Ethiopia's germplasm resources will be strengthened and expanded to cover a broad range of agro-ecological conditions and strategic sites. The process is a challenging one, demanding a comprehensive knowledge of the country's vast resources and the diverse systems under which they have been maintained. Drawing from existing experience, the rationale for projecting on a wider and more comprehensive network of *in situ* conservation of land races is already fairly well established, and in many cases is similar to those established for other regions with a similar background (Altieri and Merrick, 1987; Brush, 1991a). *In situ* conservation is considered a viable and vital component of the national overall conservation strategy, complementing the existing off-farm (*ex situ*) conservation practice; it is participatory, involving farmers and their long-established skills and knowledge of land races; it is dynamic, allowing continual evolution and generation of useful germplasm; it is relatively inexpensive for the amount of potentially useful material preserved; and, together with *ex situ* conservation, it would provide a mechanism by which Ethiopia's germplasm resources are protected and more effectively utilized on a long-term basis.

The land races of any of the crops considered in such a programme may not be maintained with adequate safety in the few areas where they are at present being established. With each crop species farmers spread their risk across time, space and the diversity of the material they grow and this occurs at the levels of the farm household, communities and regions where they exchange or diffuse their material and information about their seed, which may account for the wide range of adaptability as well as the plasticity inherent in these materials. It is essential to plan a correspondingly wide network of *in situ* conservation sites, taking all these factors into consideration, supported by more extensive research relating to the genetic, ecological and social dynamics of land races.

At least in the long run, *in situ* conservation work will be expanded to conserve wild plant species in their natural habitats. Plans are already under way to conserve *in situ* wild relatives of cultivated crops and wild plant species of potential value at strategic sites in areas where diversity exists. This may be undertaken as part of a community grazing land management programme, in areas surrounding farms, where such materials still exist but are progressively diminishing due to changes in land use or ploughing under. Such a programme might also include several trees, shrubs and grasses that grow wild and are traditionally used by communities for food, medicine and fuel.

Similarly, as part of the national coffee conservation programme, a special effort is being made to conserve the semi-cultivated coffee (*Coffea arabica*) on peasant farms in Kefa, Illubabor and parts of the Wellega region, in areas where the forest coffee occurred spontaneously. This will complement the field collection now being maintained at Chochie, Kefa (PGRC/E, 1992). More details of future projects can be found in other relevant PGRC/E activity reports and project documents.

18.8 LINKING LAND RACE CONSERVATION TO UTILIZATION

The value of land races to farmers in a developing country like Ethiopia lies in their utility as a dependable source of planting and breeding material. It is, therefore, important that locally adapted/enhanced seeds are multiplied for distribution to farmers whose requirements have not been adequately met by modern, high-input cultivars. It may otherwise make very little sense to conserve land races or may even be difficult to convince farmers to do so unless the land race conservation activity is oriented towards supporting sustainable production.

The best way to achieve this is probably through community-based seed production or marketing and distribution systems operating in networks. These could possibly be developed by enhancing or further organizing the traditional networks that were described earlier. Through this approach, farmers will be able to control the choice of crop types and cultivars to grow and will have ready access to planting material adapted to local growing conditions. They will also be in a position to evaluate critically the relative merits of a wide range of cultivars, thereby limiting undue expansion of exotic cultivars that are costly and poorly adapted.

An example of such a network which is now being developed in Ethiopia is provided in Figure 18.2, illustrating the various components of communal level genetic resource conservation, enhancement and utilization, and PGRC/E's role in coordinating such a concerted activity involving farmers, breeders and extension workers.

The Community Seed Bank (CSB) is a low cost and low technology system that will be owned and managed by local communities involving existing community service cooperatives. It comprises two major components – a seed store and a germplasm repository – for local crop improvement, complementing the gene bank at PGRC/E. The seed store

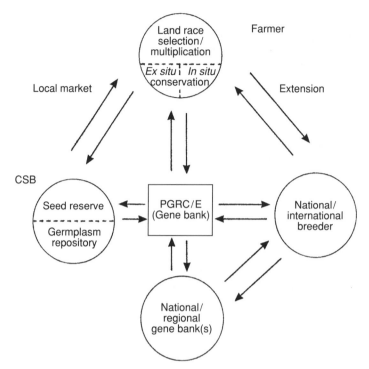

Figure 18.2 A network of seed conservation, selection (enhancement), multiplication and utilization activities in Ethiopia. CSB = Community Seed Bank; PGRC/E = Plant Genetic Resources Centre/Ethiopia.

represents a seed reserve system (largely represented by land race materials developed or multiplied contractually by the farmer) that will provide a back-up to the local (informal) market network, where farmers traditionally exchange seeds and information. The seed reserve that the CSBs maintains becomes crucial to ensuring a sustained supply of adapted seeds to farmers, channelled through the informal market system, thereby averting the risk of losing diversity.

The small traditional storage units – clay pots, rock hewn mortars, underground pits, etc. – form an integral part of the traditional seed storage systems, also representing actual *ex situ* conservation. Improved or enhanced versions of these small units may be established within a network of farm households to complement the *in situ* networks as well as the CSB.

18.9 INSTITUTIONAL BACK-UP AND COORDINATION

The institutional links and collaborative roles pertaining to the land race conservation programme in Ethiopia have been referred to already. Most of the existing activities are now well coordinated with the necessary guidance, scientific inputs and monitoring provided by PGRC/E, which has a national mandate to undertake all genetic resource activities in the country. The Centre has also established a department for community genetic resources conservation activities which coordinates the land race conservation programme, establishing links at national, regional and global levels.

The field programme is implemented by the Seeds of Survival Programme/Ethiopia (SoS/E) which represents PGRC/E's land race conservation strategy. It is supported by USC/Canada as part of its land race conservation, development and utilization programme for small-scale farms in Africa and other regions. The success of this programme in Ethiopia has inspired new global inputs emanating from, among others, the Global Environment Facility (GEF) which will support a land race conservation programme through capacity building (training and infra-structure), a network of community gene banks and *in situ* conservation activities in the country. The Community Biodiversity Development and Conservation (CBDC) Programme, another global initiative, will provide in Africa appropriate inputs for the land race improvement and utiliza-tion aspect. The African Biodiversity Network (BIDNET) of the African Ministerial Conference on the Environment (AMCEN) also seeks to establish such an activity throughout Africa, based on the Ethiopian experience. In Ethiopia, PGRC/E has been playing a key role in the initi-ation, formulation and implementation of these various regional and global initiatives, also serving as the lead agency that coordinates all these programmes for the African continent.

19

Peruvian *in situ* conservation of Andean crops

R. Ortega

19.1 INTRODUCTION

Most of the Andean crops are indigenous to that area and have become adapted to a considerable range of ecological zones, in terms of altitude as well as rainfall and soil conditions. There is a very large number of crops and also each crop comprises an astonishingly great genetic diversity.

Up to now farmers have understood and conserved this genetic diversity both as individuals and as thousands of regional communities. This has been accomplished in spite of competition from introduced crops and western technology. In this sense, *in situ* conservation is a viable option for the present and the future, and *ex situ* and *in situ* techniques should be considered as complementary ways of conserving genetic resources in the Andes.

19.2 THE ANDEAN ECOREGION

This region comprises the whole Andean chain from Venezuela, through Colombia, Ecuador, Peru, Bolivia and Argentina to Chile, with a length of 7000 km and from latitude 10°N to the southern tip of Tierra del Fuego. The varying altitudes from sea level to eternal snows and varying rainfall according to exposure and slope provide an extraordinarily wide range of conditions for plant growth.

The human population of this region has been estimated at about 30

Plant Genetic Conservation.
Edited by N. Maxted, B.V. Ford-Lloyd and J.G. Hawkes.
Published in 1997 by Chapman & Hall. ISBN 0 412 63400 7 (Hb) and 0 412 63730 8 (Pb).

million, of which 40–50% are involved in activities related to agriculture (CONDESAN, 1993). This is more than other mountainous regions such as the Himalayas (Bowman, 1980).

19.2.1 The Peruvian Andean ecosystems

The Peruvian Andean ecosystems cover some 40 million ha, representing 28–30% of the total Peruvian territory, and comprising an altitude range of between 1500 and 5000 m above sea level (Figure 19.1). This is the 'Sierra' region, which can be divided into natural ecological areas such as inter-Andean low dry valleys, high Andean valleys, the altiplano, the puna and the high cordillera. The annual rainfall varies from 200 to 1200 mm, and the average temperature from 2 to 16°C. There is a considerable amount of soil erosion in the strongly sloping fields in many regions. More than 7 million inhabitants live in this region (Valladolid, 1988; Torres, 1992).

19.2.2 Andean agriculture

The high Andean central region of South America is considered to be one of the principal centres of crop evolution (Vavilov, 1951, in Hawkes, 1983), where agriculture developed independently some 8000 years before the present. León (1964), Kaplan *et al.* (1973) and Valladolid (1988) all consider that agriculture began in the high Andes and not on the coast of Peru.

Thus the tuber crops (potato, oca, olluco, maswa), grain crops (quinoa, kañihua, achita, tarhwi), root crops (yacón, achira, arracacha, etc.) and much later maize constituted the basis of agriculture in the central Andes, which formed the foundation of the Andean civilizations, whose architectural remains are seen and admired to the present day.

19.2.3 The characteristic features of Andean agriculture

As in all regions of agricultural development, the plants and the land are only one part of the agricultural systems, which depend also on human involvement and traditions developed over many thousands of years. In this case, particularly, a communal organization for building retaining walls, aqueducts, terraces and soil improvement was an essential part of husbandry. Amelioration of micro-climates, especially in the Inca period, and the organization of communal agro-ecosystems provided food for all. Unfortunately, many of these systems, built up and prized by the Incas, are now in disarray. Even so, the times for ploughing, sowing and harvesting follow the traditional pathways of knowledge handed down from generation to generation. For this reason it has been said that the

Figure 19.1 Area covered by the Peruvian Andean ecosystems.

Andean man was truly an Ecological Man (Hart, 1974). The selection of fields for sowing or planting crops was guided by forecasts of whether the growing season was likely to be wet or dry; this was followed by irrigation, weed clearing, rotation, fertilizing and other appropriate activities of the farming year.

In the Andean agricultural systems it was held to be necessary, because of the extreme climatic and soil diversity, to select and maintain the greatest genetic diversity possible, and this was at two levels:

- **interspecific diversity**: the planting of different crops in the same field; for example, maize with rows of quinoa (*Chenopodium quinoa*), tarhwi (*Lupinus mutabilis*), etc.;

- **intraspecific diversity**: the planting of different varieties of the same crop; for example, a mixture of species and varieties of potato.

In native agriculture the sowing of pure species or varieties did not exist, because it was the custom to grow crops associated with one another in space and in time.

In connection with agricultural activities certain rituals are observed, in which libations are made to Pachamama (mother earth) and the Apus (tutelary spirits of the mountains) (Valladolid, 1988).

Plant and animal husbandry are linked together, so that most of the communities are self-supporting, even though surpluses are taken to market and traded for other commodities. The traditional agricultural system practised by Andean farmers is a type of reciprocal loan of services amongst community farmers, known as *Ayni*. Thus a person or family will solicit one or more days' work without pay, on the understanding that this will be repaid later by the same amount to the neighbour who provided it. The word *Faena* is used for collective work by the whole community for the common good. This is a moral obligation within each community.

19.2.4 Andean crops

Archaeological research has shown that pre-farming communities used about 500 plant species for food and later on domesticated some 70 different species (Minka, 1987). Outstanding among these were tubers, roots, corms, rhizomes and grains. These, together with certain legumes and fruit trees, form the well-known system of Andean crops and may be considered as the invaluable plant genetic resources of the Andes. In pre-Spanish days these crops had undoubtedly been subjected to intense selection for yield, palatability, adaptation to soils and climates, etc., resulting in the genetic richness which we know today.

After the Spanish conquest and colonization many Andean crops were to a certain extent replaced by wheat, barley, oats, peas, broad beans, etc. These displaced the traditional crops, especially on the better lands. The native animals such as the llama and alpaca were similarly pushed into higher grazing areas unsuitable for old-world cattle, sheep and goats.

In the more tropical regions the yuca (*Manihot*), camote (*Ipomoea*) and pallar (*Phaseolus lunatus*) were often displaced by rice and sugar cane (Blanco, 1988). Other plants which were used in religious rites were completely prohibited (Fano and Benavides, 1992) whilst many areas became depopulated, and the people and their crops were forced into marginal agricultural areas. It is highly likely that these processes caused what is now known as genetic erosion – the diminution of genetic diversity. In spite of these processes, many community farmers still grow their traditional crops, due to their persistence, creativity, necessity and a strong sense of tradition and continuity.

19.2.5 Natural agro-ecological regions and zones of native and introduced Andean crops

Although there are various ways of classifying the regions and zones in Peru we give here one based on Pulgar Vidal (1987) (Table 19.1). These regions, even if considered controversial, provide a strong base for synthesizing geographical and ecological concepts. The distribution of Andean crops by altitude is also of considerable interest and importance (Table 19.2). Systems of cultivation differ from region to region and according to type and use of the land, ancestral customs, locality and community (Sanchez, 1993). The best-known systems are various forms of rotation.

19.2.6 Field rotation

Field rotation is the usual practice amongst the Andean communities and includes the use of resting or fallowing the land. It is recognized by farmers as beneficial for the restoration of nutrients and recuperation in terms of plant health. The rotational details are agreed collectively but the individual farmer decides on which crops to plant.

Various names are given to this process in different parts of the Andes, such as *Suyo* (Lake Titicaca), *Aynoqa* (Aymara region), *Manda* (Puno region), *Laymi* or *Muyuy* (Cuzco Department), *Yacón* (Cerro de Pasco) and *Turno* (Junín Department). The resting or fallowing period may be prolonged for 5 to 20 years, according to soil, altitude, etc. In any case, potatoes are always the first crop grown after fallowing.

19.3 GENETIC DIVERSITY OF ANDEAN CROPS

Numerous agro-ecosystems occur in the Andes due to the diverse geographical and climatic factors mentioned above. These and, above all, the Andean farmers themselves have been instrumental in creating the wide range of genetic diversity that we see today. The small farmer considers this diversity, not as an imposition, but a part of the broad repertory available to exploit the environment for his existence. This is linked with the folklore of varietal names and traditions that form part of the Andean cultural heritage (Brush *et al.*, 1981).

19.3.1 Relic areas for Peruvian crops

(a) Potatoes

Certain areas can be considered as repositories of genetic diversity for potatoes, in terms of the rich diversity of tuber morphology such as form, colour, eye depth and also cooking quality, plant morphology, growth habit, yield, resistance and tolerance to pests, diseases and environmental stresses. These areas occur mainly in southern Peru, in the departments of Cuzco (Paucartambo, Calca, Urubamba, Grau, Aymaraes), Apurímac and Puno (Carabaya, Sandia, Huancané-Moho); also in Central Peru in the departments of Junín (Tarma, Concepción), Huánuco (Huamalies, La Unión) and Cerro de Pasco (Pasto, Daniel A Carrión) and in the northern department of Cajamarca (Calendín).

(b) Oca (Oxalis tuberosa)

The areas of greatest concentration of oca ecotypes, differentiated by tuber form, colour, flavour (sweet or bitter), maturity period, growth habit and other features, are to be found in Puno (Sandia province, Cuyo-Cuyo).

(c) Olluco (Ullucus tuberosus) and maswa (Tropaeolum tuberosum)

Significant diversity of these crops is found along the eastern ranges of the Peruvian Andes. They include yacón (*Polymnia sonchifolia*), achira (*Canna edulis*), arracacha (*Arracacia xanthorrhiza*), uncucha (*Xanthosoma sagittifolium*), ahipa (*Pachyrrhizus tuberosus*), dale-dale (*Calathea allouia*) and chichi (*Maranta arundinacea*). These are richest in the Quechua low valleys of the inter-Andean tropics and subtropics. Genetic diversity does not seem to be so great in these species as in potatoes and oca.

(d) Quinoa (Chenopodium quinoa)

Quinoa exhibits two diversity groups in Peru. One is in the valleys, characterized by high growth (to 2.5 m) with many branches and lax inflorescence, as well as high tolerance to *Peronospora farinosa*. The other is on the altiplano, with lower growth (to 1.5 m) with terminal panicle and with some ecotypes susceptible to *P. farinosa* (Tapia, 1980).

(e) Kañiwa (Chenopodium pallidicaule)

Kañiwa exhibits greatest genetic diversity in the high provinces of Cuzco Department, expressed in growth, plant colour (pink, red, yellow,

Table 19.1 Natural agro-ecological regions and zones of native and introduced Andean crops

Region	Agro-ecological zone	Tubers	Roots, corms and rhizomes	Grains	Other
Suni and Puna 3700–4500 m	Hillsides and plateaux	Potatoes: bitter varieties	Maca		
Suni and Puna 3700–4500 m	Hillsides on high plateaux (Altiplano)	Potatoes: bitter and sweet native varieties Oca Olluco Maswa		Kañiwa Quinoa Barley (forage) Barley (thrashed grain) Oats (forage)	
Suni	High plateaux	Potato: bitter varieties		Kañiwa Oats (forage)	
Suni	Low plateaux	Potato: sweet native varieties		Barley (grain) Broad bean Tarwi Oats (grain)	
Quechua 2300–3500 m	Inter-Andean high valley slopes	Potato: native varieties Dulces: sweet varieties Oca Olluco Maswa		Broad bean Quinoa Kiwicha Tarwi Barley (grain) Oats (grain) Pea Wheat Maize	
Quechua 2300–3500 m	Inter-Andean high valley bottoms			Maize Quinoa Pea Wheat Kiwicha	Cabbage (1) Onion (2)

Zone	Location			
Riverine Yunga 500–2300 m	Slopes and bottoms of low inter-Andean valleys	Yacón Achira Arracacha Uncucha Yuca	Maize Rice	Squash (3) Chilli pepper (4) Banana (5) Peanut (6) Tree tomato (7)
High Selva 400–1000 m	Slopes and bottom of very low valleys	Uncucha Sweet potato Cassava Palillo Papa Magona	Maize Beans	Peanut Coffee (8) Cocoa (9) Banana
Low Selva 80–40 m	Amazon plain	Sweet potato Uncucha Cassava Papa Magona Ahipa Chichi Dale–Dale	Maize (hybrid) Rice	Beans

(1) *Brassica oleracea* (2) *Allium cepa* (3) *Cucurbita maxima* (4) *Capsicum pubescens* (5) *Musa paradisiaca*, (6) *Arachis hypogaea* (7) *Cyphomandra betacea* (8) *Coffea arabica* (9) *Theobroma cacao* (10) *Curcuma domestica*

Achira	*Canna edulis*	Oats	*Avena sativa*
Ahipa	*Pachyrrhizus tuberosus*	Oca	*Oxalis tuberosa*
Arracacha	*Arracacia xanthorrhiza*	Olluca	*Ullucus tuberosus*
Barley	*Hordeum vulgare*	Palillo	*Escobedia scabrifolia*
Broad bean	*Vicia faba*	Papa Magona	*Dioscorea* spp.
Camote/sweet potato	*Ipomea batatas*	Peas	*Pisum sativum*
Chichi	*Maranta* spp.	Pigeon pea	*Cajanus indicus*
Dale-Dale	*Calathea allouia*	Quinua	*Chenopodium quinoa*
Kañiwa	*Chenopodium pallidicaule*	Rice	*Oryza sativa*
Kiwicha	*Amaranthus caudatus*	Tarwi	*Lupinus mutabilis*
Maca	*Lepidium meyenii*	Uncucha	*Xanthosoma sagitifolium*
Maswa	*Tropaeolum tuberosum*	Wheat	*Triticum vulgare*
		Yacón	*Polymnia sonchifolia*
		Yuca	*Manihot esculenta*

Table 19.2 Distribution and altitudinal adaptation of Andean and introduced crops

Altitude (m)	Crop		Altitude (m)
	Tubers and roots	Grains	
3700–4500	Bitter potato		
2000–4200	Native potato Sweet and bitter varieties	Kañiwa	3500–4200
		Oats (forage)	3600–4000
		Barley (forage)	3600–4000
		Quinua	1800–4000
3300–3800	Maswa		
3200–3800	Oca	Tarwi	3100–3800
		Oats (grain)	2500–4000
		Broad bean	2700–3800
		Barley (grain)	2600–3800
3200–3800	Olluco	Peas	2600–3700
		Wheat	2000–3800
		Maize*	80–3600
		Kiwicha	2400–3500
2200–3100	Yacón		
1900–2700	Achira Arracacha		
Less than 1000	Camote Yuca Ahipa	Rice Pigeonpea	Less than 1000 Less than 1000

* From the low tropics up to 3600 m.

orange, purple) and colour of grain at harvest (light purple, dark purple and black) (Tapia, 1980).

(f) *Kiwicha* (Amaranthus caudatus)

Genetic diversity in kiwicha is restricted to the temperate and cool temperate 'deep' valleys. Ayacucho, Apurímac and Cuzco (Paruro) vary in both inflorescence form and colour (Sumar, 1980).

(g) *Tarhui* (Lupinus mutabilis)

Genetic diversity of this crop is extraordinary in terms of plant size, morphology, flower colour (blue, purple, sky blue, lilac, pink, white, cream), yield, maturity, content of protein, fats and glucosides, as well as resistance and tolerance to pests and diseases. This diversity is concentrated in small plots or the borders of other crops in the northern, central and southern Peruvian departments (Blanco, 1988).

(h) Maize (Zea mays)

The greatest genetic diversity in maize is maintained in tropical, temperate and cold zones up to 3600 m or even higher. The variability is enormous (Grobman, 1961).

19.4 MAINTENANCE OF GENETIC DIVERSITY

There is no doubt that the peasant farmers possess a detailed knowledge of their crops and preserve their diversity from year to year, identifying and conserving new forms, mutations and natural hybrids which may occur from time to time. Even crosses with related or ancestral wild species which occur naturally are also conserved. This introgression of genes from wild relatives is important in adding to the total gene pool, as well as the natural hybridizations between genotypes in the crop itself (Blanco, 1993).

The potato, as shown by Ortega (1989), is one of the best crops to demonstrate the growth of genetic diversity for very long periods of time. Its reproduction by tubers, vegetatively, and by outcrossing, provides sources of stability on the one hand and change on the other.

19.5 GENETIC EROSION

This has largely taken place due to the introduction of European crops, particularly barley, which has occupied many of the niches in which indigenous crops were grown. To a lesser extent broad beans and peas, as well as other crops, have caused some genetic erosion in Andean crops. The introduction of improved potato varieties has also given rise to genetic erosion in that crop in many parts of Peru. This process has been minimal for the other indigenous crops under discussion but, whether one agrees with it or not, the peasant farmers of the Peruvian Andes have been under pressure from market forces which have undoubtedly reduced the indigenous crops to second place. Other factors such as cheap food imports, the replacement of crop-growing by cattle pastures, destruction of forests and many other pressures have all contributed to the genetic erosion of the indigenous Andean crops.

This has been pointed out also in respect of potatoes, where a diminution of varieties is observed now, in contrast to Russian publications which listed far more in 1933. Other evidence points in the same direction. Also the potato blight (*Phytophthora infestans*), which was not known in Peru in the past, has now decimated many potato fields in areas where great genetic diversity was once found.

19.6 *IN SITU* CONSERVATION

In the words of Frankel (1970), *in situ* conservation of genetic resources is 'the continued maintenance of a plant population in the community to which it belongs, within the environment to which it is adapted'. This activity started when humans first came into contact with plants thousands of years ago and has continued to this day. The diversity that developed during this process was maintained in the same way by the peasant farmers even after the Spanish Conquest some 500 years ago (Brush, 1991b). In establishing techniques for *in situ* conservation the traditional techniques of the peasant farmers must be taken into account. The following section takes the potato as a case study.

19.7 FARMING STRATEGIES FOR POTATO *IN SITU* CONSERVATION

The Peruvian peasant farmer is not so much looking for high financial returns but is basically more concerned with the greatest amount possible of varied food for the whole year, considering also the effect of climate and other physical factors on total yield. This philosophy directs the farmer less towards uniform commercial crops than to the mixture of genotypes in small plots that will ensure the survival of at least some of them under various environmental pressures. This system does not ensure high yields, but nor does it lead to total crop failures (see Ortega, 1991, on a project in Yauri, Cuzco). This diversity is related also to the particular use to which each variety is put – for example, *sancochado* (boiled), *huatia* (oven-baked, roasted, etc.), *chuño* (freeze-dried) or *moraya* (freeze-dried and washed in a stream), whilst others serve for particular rituals which still remain in certain communities.

Certain genetic conservation strategies are worth mentioning here:

- The first is vertical control of ecological zones. Native varieties are on the whole imperfectly adapted to lower altitudes, but do well in the higher cool zones where they resist cold remarkably well. In certain areas of Paucartambo province, Cuzco, such as the community of Carpapampa, potatoes are grown at three distinct levels or ecological zones, according to varietal adaptations (Table 19.3). This stratification enables mixtures of varieties of different levels of frost resistance to be grown most efficiently in the farmers' small plots, so avoiding the erosion or loss of various genotypes.

Table 19.3 Potato varieties at different altitudes in Paucartambo province, Cuzco

Height above sea level (m)	Region	Type of potato
4300–4700	Very high	Ruk'i Allpa
4000–4300	Intermediate	Kus'i Allpa
3600–4000	Lower	Queshua Allpa (sweet varieties for direct consumption)

- Fallow land (called *muyos*, *laymes* or *turnos*) is brought back into use after some seven years and provides good yields, especially when the soil is black and rich in nutrients.
- The production cycle in non-irrigated land is as follows. The first preparation of the soil is carried out in February to March up to the first fortnight of April and the final stages take place in September to November. The first sowings give better yields than the second ones, and each native variety is sown in two distinct areas, so that if one is destroyed by frost, for instance, the other may survive. The farmers also practise 'horizontal crop control', using different varieties. Thus in the province of Espinar-Cuzco where bitter varieties belonging to *Solanum juzepczukii* and *S. curtilobum* are sown, different varieties of the former are mixed with the scarcer ones of the latter. Tubers of the former species are smaller but more resistant, whilst those of the latter are larger but less resistant to frost. The farmers in fact never sow the varieties of these species apart but always together.
- In certain regions farmers refuse to grow newly bred improved varieties because they are convinced that their traditional varieties are better flavoured, whilst the new ones are said to be watery and only fit to put into soups, and that they do not keep well, turning black inside after a short while. The native varieties keep better, especially those named 'Mestiza', 'Mactillo', 'Chimaco' and others.
- It is important to note that the peasant farmer largely grows the native varieties for special events such as family birthdays, religious feasts and other events. Guests, however, are served potatoes of lower quality, such as recently bred varieties, which are generally watery or tasteless.
- It seems that when a tuber is 'stolen' from a farmer, this is often by a neighbour, who quickly sows it in his own ground, thus paradoxically promoting its spread and conservation.
- On the whole the native varieties are used for home consumption and the breeders' varieties for selling in the markets. This helps to conserve the indigenous varieties. Even so in certain communities the variety 'Mariva', introduced 25 years ago, and 'CICA' (20 years ago) have been adopted.

- The native potato varieties fetch a lower price in the market, because the stall-holder is looking for uniformity and thus prefers the improved varieties; this applies also to potatoes that are sold in the large cities. Thus this process works towards conservation of the native varieties, which fetch a lower price and hence are not taken to market except in small quantities.
- The most usual marketed varieties are 'Huayro', 'Peruanita', 'Ccompis' and 'Bole', of which 'Ccompis' is most liked, but the medium-size and small tubers are left for eating and seed sowing by the farmers. It is also observed that some native varieties from one region are brought to the Cuzco market and sold to market women from another region. In consequences the potatoes grown in the plots of the latter are kept for the next sowing, thus avoiding the genetic erosion of their own area.
- Although in general only the smaller tubers are used for sowing, this does not seem to diminish the viability or number of genotypes conserved. On the contrary, it may help to increase the diversity by introducing small tubers from the first generation growing naturally from true seed derived from natural outcrossing (and sometimes with wild species). If yields are very bad due to abnormal weather conditions the farmers prefer not to eat, rather than consume their seed stocks. To solve this problem they may migrate temporarily to other regions to find work or will eat the produce from other crops. In no way will they eat their own seed potatoes, thus preserving the genetic integrity of their crops.
- Peasant farmers have some doubts about identifying the varieties when they are in leaf or flower, but they will always and without hesitation identify them from the tubers with 100% accuracy. This clearly helps to maintain their stocks, because if one variety (phenotype) is missing they will immediately try to obtain it from other farmers in the community. If it is unobtainable by this means the farmer will try to obtain it from other bordering communities by buying, exchanging or even working to obtain the means for buying it.

In summary, it is clear that the central or fundamental aspect of *in situ* conservation practised by peasant farmers in southern Peru is the continuing flow of genetic resources in terms of genes and gene complexes from each generation to the next, no matter what may be the methods and origins of these materials.

ACKNOWLEDGEMENT

I would like to thank M. Sawkins for preparation of the map.

20

Central Asian *in situ* conservation of wild relatives of cultivated plants

N. Lunyova and T. Ulyanova

At the present time, taxonomists, historians and botanists tend to regard Central Asia and Kazakhstan as one cohesive region from the points of view of natural history and economy, and thus they refer to the Kazakhstan–Central Asian Region, one of the largest natural regions in the world (Tikhonov and Gerasimova, 1990). Unique ancient relict plants, endemic species and even fragments of ancient landscapes are represented in abundance in Central Asia and Kazakhstan. At the same time, this region is one of the richest natural foci of plant genetic diversity, with over 8000 species.

In his numerous works, N.I. Vavilov repeatedly paid special attention to the richness of the Central Asian flora, particularly in the so-called wild relatives of cultivated plants. Vavilov (1965) also stressed that the search for new useful plants should be carried out in that region in the first place. Some wild Central Asian species are known to have directly participated in the development of cultivated plants, as for example *Aegilops* species have in the spontaneous formation of tetraploid and hexaploid wheats.

The N.I. Vavilov Institute of Plant Industry has compiled an inventory of the wild relatives of cultivated plants in the flora of the former Soviet Union (European part, Crimea, Caucasus, Far East, Central Asia). The compiled list includes 613 species belonging to 130 genera of 36 families (Brezhnev and Korovina, 1981). Of these, 304 species (i.e. 50%) of the

Plant Genetic Conservation.
Edited by N. Maxted, B.V. Ford-Lloyd and J.G. Hawkes.
Published in 1997 by Chapman & Hall. ISBN 0 412 63400 7 (Hb) and 0 412 63730 8 (Pb).

wild relatives of cultivated plants come from Central Asia and Kazakhstan. It should be noted that the specific diversity is very unevenly distributed in the region. Thus, the highest concentration of species has been recorded for the foothills and mountainous regions of Tien Shan, Kopet Dagh and the Pamir-Alay mountains.

For instance, the flora of Turkmenia is outstandingly rich and unique, especially Kopet Dagh, the southern mountains of which contain a multitude of species which have never been encountered anywhere else in the world. Kopet Dagh is known to occupy the first place among other mountain systems of Central Asia by the number of wild fruit species present, and south-western Kopet Dagh (dry subtropics) is one of the world foci of type-formation of the most valuable fruit crops characteristic of the dry subtropics. It should be mentioned that the wild *Vitis vinifera* L., which occurs here along rivers and near springs, and the wild *Vitis sylvestris* C.C.Gmel. have been spontaneously crossing with the cultivated varieties that have existed since the times of the Parthian kingdom (250 BC to AD 229), and which have eventually produced the large diversity of forms that currently exists in nature.

The native flora of this region contains four species of pear (*Pyrus boissiariana* Buhse, *P. communis* L., *P. turcomanica* Maleev, *P. regelii* RCHD); the endemic apple species, *Malus turkmenorum* Juz. et M.Pop.; common quince, *Cydonia oblonga* Mill.; six species of almond (*Amygdalus communis* L., *A. scoparia* Spach., *A. bucharica* Kersh., *A. turcomanica* Lincz., *A. vavilovii* M.Pop., *A. spinosissima* Bunge); and seven species of sour cherry, including such rare ones as *Cerasus erythrocarpa* Nevsky and *C. turcomanica* Pojark.

The foothills, the lower mountain belt and most canyons of southwestern Kopet Dagh are rich in *Punica granatum* L. Hundreds of forms that have been found here differ by sugar content, fruit colour, cold tolerance and disease resistance. All this genetic diversity serves as a basis for breeding new crop varieties.

As well as subtropical and other fruit plants, the closest relatives of the cultivated cereal and forage plants occur in Kopet Dagh Mountains. These are *Aegilops cylindrica* Host, *Ae. tauschii* Coss., *Ae. crassa* Boiss., *Ae. juvenalis* (Thell.) Eig, *Ae. triuncialis* L., *Ae. biuncialis* Vis., *Ae. ovata* L., *Secale sylvestre* Hust, *S. segetale* (Link.) Roshev., *Hordeum brevisubalatum* (Trin.) Link., *H. bogdanii* Wilensky, *H. bulbosum* L., *H. geniculatum* All., *H. leporinum* Link., *H. spontaneum* C.Koch, *H. lagunculiforme* (Bacht.) ex Nikif., *Avena clauda* Durieu, *A. eriantha* Durieu, *A. barbata* Pott. ex Link., *A. meridionalis* (Malz.) Roshev., *A. fatua* L., *A. ludoviciana* Durieu, *Sorghum halepense* (L.) Pers., *Vicia fedtschenkoana* V.V.Nikit. (an endemic species), *V. hyrcanica* Fisch. et Mey, *Onobrychis transcaspica* V.V.Nikit. (an endemic species), etc.

The flora of Kopet Dagh is extremely rich in ornamental plants

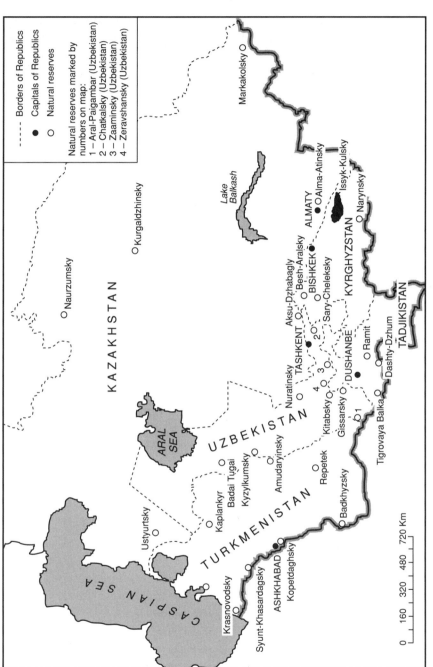

Figure 20.1 Natural reserves of Central Asia and Kazakhstan.

belonging to such genera as *Tulipa* L., *Fritillaria* L., *Eremurus* Bieb., *Hyacinthus* L., *Colchicum* L., etc. The south-western Kopet Dagh is the only place where *Mandragora turkomanica* Mizg., a very rare endemic species, occurs on gravel and stony slopes in the lower part of the mountains.

At present, 30 natural reserves exist in Central Asia and Kazakhstan (Figure 20.1). Most of them (Table 20.1) are located in Uzbekistan, Kazakhstan and Turkmenia. Their numbers are significantly smaller in Kyrgyzstan and Tadjikistan.

Table 20.1 Natural reserves in Central Asia and Kazakhstan

Region	Reserves
Uzbekistan	Badai Tugai
	Kyzylkumsky
	Nuratinsky
	Zeravshansky
	Kitabsky
	Aral-Paigambar
	Gissarsky
	Zaaminsky
	Chatkalsky
Kazakhstan	Ustyurtsky
	Barsakelmes
	Naurzumsky
	Kurgaldzhinsky
	Aksu-Dzhabagly
	Alma-Atinsky
	Markakolsky
Turkmenia	Krasnovodsky
	Kaplankyr
	Syunt-Khasardagsky
	Kopetdaghsky
	Amudaryinsky
	Repetek
	Badkhyzsky
Kyrgyzstan	Besh-Aralsky
	Sary-Cheleksky
	Issyk-Kulsky
	Narynsky
Tadjikistan	Tigrovaya Balka
	Ramit
	Dashty-Dzhum

These reserves feature relatives of many cultivated cereal and forage species, including numerous species of *Secale* L., *Hordeum* L., *Avena* L., *Sorghum* Pers., *Onobrychis* Mill., *Vicia* L., *Medicago* L., *Lotus* L., etc. The reserves are also extremely rich in wild vegetable plants, especially in *Allium* L. species, many of which are endemic ones (e.g. *Allium pskemense* B.Fedtsch., *A. suvorovii* Regel., *A. vavilovii* M.Pop et Vved., etc.), as well as in *Beta* L., *Spinacia* L., *Rheum* L. species, and many others.

Horticultural plant breeders of the world are deeply interested in fruit plants of the region, for instance in the species of *Ribes* L., *Amygdalus* L., *Armeniaca* Scop., *Cerasus* Mill., *Crataegus* L., *Cydonia* Mill., *Fragaria* L., *Malus* Mill., *Mespilus* L., *Punica* L., *Eleagnus* L., *Diospiros* L., *Ficus* L., *Vitis* L., *Hippophae* L., *Pistacia* L., etc., most of which grow in existing natural reserves. In addition, the mountainous forests of the Central Asian natural reserves, especially in the Sary-Cheleksky reserve (Kyrgyzstan), are rich in *Juglans regia* L.

It has been argued that plant conservation *ex situ* leads to the loss of characteristic features in storage and significant input is required to safeguard them; therefore botanists and plant breeders throughout the world are now striving to maintain plant global diversity *in situ*, especially that of the wild relatives of cultivated plants. It is quite natural that, with the ever growing anthropogenic influence on nature, humanity should place stronger emphasis on plant conservation in natural reserves.

The problem of preserving natural reserves and maintaining genetic diversity in Central Asia and Kazakhstan attracts the interest of scientists from many countries and ranks among the major priorities in botanical investigations. The acuteness of the problem is determined by the vulnerability of mountainous and desert bio-complexes. The problem may be solved only through combining ecologically reasonable economic activities with nature protection measures that would include the development of a wide network of protected territories, as well as monitoring the ecological processes taking place in the arid zone. In the pursuit of this objective, natural reserves are to play a part of paramount importance. At present, however, the general status of nature protection activities in the region is depressing. The number of natural reserves, just 30, is wholly inadequate for such a vast region: they cover less than 1% of the total territory; they are located unevenly through the territory; their location has not been scientifically justified; and the areas occupied are unacceptably small. Procrastination with setting up new natural reserves may bring about the total loss of a significant number of wild plant species and irreversible destruction of ecosystems (Dedkov, 1990).

In 1986, botanists from Central Asia formulated major principles of genetic resources protection in the Asian arid zone. For instance, when organizing new natural reserves it is necessary to proceed from the fact that the number of species and their diversity tend to increase from plains towards the mountains, and therefore the number of protected territories should increase accordingly. Special attention should be paid to setting up new reserves in the Turkmenian-Khorasan, Pamir-Alay and Tien Shan mountains as these are centres of origin of many fruit and cereal crops, as well as foci of ongoing species formation, i.e. the 'Vavilov centres'. Unfortunately, when creating a network of natural reserves in the region, this consideration has been little regarded, if at all, and therefore many Vavilov centres were left outside the protected territories.

In compliance with the second principle, it is very important to ensure in the potential protected territory the presence of rare and endemic species, and of those with limited areas of distribution. Quite often, rarity of many species in the region is a result of either uncontrolled felling of forests or heavy overgrazing. Therefore, the goals of natural reserves must be to preserve plant genetic resources; that is, restoration of populations of rare plant species on the protected territories and adjacent areas, and creation of seed banks for these species (Tikhonov and Gerasimova, 1990). A newly created natural reserve should cover an area sufficient for the normal life of ecosystems. Only through adherence to these principles in setting up new reserves and arranging their internal structure would it be possible to preserve the specific diversity in the region. In addition to the existing reserves, most of which need a considerable increase in terms of size, it is necessary to establish 25 new reserves both in the mountains and on the plains (Gounin *et al.*, 1990).

A priority trend in research activities in natural reserves is the monitoring of ecosystem dynamics with the aim of controlling, evaluating and predicting the state of the environment in order to be able to undertake timely nature-protective measures. At present, natural reserves of the world are involved in the implementation of the 'Man and Biosphere' International Programme, and part of the research plans of the Central Asian reserves becomes the search for tests that would reveal disturbances in ecosystems caused by natural processes and/or human activities (Dedkov, 1990).

In general, most scientific programmes under implementation within natural reserves in Central Asia and Kazakhstan are far from being perfect. The current research is concentrated on very narrow, disconnected problems of an applied character. The Turkmenian natural reserves, however, constitute an exception, as they are better organized and coordinated with clear research aims concerned with preserving and restoring natural ecosystems (Gounin and Neronov, 1986).

The results of wide-scale ecological investigations carried out in several biosphere natural reserves in the moderate belt of the desert zone, including Repetek in Turkmenia, have provided a basis for devising a special programme for all natural reserves in the Central Asian desert zone (Gounin and Neronov, 1986). This programme has formulated the major objective for natural reserves; namely, organization of research aimed at developing measures that would ensure protection of genetic diversity. The planned research will include the development of ecological grounds for preserving and restoring rare species and ecosystems, and studying their role in preserving genetic diversity, as well as the development of ecologically based mechanisms of control of the abundance of common species and the maintenance of common ecosystems. A study on the adaptability and evolution of organisms in the intact and disturbed ecosystems along with the monitoring of desertification form an important part of this programme. A new trend in natural reserves research has been proposed with the aim of the development of regional plans for nature protection and reasonable utilization of natural resources. This trend should cover such aspects as the evaluation of the economic potential of ecosystems in the arid zone, compiling of regional medium-scale maps, elaboration of the economic development pattern for the territories in the region, as well as planning the rational use of natural resources.

When the programme was published in 1986, even its authors were far from thinking that it could be implemented in all the reserves. However, it was expected that promulgation of the programme would help natural reserves in the long-term planning of scientific activities and in broadening their international collaboration.

Analysis of literature sources regarding this matter for the recent years (including a monograph, *Natural Reserves of Central Asia and Kazakhstan*; Anon., 1990) has shown not only the complete absence of wide-scale investigations within the framework of the proposed programme, but also a sharp decline in nature protection activities due to insufficient financing and lack of trained staff, laboratory and field equipment, etc.

The human advance on the desert has turned many natural reserves into no more than small islands of wild nature surrounded by cultivated lands and absolutely defenceless against powerful anthropogenic influences. In many natural reserves of the region, the nature protection status is being violated and human economic activities still continue on the protected lands. All this stems from the misunderstanding of true nature-protecting objectives of natural reserves, deep-rooted in the local population at large. To worsen the situation, neither a coordinated plan of developing a network of natural reserves nor a regional research programme has been elaborated so far. Besides, participation in such programmes, as well as in those under the aegis of UNESCO dealing with changes in the environment and with ecological monitoring, is possible only provided the natural reserves will be better staffed and financed, and would have better material and technical support.

In general, the status of genetic diversity protection in natural reserves in the arid zone of the former Soviet Union is at such a low level that procrastination with regard to improving the existing natural reserves and setting up the new ones may lead to total loss of a significant number of species and irreversible degradation of natural ecosystems. It is not only the Central Asian natural reserves; the whole system of nature protection and utilization of natural resources in the region badly needs crucial transformation. Otherwise the destruction of natural bio-complexes cannot be stopped.

ACKNOWLEDGEMENTS

We thank S. Shuvalov for preparation of the map and translation of the text from Russian to English.

21

Plant conservation *in situ* for disease resistance

A. Dinoor and N. Eshed[†]

21.1 INTRODUCTION

It is well appreciated that *in situ* conservation of germplasm resources for disease resistance is needed for wild germplasm as well as for land races. There is no doubt that diversity of sources of resistance is necessary regardless of the protective management systems adopted. It is well documented that a wealth of resistance genes and resistance mechanisms is found in wild plants and land races (e.g. Frankel and Bennett, 1970; Watson, 1970; Harlan, 1977; Segal *et al.*, 1980; Dinoor and Eshed, 1987; Burdon and Jarosz, 1989; Williams, 1989; Hawkes, 1991b; Lenné and Wood, 1991; Alexander, 1992a; Parker, 1992). There is, though, an appreciable controversy about very important issues such as the designation of promising populations and the methodology of sampling, evaluation and exploitation. Some basic concepts in host–parasite relationships are sometimes disregarded through the zest of postulating generalized theories. Descriptions of short-term experimentation (Segal *et al.*, 1980; Alexander, 1992a) are sometimes recruited for the composition of attractive conclusions.

This chapter will present differing approaches and conclusions and examine their basic assumptions in view of basic concepts. It then leads to the discussion of the following conclusions:

1. The performance of germplasm resources *in situ* depends on environmental factors no less than on genetical factors.

Plant Genetic Conservation.
Edited by N. Maxted, B.V. Ford-Lloyd and J.G. Hawkes.
Published in 1997 by Chapman & Hall. ISBN 0 412 63400 7 (Hb) and 0 412 63730 8 (Pb).

[†] Deceased

2. Promising populations should not be *a priori* determined by the physical conditions and/or other non-related markers (e.g. allozymes).
3. One of the most important key factors in choosing appropriate populations is the relevant pathogen selectors to be used.
4. Sites with mosaics of environmental conditions are the most suitable for *in situ* conservation.
5. Appropriate management systems need to be developed for optimal maintenance of target host plants in genetic reserves.
6. The optimal utilization of genetic reserves for breeding resistant crops would be through random collections and centralized targeted evaluation, using relevant selectors.

21.2 PHENOTYPES AND GENOTYPES OF PLANTS AND PATHOGENS IN AND FROM NATURAL COMMUNITIES

Review of the massive literature on resistance of wild plants to disease has convinced the authors that, before any analysis is carried out about the expectations from and benefits of *in situ* conservation, some basic concepts have to be briefly reviewed, to avoid misconceptions.

Resistance of plants and pathogenicity of parasites are genetically controlled. There is ample evidence for the monogenic control of complete, absolute or vertical resistance and pathogens virulence. There is also well-documented genetic evidence for gene-for-gene relationships that operate in these types of resistance and virulence (e.g. Flor, 1942; Person, 1959; Burdon, 1987a; Parker, 1992). The genetic control of quantitative, partial or horizontal resistance of plants and the aggressiveness of the parasites has also been analysed in several systems. The genetic relations between organisms, though, are not very clear, and there is a tendency to assume that clear-cut genetic relationships do not exist.

Since most of the documented evidence for resistance of wild plants to disease deals with monogenic or oligogenic resistance (Parker, 1992), it should be borne in mind that the behaviour of natural communities of hosts and parasites abides by the principles of the gene-for-gene concept (Flor, 1942; Person, 1959). Therefore the distinction between phenotypes and genotypes, regarding resistance and virulence, has much wider scope than in other traits, where dominance, synergism, antagonism and epistasis may play a role. In host–parasite relationships, a phenotype, being as clearly defined as it can be, cannot disclose the genetic background of either the host or the parasite. A resistance phenotype in this case is determined by the presence of at least one resistance gene in the host versus a specific gene for avirulence in the parasite. According to the phenomenon of epistasis in a gene-for-gene system, one such pair of genes for incompatibility will determine the phenotype of resistance regardless of other, compatible gene pairs, that would have brought about an expression as a susceptible phenotype in the absence of that pair for incompatibility. It is also not known how many other pairs of incompatible interactions are involved in the same host–pathogen interaction. Even when the non-genetical factors do not limit disease development, the genotypes underlying the phenotypes cannot be identified in natural surroundings. The genetic background may be resolved by several different methods only under artificial inoculation with single separate isolates (Loegering *et al.*, 1971; Dinoor and Peleg, 1972).

Quantitative or partial resistance is not categorically non-specific. In those cases where some specificity is involved, the predicted performance will probably be akin to that of vertical resistance. However, in those cases where non-specificity is claimed, the effect of the environment will markedly interfere in most cases, and pathogen aggressiveness will not be equally expressed in every location. Therefore, *in situ* performance of non-specific quantitative resistance cannot be indicative of its performance *ex situ* (Dinoor and Eshed, 1987, 1990).

21.2.1 *In situ* performance

In natural communities, several different host genotypes interact with several pathogen genotypes. Each individual host may be parasitized by several genotypes of the parasite. A plant will be diseased even if only one pathogen genotype will be compatible with it. Also, incompatibility in a host, due to several resistance genes, will not be phenotypically evident even if a single compatible isolate develops. This phenomenon is a 'masking' of one phenotype by another. A plant will be healthy if all the pathogen isolates that have access to it are avirulent. Therefore the potential performance (phenotype) of any particular plant in other locations cannot be predicted from its performance in one location. The pathogen's population in the other location may differ in genotype composition and will interact with the host accordingly, producing a different phenotype (Dinoor and Eshed, 1990).

21.2.2 *Ex situ* performance

Plants removed from their natural surroundings may be exposed under field conditions either to the local population of the parasite in the new location, or to mass inoculation of a specific isolate, or a group of isolates. In this type of testing the use of a mixed inoculum is bound to involve the ambiguity of the masking expression and underlying resistance genes will not be identified. Only when exposed to massive inoculation with a specific isolate or isolates will the performance provide more accurate and specific information applicable wherever these strains are found (Dinoor and Eshed, 1987, 1990; Dinoor *et al.*, 1991). Evaluation under isolation will be much more accurate, and will provide information about protection from each of many specific isolates (section 21.4). This type of information would be of much wider benefit since a researcher in one geographical location, in need of resistance, could choose germplasm which will be useful against the specific isolates of that particular region.

21.3 DISTRIBUTION MAPS OF RESISTANCE AND VIRULENCE

Distribution maps have been constructed from several types of data collected: presence of specific resistance (Dinoor, 1970); presence of pathogen races, spatial distribution of diseases and of reactions to diseases along transects (Segal *et al.*, 1980); or in three dimensional fashion (Dinoor and Eshed, 1987, 1990). These maps may be quite useful in gaining a visual perspective of distribution of any mapped trait. They could be quite useful when potential areas for genetic reserves or *in situ* conservation need to be identified. In such forms they are not meant to provide an insight into population structures. If and when distribution maps of several genes for resistance or virulence are constructed for the same populations, only then will an insight into population structures be obtained. It can be envisaged that distribution maps of resistance and virulence for any plant and pathogen population, and especially for genetic reserves and sites of *in situ* conservation, will be very helpful in future utilization of these resources by the international community.

21.4 STRUCTURE OF POPULATIONS AND COMMUNITIES

Populations of progenitors of several different crops have been the main subject of many studies (Nevo *et al.*, 1984, 1985; Manisterski *et al.*, 1986; Burdon, 1987a; Dinoor and Eshed, 1987; Segal et al., 1987; Moseman *et al.*, 1990; Burdon and Jarosz, 1991; Dinoor *et al.*, 1991). Attempts were made to describe their structures as far as types and specificities of resistance are concerned. There is a range of descriptions from resistance to a single specific pathogen biotype, to a bulk of biotypes and even several pathogens. The structures of communities have been of more concern to plant demographers and are not very often described in the context of disease resistance. The issue of resistance-gene deployment has been intensively studied in successful attempts to diversify crops in their disease resistance. In wild plants, this aspect has not been adequately treated. Deployment of resistance types has been described for members of single species populations (Segal *et al.*, 1980; Burdon and Jarosz, 1991;), scattered among other components of the plant community, but not for the community as a whole, where hosts and non-hosts of a specified pathogen are intermingled. More than that, some studies, while attempting to evaluate the impact of population structures on population protection, have reconstructed a composite natural population in a field trial and have actually discriminately miniaturized the original community, roguing out all other members of the community and filling the gaps that existed in nature between members of a population (Segal *et al.*, 1980, 1987). A distance of 1–2 m between samples, with many intervening species and genotypes in between, was miniaturized to 20–30 cm, excluding all the intervening species and genotypes. It would have been interesting to see values of disease incidence of the whole community.

Descriptions of population structures should have details about gene frequencies, gene combinations and genotype frequencies (Burdon, 1987a; Dinoor and Eshed, 1987). An inherent problem in accumulating this information, in particular for disease resistance and pathogen virulence, is the identification of those genes. Genes for resistance are identified by genes for avirulence and genes for virulence are identified by genes for resistance, with the limitations of epistasis to be taken into account. Therefore an appreciable array of pathogen cultures and host lines needs to be available and used separately from one another. Many of the descriptions are only indications of the proportion of resistant plants to one or two or even a mixture of cultures (Dinoor, 1970, 1975; Segal *et al.*, 1980, 1987; Wahl and Segal, 1986). Very few descriptions go into details of genotypes, their reactions to several cultures separately and their frequency (Burdon, 1987a; Burdon and Jarosz, 1991; Eshed *et al.*, 1994). Much more than that is needed for a comprehensive description of population structures.

Another very vague issue relates to the types of pathogen cultures and host lines to be used in determining the genotype components of populations. In the national project in Israel (Anikster *et al.*, 1988) devoted to gaining insight on future *in situ* conservation of wild wheat, resistance to diseases was also investigated (Dinoor *et al.*, 1991). Two hundred and fifty wheat lines established from seed collected along transects were tested for resistance to powdery mildew. Twelve mildew cultures were used: three from local isolates, three from wild wheat elsewhere, three from durum wheat and three from bread wheat. No line was resistant to each of the cultures derived from wild wheat, while 40–60% of these lines were resistant to the mildew from bread wheat (Eshed *et al.*, 1994). This is a major difference, reflecting the importance of the choice of cultures for evaluation.

Populations of plants with long associations with parasites have gone through a pathway of coevolution leaving behind a trail of 'defeated genes'. Genes for resistance were defeated by mutant pathogen strains newly established from time to time through history. On the other hand, there are probably also defeated pathogen genes. We assume that these 'defeated' pathogen genes are avirulence genes that were harmless to the parasite as long as the corresponding host-resistance genes did not exist in the host population. Once the host-resistance genes appeared, by mutation or migration, they curtailed the pathogen population. The defeated genes of both host and pathogen are not only milestones in the mutual struggle for survival (Burdon and Marshall, 1985); they are also an important reservoir of genes that can be used by humans (Watson, 1970; Dinoor, 1981; Dinoor and Eshed, 1983, 1987, 1990; Wolfe, 1993).

In the host, following the trend of using mixed host crops (multi-lines and/or varietal mixtures), hopefully with rotation of genes, there will be benefit from additional genes added to the arsenal. These genes, locally defeated, may not yet have been encountered by pathogens in other regions, and will provide resistance there at least for a while (Dinoor, 1975, 1981; Dinoor and Eshed, 1983, 1987).

In the parasite, defeated genes are genes for avirulence not matched before by the corresponding resistance genes of the host, that were non-existent at that time. These genes for avirulence, which have triggered the selection and establishment of genes for resistance in the host, will restrict the number of host genotypes that may be parasitized by those members of the pathogen's population. In sites of *in situ* conservation the diversity of the parasites will also be preserved (Dinoor, 1981; Dinoor *et al.*, 1991). This would be an important bonus, not only for the maintenance of host diversity but also as a reservoir of cultures to be used for the identification of resistance genes from the same site and from other sites. Alleles both for avirulence and for virulence are necessary tools for identification of resistance genes, differentiating these genes from each other and following them up in breeding programmes.

The term 'defeated genes' cannot be discussed without reference to what is being termed 'cost of resistance' (Smedegaard-Peterson and Stolen, 1981; Burdon and Muller, 1987) and the cost of virulence. This concept, adapted from genetic studies of other traits and integrated in studies of pest resistance, deserves a different approach when disease resistance is concerned. Burdon and Muller (1987) actually tried to measure experimentally the cost of resistance, but they could not come to any conclusions due to apparent association of resistance with some other traits which affect competitive ability and survival. Is there really a cost associated with disease resistance? Resistance to pests involves the constitutive production of metabolites by the resistant plant; this could well involve some cost to the plant when the protection is not needed, and this cost would lower the net productivity of the plant. Many of the mechanisms of disease resistance, especially against obligate parasites (which are the subject of most studies so far on resistance in wild relatives), are induced rather than constitutive. Therefore, if resistance genes are not needed, or not used by the plant, no cost should be involved with their existence. Burdon and Muller (1987) and Burdon and Jarosz (1991) have argued that these resistance genes, that are being maintained in the population without crucial use, have already been selected as 'low cost' or 'no cost' genes. This could well be the case but our assumption is that resistance genes for induced mechanisms have no cost to start with.

The situation in the parasite is different: virulence is associated in most cases with a mutation from dominant avirulence to recessive virulence. The mechanism is of avoidance; the mutant genes for virulence probably do not produce some specified products that the gene for avirulence does. This product, probably of some value to the pathogen, is used by the resistance gene of the host to recognize the presence of the pathogen and trigger the resistance mechanism. The virulent strain does not produce the product and therefore does not trigger the inducible resistance mechanism. This would involve some cost to the virulent pathogen, and strains possessing unnecessary genes for virulence might be less competitive. Some early experiments by Watson (in Van der Plank, 1963) have demonstrated such losses.

In some studies the low frequency of strains with wide virulence has been attributed to low competitiveness. M.S. Wolfe (personal communication) suggested that most of the low-frequency occurrence of the more virulent strains is only due to probability. The probability of several genes being combined is the product of the frequency of occurrence of each gene separately. One could thus claim genuine low competitiveness when the frequency observed is lower than the frequency predicted on the basis of the frequencies of the single genes.

Since most of the local pathogen isolates are virulent on most of the local host lines (Burdon, 1987b; Dinoor and Eshed, 1987, 1990; Burdon and Jarosz, 1989; Dinoor *et al.*, 1991; Alexander, 1992b) many alien cultures should be added for determining resistance. By the same token, many alien host lines should be added for determining the virulence of the local pathogen cultures. A matrix of host reactions to pathogen isolates could then be analysed by a computerized system, based on the gene-for-gene concept (Loegering *et al.*, 1971; Dinoor and Peleg, 1972). To be more accurate and practical, two sets of data should be separately analysed: components of the host population reacting to a few local and to many alien pathogen cultures; and components of the pathogen population performing on a few local and many alien host lines or entries.

Association of disease resistance with allozymes (Nevo *et al.*, 1984, 1985; Moseman *et al.*, 1990) seems very attractive for more general and convenient descriptions of population structures. Nevertheless, resistance is very specific: some plants might be resistant to certain pathogen strains and susceptible to others and other plants might be *vice versa*. Does this association with allozymes pertain to any resistance, or to certain specific resistance? Burdon and Jarosz (1989) have also expressed their doubts whether these associations are really only occasional ones.

Plant communities are for themselves an odyssey of struggle among species and continuous dynamic change towards a climax (Harper, 1977). Management systems should be adapted for maintenance of the target plant species. Lack of intervention and complete protection may lead to gradual decline of diversity leading to elimination of the target plant species (Dinoor and Eshed, 1990).

Future prospects for *in situ* conservation of resistance will largely depend on the choice of appropriate locations and populations. Documentation of the types of resistance present in candidate populations is therefore very important. It has been demonstrated, in selected locations, that monogenic resistance, quantitative resistance and tolerance are common (Segal *et al.*, 1980, 1987). Evaluation methods need not affect the identification of quantitative resistance and tolerance, but they may have a crucial impact on the identification of existing monogenic resistance.

Natural surroundings are conglomerates of mosaics of ecological niches. Therefore, the transect sampling method (Segal *et al.*, 1980, 1987) as a sole sampling of an area has its limitations, in missing out components of the population. Mosaic sampling, ecologically oriented (Dinoor and Eshed, 1987, 1990) would include more components of the diverse populations.

Evaluation of resistance in the population is a key issue. Evaluation *in situ* has triple shortcomings:

- The exposure to a mixed pathogen population will result in 'masking' the resistance of individual plants to at least some members of the mixed pathogens population (Dinoor and Eshed, 1987).
- Resistance to components of the local pathogen population is rare or uncommon. The major value of the local plant population is the resistance to alien biotypes of the pathogen (Dinoor *et al.*, 1991; Eshed *et al.*, 1994).
- The major effects of the environmental conditions will lead to underestimation of the aggressiveness and virulence of the local pathogen population and much more resistance will be claimed, based on healthy plants (Dinoor and Eshed, 1987; Dinoor *et al.*, 1991; Eshed et al., 1994).

Evaluation *ex situ* is being performed either under field conditions or in isolation. Under field conditions, the environmental conditions are quite uniform, and may also be artificially accentuated (e.g. shading: Dinoor and Eshed, 1990; overhead sprinkling, timing of the growing season). The indiscriminate exposure of the local pathogen population still has the shortcoming of masking despite having the advantage of exposure to alien biotypes. Evaluation in isolation is being performed under field conditions or controlled conditions. Massive inoculation in the field will ensure that only specified pathogen biotypes are used. If single pathogen biotypes are used, the masking effect is avoided as well. The remaining limitation is space; not many single pathogen biotypes can be handled concomitantly in the field. In controlled conditions, most limitations are avoided and many pathogen biotypes may be used separately. The major factor of concern now is the choice of pathogen biotypes. These can be specifically targeted: evaluation may be performed in different problem areas using the local problematic pathogen biotypes (Dinoor, 1975, 1981) and not those isolates found in the natural population (Williams, 1989).

The recommendations put forward for *in situ* evaluation for resistance (Lenné and Wood, 1991) are in sharp contrast to the above situation and considerations. In areas of low disease incidence and severity, resistance *in situ* will hardly differ from susceptibility. In areas of medium to high disease incidence, the responses of hosts *in situ* will be to a mixture of local isolates which might, in most cases, not be relevant to the target areas for resistance breeding. In addition, under infection with a mixed pathogen population, virulent types will mask any resistance. Lenné and Wood (1991) have actually misinterpreted the data presented by Dinoor (1970) which they thought supported their recommendations. Resistance identified and reported was not from *in situ* evaluation, but rather from adult stage evaluation under mass inoculation, in isolation, by specific cultures.

There would be no limit to the scope of population structure description. The more pathogen biotypes are used, the wider will be the scope of description, and the description of diversity will be more comprehensive. This in turn, will be of much more help in assigning appropriate sites for *in situ* conservation.

21.5 PERFORMANCE OF NATURAL PLANT POPULATIONS *IN SITU*

Descriptions of diseases in natural plant populations indicate that diseases develop very often but they seldom reach epidemic proportions (Harlan, 1977; Segal *et al.*, 1980; Dinoor and Eshed, 1984; Burdon and Jarosz, 1988, 1989; Alexander, 1989; Kranz, 1990; Dinoor *et al.*, 1991). This situation was generally related to the status of dynamic equilibrium between the defending hosts and the attacking parasites, achieved after a long-lasting coexistence. Only later have researchers become more and more aware of the involvement of the environment, or the physical conditions affecting disease development and thus affecting the performance of the plants (Dinoor and Eshed, 1987, 1990; Jarosz and Burdon, 1988; Burdon *et al.*, 1989; Paul, 1990). The involvement of that third party of the triangle (host–pathogen–environment) is especially conspicuous *in situ*, when marked differences show up between adjacent parts of the population of a plant species living in clearly different but adjacent ecological niches and sometimes in niches that are probably but inconspicuously different (Dinoor and Eshed, 1987). The physical factors involved are, for example, the aspect of the slope, depth of soil, proximity of rocks, distance from a beach, and shade of trees and bushes.

Underneath a tree and around it, at least four niches can be identified (Dinoor, 1962; Dinoor and Eshed, 1990). These involve the combinations of illumination and canopy projection. In the area under the tree, heat radiation from the ground to the atmosphere is greatly reduced. Therefore, it is warmer at night, the foliage of the herbaceous cover does not cool enough for dew deposition and leaf surfaces are relatively dry. The area under a tree may also be covered by shed leaves, which might, allelopathically, affect the herbaceous undercover. The area shaded by the tree is determined by the angle of the sun irradiation and does not overlap the area under the canopy projection. The four niches will therefore be:

(a) sunny, unprotected (wet and cooler at night, warmer during the day);
(b) sunny, protected (dry and warmer at night, warmer during the day);
(c) shaded, protected (dry and warmer at night, cooler during the day);
(d) shaded, unprotected (wet and cooler during the night, cooler during the day).

In a population of wild barley, *Hordeum spontaneum*, growing in this habitat, a clear difference may be seen especially between niche types (c) and (d). In niche (c) powdery mildew dominates and reaches epidemic proportions while in niche (d) net-blotch dominates and reaches epidemic proportions. This clear differentiation is due to the ability of mildew to develop under dry conditions while the net-blotch fungus (*Drechslera teres*) needs free water on the leaf surface for sporulation, spore germination and host penetration. On the other hand, the net-blotch fungus has a shorter incubation period and, being necrophytic, it competes severely with mildew. Moisture limitations favour mildew development.

The effects of the environment on the spatial distribution of diseases can be validated by transplant experiments *in situ* (Harper, 1977). Healthy-looking plants from disease-free niches and diseased plants from infested niches, or seeds from them, may be planted side by side in both types of niche. Natural disease development would discriminate between host resistance and environmental effects. This was done for the system of *Limonium* and rust at two different locations, and the results were clear cut, pointing out the overriding effects of the environment (Dinoor and Eshed, 1987).

Harper (1990) expressed his doubts about the role of plant diseases in determining the composition of plant communities, since most associations of plants and diseases show very low disease severity and incidence. We are actually faced with a state of dynamic equilibrium among at least the three main factors involved, namely plants, pathogens and the environment. The pathogens are here, probably at their optimal state of balance with the host and the environment. A reflection of what could happen to the host population when the delicate balance is disrupted may be gained from at least two situations of imbalance:

1. **New weeds and weed control by pathogens**. Plants like *Chondrilla juncea* or *Rubus* spp. may not be considered as weeds in their homeland. Once introduced into other continents, free from their pests, pathogens and competitors, they have become very severe weeds. Rust diseases of both plants do not seem to be a devastating problem in their home countries. In the case of *Rubus* rust, it is very rare or even hardly seen in Germany, for example, where *Rubus* is a component of some forest communities. When these rusts were introduced into the new habitats, where their hosts became weeds, they were successfully established and controlled the weeds (Oehrens, 1977; Delfosse *et al.*, 1986).
2. **Invasion of wild plants into habitats and seasons beyond their optimal and prime performance**. Prosperous wild plants like *Avena sterilis* or *A. barbata*, propagating nicely and producing large amounts of seed under moderate rust epidemics, may invade other habitats like fields and gardens. In some of these situations, devastating epidemics have broken out, especially when the season was extended by water supply, and the whole population in this habitat collapsed and did not produce any seed (Dinoor, unpublished).

Removal of the pathogen or enhancing it, as suggested by Harper (1990) and Alexander (1992b), could well demonstrate a possible role of pathogens. This would need long-term experimentation, since the effects would need to be accumulated over seasons and would involve also long-term interhost genotype competition, mediated by the diseases. Since pathogens are specific and fungicides less so, enhancement of pathogens would seem more applicable (which was the case of biological control of weeds referred to above), while suppression of them by fungicides might be complicated and non-applicable, since they may affect other diseases of other plant components of the community.

21.6 PROSPECTS FOR CONSERVATION OF DIVERSITY IN HOST–PARASITE RELATIONSHIPS

Analyses of host–parasite relationships in natural populations of wild plants have come up with different explanations and theories about the dynamic equilibrium in those systems, how it was achieved and what evolutionary processes are involved. When the time comes, and the genetic reserves or parks are planned (Frankel and Bennett, 1970; Frankel, 1974; Jain, 1975; Dinoor, 1981; Dinoor and Eshed, 1984; Hawkes, 1991b;

Eriksson *et al.*, 1993), what guidelines will plant pathologists formulate to help to determine and assign the most appropriate sites for these reserves and parks? Several reviews have addressed the question of where to search for resistance. Similar reasoning would be applicable to the question of where to establish *in situ* conservation parks. These reviews came to the conclusions that the most promising locations are those where diseases are widespread, where the selection pressures can physically be observed. How does this conclusion fit with the other conclusion that disease levels are low in most wild populations? Conversely, where are the resistance genes if the source population is expected to be heavily diseased? The ecologist's view of Harper (1990) is based on the fact that the levels of diseases in natural plant populations are very low. He expressed his doubts as to whether pathogens and diseases play any important role in natural communities.

Burdon and Jarosz (1989) expressed their reservations as a result of collecting efforts in areas environmentally favourable for disease development. Our own experience with some natural systems (Dinoor and Eshed, 1987, 1990) does not conform with the generalization that the search for resistance should be directed only to highly diseased populations (e.g. Manisterski *et al.*, 1986; Burdon, 1987a). In the Ammiad project in Israel (Dinoor *et al.*, 1991; see also Chapter 15), diseases of wild wheat were very scarce or even absent, yet under *ex situ* tests the plants were susceptible to the local mildew isolates and to other mildew isolates from wild wheat (Eshed *et al.*, 1994). At the same time 40–60% of them were resistant to alien cultures from bread wheat (in that crop in which the resistance is actually needed). We disagree with the scheme, advocated by Lenné and Wood (1991), that emphasizes *in situ* evaluation for resistance.

There are five main steps that we would suggest for urgent operational projects, to rationally determine appropriate sites for optimal *in situ* conservation for disease resistance:

1. Representative sampling of the progenitor plant species from natural populations (Allard, 1970; Zohary, 1970; Marshall and Brown, 1975; Chapman, 1989). Priority populations should be determined by the following considerations:
   - sites which could most likely be set aside for gene parks;
   - sites with a mosaic of ecological niches, concentrated in relatively small areas (Burdon *et al.*, 1989; Robinson and Quinn, 1992);
   - sites at risk of destruction;
   - sites where vegetation management will be in accord with other potential uses (controlled grazing, nature reserves, national parks, recreation areas, etc.).
2. Propagation of the plant material.
3. Evaluation of the propagated material in relevant research centres,

scattered worldwide, using separately, in each place, local biotypes of the pathogen concerned. A standard set of plant lines with known resistance should accompany the tests.

4. Integration of the results from the different testing centres, into profiles of population structures.

5. Determination of site priorities based on the criteria used in step 1 and the diversity and specifications described in step 4.

For long-term activities, not pressed by urgent rescue operations, more comparative research is needed. Population structures should be more comprehensively determined for populations in different ecological niches, for different population management systems and regimes and for plants having different propagation and dispersal systems.

In memoriam to Dr Nava Eshed

Dr Nava Eshed was a daughter of an agricultural pioneer family. She was herself a pioneer, among the founders of an agricultural community (kibbutz) in the desert. From a vegetable grower she turned into a student of agriculture, specializing in plant protection and field crops. She then joined the staff of the Department of Plant Pathology and Microbiology where she taught the full specialized course 'Ornamental Plant Pathology' and the laboratory course 'Introductory Plant Pathology' for which she also produced the detailed, illustrated laboratory manual (in Hebrew). Her research embraced several topics: the genetics of physiological specialization into physiologic forms; diseases and resistance in natural plant communities and biological control of weeds by pathogens.

Dr Eshed concentrated in her theses on the complicated topic of physiologic specialization of pathogens in wild plants. She showed for both powdery mildew (*Erysiphe graminis*) and crown rust (*Puccinia coronata*) that specificity of pathogens is very strict in cultivated plants but very vague in wild plants. While being specific on cultivated species, strictly specific physiologic forms of pathogens have several common hosts in the wild. This opens up new frontiers for recombination and adaptation to more hosts. Through a very thorough and fundamental genetic study of crown rust forms, she managed to produce new recombinant pathogens with a wider host range. Such hybrid pathogens are a potential theoretical threat to novel ideas in breeding for disease resistance. However, Dr Eshed also showed that these new types, exhibiting a wider host range, are less competitive than their narrow host range parents. Her genetic work with the physiologic forms of crown rust was unique: no one else has succeeded in successfully conducting this type of work on rust genetics.

Dr Eshed as a person is an interesting and admirable story. Apart from her extreme dedication to research and teaching, she intensely loved nature and people. Travelling countrywide and worldwide did not satisfy her love of colours, shapes and different modes of behaviour. She was always excited by sunsets, spring and autumn colours, and sought out partners to join in celebrating the beauty of nature. With all that, she always found time not only to pay attention to the people around her but also to be come involved with their problems, doubts, joys and sufferings. She offered her help to everybody, without having to be asked to do so.

May her memory be blessed.

Part Four

Discussion

22

A practical model for *in situ* genetic conservation

N. Maxted, J.G. Hawkes, B.V. Ford-Lloyd and J.T. Williams

22.1 INTRODUCTION

Having discussed specific aspects of genetic reserve and on-farm conservation and provided case studies written by those actively engaged in conserving genetic diversity *in situ*, we would like to draw some overall conclusions concerning the practical application of this conservation strategy. Of the two basic strategies for genetic conservation discussed in Chapter 2, the majority of research activity has been focused on *ex situ* genetic conservation; relatively little progress has been made in developing scientific principles appropriate for *in situ* genetic conservation. Since the mid 1980s, when *ex situ* conservation of world crops had been largely achieved, attention began to switch toward the *in situ* conservation of plant genetic diversity.

Increased professional and public interest in the conservation of biodiversity since UNCED and the legalities concerning sovereign rights over genetic resources means that the time has long passed for debate and academic discussion. There is now an urgent need to formulate sound scientific principles for the conservation of populations in their natural surroundings and, where appropriate, in their agricultural systems. We need to provide the scientific basis for *in situ* conservation of wild species which are related to a greater or lesser degree to crops, and which may well be needed by humankind in the future. To conserve genetic resources *in situ* efficiently, we need to build on the experience of

Plant Genetic Conservation.
Edited by N. Maxted, B.V. Ford-Lloyd and J.G. Hawkes.
Published in 1997 by Chapman & Hall. ISBN 0 412 63400 7 (Hb) and 0 412 63730 8 (Pb).

ecologists and population geneticists, but first the biological processes involved must be clearly defined and the procedures modelled.

22.2 DEFINITION

Definitions in the literature and even in the Convention on Biological Diversity are unclear, because they confuse two distinct techniques: genetic reserves and on-farm conservation. In part this lack of clarity is due to *in situ* conservation being applied to wild species on the one hand and to domesticates on the other. Also the use of genetic reserves as a term is not followed by all; synonymous terms include genetic reserve management units, gene management zones, gene or genetic sanctuaries, crop reservations and other terms. The basic unit of conservation can be understood by reference to Jain (1975) and also Williams (1991). The essence of *in situ* conservation is that germplasm is conserved in the locality where it is currently found, either where it is naturally located or where it has developed distinctive traits under cultivation. In this context the definition does not include transportation of the germplasm to a distant location to facilitate conservation. If we are to avoid confusion, however, it is important to distinguish between these two distinct *in situ* techniques. The following working definitions are proposed:

- **Genetic reserve conservation** – the location, management and monitoring of genetic diversity in natural wild populations within defined areas designated for active, long-term conservation.
- **On-farm conservation** – the sustainable management of genetic diversity of locally developed traditional crop varieties, with associated wild and weedy species or forms, by farmers within traditional agricultural, horticultural or agri-silvicultural cultivation systems.

Of these two basic techniques, on-farm conservation is less directly under the scientific control of the conservationist. The conservationist's responsibility is to promote and preserve the conditions in which the traditional farmer can maintain genetic diversity in land race varieties and related crop weeds within the traditional production systems employed. However, conservationists do not actively intercede in the act of conservation, although they need to keep a watching brief on the process. In contrast, genetic reserve conservation involves a more active role for conservationists, who positively intervene to promote the conservation of the target taxon. It is the details of this latter intervention that currently require urgent clarification.

A distinction can be made between 'active' and 'passive' *in situ* conservation. Plant species are undoubtedly conserved in numerous environments unlikely to be considered genetic reserves, such as areas of wasteland, field margins, primary forest, even national parks, but in each of these cases the existence of particular species is coincidental, therefore passive and not the result of active conservation management by humans. These populations are not actively monitored and, as such, are more vulnerable to extinction, i.e. it is unlikely that any deleterious environmental trend would be noted and counter-measures adopted. In this sense active conservation requires positive action to promote the sustainability of the target taxa and the maintenance of the natural or artificial (e.g. agricultural) ecosystems which contain them, thereby implying the need for associated habitat monitoring, management and protection.

22.3 TYPES OF ON-FARM CONSERVATION

It should be noted, as already proposed in Chapter 2, that the conservation of genetic diversity on-farm in traditional cultural systems involves various distinct types of germplasm, each of which is associated with a different agro-environment. In fact, there are at least three types of on-farm conservation. Two of these are associated with the genetic resources of domesticates and one with the wild and weedy species related to crops or ones from which the crops themselves were almost certainly derived.

22.3.1 On-farm conservation of seed crop species

These exist and have evolved in a human-controlled environment and consist of seed and grain crops, vegetables, forages and fodder species. These, of course, have been enhanced to a degree by farmers and further improved by plant breeders, but the strains we are concerned with here are what used to be called 'primitive forms' but are now generally referred to as 'land races'. These were not derived from purposeful breeding but are largely the result of repeated unconscious selection. Such land races are mega-populations (often composing numerous individual populations), containing great genetic diversity in some cases, less in others. They exist under human and environmental selection; they may or may not be very old, but are evidently maintained in balance with environmental and human needs, as well as with insect and other pests and viral, bacterial and fungal pathogens.

It is essential here to collaborate with farmers and their families so as to ensure that the traditional varieties or land races are conserved in perpetuity. Some sort of convention or written agreement may be needed, particularly if the farmers or communities are at the same time growing new highly bred varieties. A linked but distinct activity amongst the farmers is that of breeding special land races or selection of particular races from those grown originally. Government officials concerned with crop diversity conservation should be appointed to inspect fields and harvests to ensure that the traditional farming practices are maintained and to pay compensation where needed (Chapter 18).

22.3.2 On-farm conservation of vegetatively propagated crops

These also have evolved in a human-controlled environment and include vegetatively propagated staples such as potato, sweet potato, yam, cassava, taro, *Xanthosoma* and a range of other minor crops, such as *Curcuma*, Jerusalem artichoke, yam-bean and various indigenous tuber crops in the Andes of South America, as well as one (*Plectranthus*) in Ethiopia. We should also include here *Musa ensete* and a range of mostly tropical plants that are maintained as vegetatively propagated cultivars, especially fruits.

As with seed crop on-farm conservation, it is essential to collaborate with farmers and their families to ensure that the traditional genotypes are conserved. With this type of material detailed descriptions and photographs might well be made in order to verify that the same genotypes are continuing generation by generation. Whole villages and communities might well be involved in the maintenance of a larger range of genotypes than would be possible for one single farmer. A certain amount of duplication of genotypes might be advisable in most instances (Chapter 19).

The maintenance and propagation of these materials is probably well known in each individual case. However, we are not so clear about how to deal with them on a scientific basis. They are certainly not populations or land races in the sense that the seed and grain crops are, but it would seem that they are populations in a wider sense or relics of these. Many of these species retain sexual fertility (e.g. potatoes in the Andes) so that new recombinants can occur. Farmers in many regions maintain a whole range of 20 to perhaps 50 or more clones, all in one field or a series of fields, each clone being given a distinct name. Although at first sight they would seem to be very different from the seed and grain crops, evidence now available shows that these vegetatively propagated crops are also maintained by the farmers in more remote regions under the pressure of environmental selection (frost, heat, drought, etc.) and by their pests and diseases, with which they maintain an equilibrium of resistance or at least a tolerance at the levels needed for their survival. We might introduce the idea of conservation of vegetatively maintained crops by whole villages or communities, rather than by individual farmers. This has already been described at a recent conference of potato breeders in Bolivia (Hawkes, unpublished) and was, in August 1995, inherent in a decision at a south-east Asian regional symposium to focus on integrated community-based management of genetic resources.

22.3.3 On-farm conservation of wild or semi-cultivated crop relatives dependent on man-made habitats

The importance of these crop relatives was realized by Vavilov even in the early years of the twentieth century, particularly in their ability to exchange genes with the crops themselves. In some instances species closely related to crop species may represent ancestral forms from which the crops evolved; in others they may be closely related weedy forms or other related species, but not the ancestors of crop plants. It may often be difficult to distinguish which category such forms belong to, and it may not be of great importance in terms of their practical conservation.

These are the weedy or ruderal species which are unable to survive under natural habitat conditions and need open areas amongst crops, around dwellings and by walls, hedges, path sides and roadsides for their survival. In the general literature of plant conservation these materials, which are often classed as varieties or subspecies of the crops themselves, are omitted or considered of little value. Yet it is often to these materials that plant breeders need to turn in order to introduce valuable genetic materials into their crops. Thus, a strategy for conservation of these species must also be considered.

In this case, the plant may have an obligate relationship with a traditional agricultural system; for example, it may be a weed of traditional cereal fields. If the system changes (e.g. herbicide spraying is introduced), it will result in the decline in the local populations of the weed taxon. The objective, in this case, would be to maintain the traditional agricultural system, prohibiting herbicide spraying, thus maintaining the taxon as a weed of cereal fields or borders. Conservation of the target taxon would be by indirect on-farm methods, promoting traditional agricultural systems, rather than establishing a specific genetic reserve (Brush, 1991a). Examples of this form of indirect conservation through traditional farming systems are provided by *Solanum sparsipilum* growing in fields of *Solanum tuberosum* in southern Peru and northern Bolivia and *Vicia narbonensis* growing in fields of *Vicia faba* in southern Syria. Figure 22.1 shows a weedy field of wheat growing in Azerbaijan; this field was visited during a forage legume collection mission and was found to contain 37 cultivated or wild forage legume species. Vavilov (1962) refers to weedy rye species infesting in a similar manner local cultivated *Triticum turgidum* in Afghanistan and barnyard millet (*Panicum crus-galli* L.) in rice in Iran. These examples require the maintenance of relatively low agricultural productivity, which raises the whole question of provision of incentives.

In many cases such materials can be well conserved in *ex situ* collections and the degree to which *in situ* conservation is needed has to be determined, since in these cases these materials would be conserved

Figure 22.1 A weedy field of wheat containing 37 cultivated or wild forage legume species.

under traditional agricultural systems as weeds of existing crops or in suitable habitats round field borders and waste places.

22.4 CLARIFYING METHODOLOGIES FOR *IN SITU* GENETIC CONSERVATION

With increasing levels of genetic erosion there is currently a critical need to develop and apply detailed *in situ* conservation methodologies. However, in terms of on-farm conservation it is inherent within our definition that a single methodology cannot be developed. By and large, farmers are the ones who ultimately undertake the conservation and not the scientists, and each farmer in any part of the world must act according to his or her own traditions and practices. The degree, if any, to which scientists or politicians intervene in the process of on-farm conservation varies between different parts of the world where different national programmes and political regimes exist. Whatever the degree of such intervention, there will be a continuing moral obligation on national programmes to maintain back-up collections of *ex situ* materials to provide farmers with replacements if, for some reason, the latter lose them. It could be argued (Chapter 17) that the more intervention which takes place, the greater the danger of disrupting the traditional cultural practices and the natural processes of conservation. This would appear to be more likely, given our general lack of understanding of what goes on in traditional agricultural systems in terms of genetics and evolution. Our scientific lack of understanding has recently been underlined by population dynamic and genetic studies undertaken by the International Rice Research Institute on rice land race populations in non-intensive agricultural systems in south-east Asia and potato and barley land races elsewhere (Brush *et al.*, 1995).

It is difficult to propose an overall methodology for on-farm conservation because the conservationist is not actively involved: it is the farmer who undertakes the conservation and the mode of conservation used by traditional farmers is poorly understood. It is obviously true, however, that if traditional agriculture disappears from regions with high concentrations of crop relatives, then the crop relatives dependent on the traditional cultural system will also disappear. Therefore any general methodology for on-farm conservation would involve the maintenance and promotion of traditional cultural systems and the continuing conservation of the germplasm conserved within those systems.

22.5 A METHODOLOGY FOR GENETIC RESERVE CONSERVATION

It is somewhat easier to propose a generalized methodology for conservation in a wild species genetic reserve. Any methodology would involve the conservationists using their knowledge of genetics, ecology, geography, taxonomy and other disciplines to locate, monitor and manage the genetic diversity they wish to conserve. But it is the application of these disciplines to the maintenance of natural wild populations that must be made explicit, if we are to develop and apply a detailed methodology for the conservation of genetic diversity within a reserve.

The conservation of germplasm of a particular taxon, whether *in situ* or *ex situ*, necessarily involves only a section of the total gene pool, as the practical resources available for conservation are always finite. It is therefore impossible to conserve the entire gene pool unless it is very small indeed. The conservationist's objective is to ensure that the maximum possible range of genetic diversity is represented within the minimum number and size of *in situ* genetic reserves. This is, however, a complex goal to achieve because detailed information on the amount of genetic variation, population structure, breeding system, habitat requirements and the geographical distribution of the target taxon are required and these data are commonly unavailable for even the most well studied groups of crop relatives. The process of finding optimum locations for genetic reserves is further complicated, because reserves will seldom be set up with the aim of conserving a single species. The target will often be a group of species, each potentially with differing ecogeographic requirements. The broader the target (genus, tribe or even family), the more numerous the taxa included. With the application of niche theory, the survival of each taxon is likely to depend upon the availability of diverse ecogeographic niches. If diverse taxa are to be conserved together then there is even less likelihood of the required background biological data being available to formulate an effective conservation strategy for all the species included.

Considering these problems and in conjunction with the conservationist's central objective defined above, the primary goal of a general methodology for genetic reserve conservation would be to locate, monitor and maintain diverse populations of the target taxon within specifically designated wild habitat. Hawkes (1991a) outlined some basic requirements for establishing a genetic reserve for crop plant relatives. These requirements have been developed into the model for genetic reserve conservation proposed in Figure 22.2. It should be noted that this specific model for genetic reserve conservation fits within the general model of plant genetic conservation proposed in Chapter 2. Therefore, before a target taxon is conserved in a genetic reserve, several steps in the conservation process must have already been undertaken: it must have been decided that the target taxon is of sufficient interest to warrant active conservation, an ecogeographic survey or a survey mission must have been undertaken, and the particular conservation objectives and appropriate strategies outlined. Within the latter point it must be established that conservation in a genetic reserve is appropriate. Finally, once the gene pool is conserved in the genetic reserve, a scheme that makes the material available for current and future utilization must also be approved.

The model for genetic reserve conservation is divided into three phases: reserve planning, reserve management and reserve utilization.

22.6 RESERVE PLANNING AND ESTABLISHMENT

Before commencing the conservation procedures associated with *in situ* conservation, those mandated to carry out the conservation must select the target taxa (Chapter 3) and develop a clear strategic plan, which states the objectives of the conservation activities. As a prerequisite or as part of this, there will be a need to review the target gene pool and to undertake an ecogeographic survey or preliminary survey mission (Chapter 4). The ecogeographic survey should conclude with a clear, concise statement of the proposed conservation objectives and priorities, and should identify appropriate strategies and methods for their implementation.

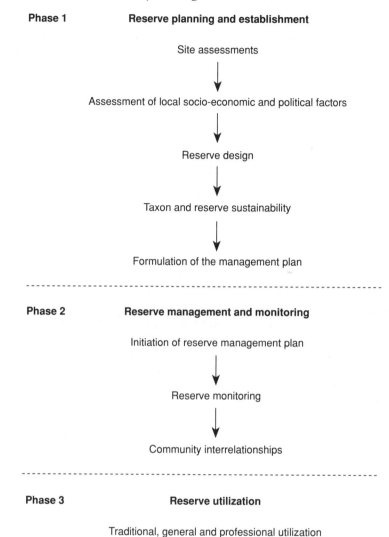

Phase 1 **Reserve planning and establishment**

Site assessments

Assessment of local socio-economic and political factors

Reserve design

Taxon and reserve sustainability

Formulation of the management plan

- -

Phase 2 **Reserve management and monitoring**

Initiation of reserve management plan

Reserve monitoring

Community interrelationships

- -

Phase 3 **Reserve utilization**

Traditional, general and professional utilization

Linkage to *ex situ* conservation, research, duplication and education

Figure 22.2 Model for genetic reserve conservation.

22.6.1 Site assessments

Although ecogeographic techniques, especially when used in conjunction with geographical information systems, will identify broad areas of genetic diversity where genetic reserves could be sited, they will not provide the precise location for reserves. The ecogeographic survey may suggest one relatively large reserve area or multiple reserve areas. To establish which is most appropriate, the potential sites will require surveying. This will commonly involve assessment of taxonomic diversity (numbers of taxa and variation within taxa) and will increasingly involve assessment of genetic diversity using molecular techniques (Chapter 12). This allows assessment of the variation present at a particular site, as well as comparison between alternative sites. The overall criterion, established in the core objective of genetic conservation, is to maximize the conserved genetic diversity of the various taxa and actual population sizes, as it will not always prove possible to determine the minimum effective population size, and hence the minimum viable population size necessary for conservation, in advance of conservation.

To facilitate security of conservation into the foreseeable and long-term future it is advisable for any target taxon that more than one reserve is established and that the reserves should be deliberately selected to complement each other. This will permit the conservation of any ecotypes present in the target taxon and enhance the conservation of the entire gene pool. Sites should be selected to encompass the widest possible range of ecogeographic conditions in which the target taxon is located.

When considering potential sites for establishing genetic reserves it would be advantageous to link these to other ongoing conservation interests. In the wider conservation movement the types of protected areas are fairly well understood as defined by the IUCN Commission on National Parks and Protected Areas and also by international networks and conventions. As far as possible, genetic reserves should be designed within this general framework. If the ecogeographic data suggest locating the reserve in a particular area, it may be that the area or a closely adjacent area contains an existing protected area, such as a nature reserve or national park. In this case the management plan could be adapted to permit genetic conservation in a portion of the wider reserve, therefore saving some of the costs of genetic reserve establishment and maintenance.

The ecogeographic survey should conclude with a clear statement of the habitat requirements of the target taxa (Chapter 4). Combined with the relative assessment of alternative sites, the habitat requirements of the target taxa will need to be reviewed and compared with those provided by the potential sites being considered. If the predictive value of the ecogeographic survey is high (and this is dependent on the quality and quantity of the ecogeographic data collated), then the habitats of the potential sites will match those required by the target taxa and the most suitable sites will contain robust populations of the target taxa.

The relative cost of reserve establishment will also affect selection of alternative sites. Faced with a choice of potentially equally suitable sites and differing establishment costs, there would be little justification for selecting a site other than the least expensive. The same logic would apply to the running costs of the reserve once established. This will necessitate the application of some form of cost–benefit analysis prior to actual reserve selection.

This is one of the largest 'unknowns' in promoting this type of conservation. Unless there are indications that materials conserved *in situ* will be used commercially, the justifications remains moral and scientific, which is unlikely to convince policy makers to fund reserves 'in perpetuity'. However, it should be noted that past experience has often shown that wild species once considered commercially 'useless' have proved on further examination to be 'useful', because they contain resistance genes (Hawkes, 1990: 206–210) or pharmaceutically active compounds, for example. Thus some economic component can be built into the commercial utilization equation of even wild species of no immediate utilization value, though it remains difficult to quantify how much. The Convention on Biological Diversity includes the concept of 'incentive measures' embracing such things as subsidies and tax exemption. In this way governments might pay landowners subsidies to manage their land in particular ways. However, this will only be feasible in the long term, as will need to be the case for genetic reserves, if benefits are expected and ultimately demonstrable – for example, that material from reserves is used in the production of disease-resistant and pest-resistant commercial varieties, or that the reserve does provide recreational value.

The reserve site must also be sustainable as a reserve for the foreseeable future. The reserve must be accessible for the reserve manager and potential germplasm utilizer and it must be secure from deleterious changes in human or natural (e.g. increased or decreased levels of fires or wild grazing animals) intervention.

22.6.2 Assessment of local socio-economic and political factors

Constraints ranging from economic to scientific and organizational will affect the implementation of the model for *in situ* conservation. The simplest way forward in economic and political terms is for countries to take action on establishing a series of national parks or heritage sites, since in this case there is likely to be some benefit to the people of the countries concerned, and in many instances the benefits may be more apparent when there is a strong forestry sector. Such areas are often managed in a minimal way, using normal forestry practices. In many cases these areas can also include genetic reserves, even when the prime designation of the area is for broad ecosystem conservation.

There are clearly major costs involved in delineating and appropriating broad reserve areas, and many reserve area networks are voluntary, while other networks do have a legal obligation. Even though moral and legal obligations may have been recognized, government policies will often need adjustment because many such policies provide incentives to mismanage the environment, especially in the agriculture and forestry sectors. At present, with widespread debate on trade protection, external debt and other international problems, developing countries will need to produce national development strategies incorporating economic and environmental objectives, while all countries are being required to promote sustainable development.

Soundly based *in situ* conservation of crop gene pools with intensive management may well show only slow progress, but two actions will be clearly helpful: firstly, in terms of economics, the availability of outside funding for the initial planning, whether from national donors or the Global Environmental Facility; and secondly, in terms of organization, the widespread viability of standard blueprints and summaries of organizational needs to national governments and for these to be part of the national biodiversity action plan.

If genetic material is going to be used it will need to be available to some extent from *in situ* reserve areas. The Convention on Biological Diversity states that access needs to be determined by sovereign governments but contracting parties should facilitate this access for environmentally sound uses, subject to prior informed consent and on mutually agreed terms. Any scientific research on *in situ* genetic resources should therefore include participation of the provider and be conducted in the providing state. Furthermore, the users of the resources should share in the results of research and development and the benefits arising from possible commercial utilization.

Many plant materials which require conservation *in situ* will exist in reserve areas that transcend the national boundaries of sovereign states. The principles related to environments and ecologies, and linkages between *in situ* and *ex situ* conservation, discussed above, are particularly relevant and need to be built into intercountry collaboration. Constraints could be overcome by promoting regional activities, and the process would be helped by organizations representing states developing the planning across national boundaries. The interdependency of nations in crop improvement makes it imperative that constraints are dealt with quickly. However, plans for *in situ* conservation with uncosted and only potential long-term benefits will not form a basis. More than ever the partnership of scientists and international organizations is needed to provide leadership in this area.

Finally, when *in situ* conservation work becomes ongoing, there will be a need for adequate curation of the resources in the *in situ* reserves. Internationally, trained human resources will need to be identified to conserve and manage the genetic materials *in situ* (these professional staff are currently in short supply or unavailable). Otherwise the availability, utilization and flow of benefits resulting from conservation will be constrained.

22.6.3 Reserve design

The previous stages of the model have highlighted how the reserve site is selected, but now having identified the potential site, we must design the reserve appropriate for the target taxon at that location. This involves consideration of various factors, such as structure, size, whether a single large or multiple smaller sites are best for the target taxon, the use of corridors, reserve shape, environmental heterogeneity and potential user communities. Each of these factors is associated with a copious and controversial literature (Chapter 8), but it is possible to draw some overall conclusions.

The current consensus view of reserve structure is that based on the Man and the Biosphere programme (UNESCO), as discussed in Chapter 8. This establishes a central core area with a stable habitat, surrounded by a buffer zone and outside this, where possible, a transition zone shielding the reserve core from general areas of human exploitation. This plan assumes that the core area is sufficiently large to accumulate 1000–5000 potentially breeding individuals. Rather than focus on the actual size of the reserve, it is more objective and appropriate to target the ideal numbers of individuals that form a viable population – the effective population size – that will ensure the effective conservation of genetic diversity in the target species for an indefinite period.

The reserve design debate is often centred on the relative advantage of single large versus several small reserves, the so-called SLOSS debate (Single Large Or Several Small). For example, is it better to have one large reserve of 15 000 ha or a network of five each of 3000 ha? Large reserves obviously enable a more ecogeographically diverse environment to be included within a single location with minimal edge effect. Alternatively, if a network of smaller reserves is established, each reserve could be sited in a distinct environment, which would better enable conservation of extreme ecotypes. So the conservation value of multiple small reserves may be greater than the sum of its individual components; however, if reserves are too small or too isolated the populations of the target taxon they contain will become inviable. The current consensus is that the optimal number and size of reserves will depend on the characteristics of the target species. Large reserves are better able to maintain species and population diversity because of their greater species and population numbers and internal range of habitats, but small multiple reserves may be more appropriate for annual plant species, which are naturally found in dense but restricted stands.

Practically, the vast majority of the wild relatives of crop plants are much more widely distributed. In such cases a single reserve (or a single cluster of sites) would not suffice. A number of reserves, located in different segments of the distribution area of the target species, would be required to cover its ecogeographic divergence and to deal adequately with the genetic changes which occur over its geographical range. If multiple small reserves are selected, their potential conservation value can be enhanced by using habitat corridors to link individual reserves, thus facilitating gene flow and migration between the component reserves. In this way individual populations can be effectively managed at a meta-population level.

The edges of a reserve should be kept to a minimum to avoid deleterious micro-environmental effects including changes in light, temperature, wind, the incidence of fire, introduction of alien species, grazing and deleterious anthropogenic effects. A round reserve will have the minimum edge to area ratio. Fragmentation of the reserve by roads, fences, pipelines, dams, agriculture, intensive forestry and other human activities will necessarily fragment, diminish and limit the effective reserve size and multiply the edge effects, and may leave populations in each fragment unsustainable.

The concept of environmental heterogeneity should be built into the design. When selecting sites for establishing a genetic reserve, sites with spatial or temporal heterogeneity (e.g. taxa contained in the reserve have different flowering times) should be given priority over homogeneous areas. The wider the range of habitat diversity in a potential reserve site, the better. This will ensure that the target species will preserve the various genes and genetic combinations associated with any ecotypic differentiation.

As will be discussed in more detail in section 22.8, the ultimate rationale behind conservation is potential human utilization; therefore the user communities must be considered when designing the reserve, whether in terms of permitting sustainable exploitation within the buffer or transition zone by traditional farmers, or building appropriate facilities for ecotourists or scientific visitors. Each user community will have a different view of the reserve and a different set of priorities. The requirements of each group of users should be surveyed before the reserve is established and their needs met as part of the management regime.

22.6.4 Taxon and reserve sustainability

Sustainability in the sense of continuance is a fundamental concept for genetic reserve conservation. It is not an inexpensive option compared with *ex situ* conservation, because the reserve, once designated and established, will require active and consistent population monitoring, habitat management and site security for a substantial period of time. This will necessitate the commitment of substantial levels of resources for a similar time period. *In situ* conservation is unlike *ex situ* conservation, where if the original collection from one gene bank is lost the curator may either obtain a sample of the duplicate collection placed in another gene bank or possibly return to the original collection site and re-collect from the original or a similar population, both of which options may be associated with relatively minor costs. If the material is lost from an *in situ* reserve, however, unless there is a back-up reserve somewhere else, the large quantity of resources expended on establishing the reserve would have been wasted and the cost of rehabilitating populations using materials stored *ex situ* would have to be considered. The latter option is commonly expensive and may require extensive research to ensure that the reintroduced materials do not likewise go extinct. Therefore it is important that both the target taxon and the reserve are sustainable.

'Taxon sustainability' means that the taxon is suitable for conservation over an extended period in an *in situ* genetic reserve. A highly mobile species or a weed associated with human disturbance, such as many crop relatives, may not be suitable for genetic reserve conservation. These may be better conserved as part of an on-farm conservation regime. The reason for this is that although large populations may be initially found in the reserve site, over a period of 10, 20 or 100 years the population may have migrated due to changes in the associated plant and animal populations or environmental conditions. Concomitant with this, the management plan for a genetic reserve can probably often be reviewed only in the short or medium term. Sites and therefore population stability can never be fully guaranteed indefinitely, because of possible unforseen or unavoidable long-term influences, even if that is the goal. Taxon sustainability is characterized as population viability over an extended period of time. IUCN has defined a viable population as one which maintains its genetic diversity, maintains its potential for evolutionary adaptation and is at minimal risk of extinction from demographic fluctuations, environmental variations and potential catastrophe, including over-use (IUCN, 1993).

In terms of reserve sustainability, a large investment of resources is required over a considerable period of time. There would therefore be little value in establishing a genetic reserve unless the reserve was unlikely to be affected by any form of human development project (Ingram and Williams, 1984). Reserves near settlements may be encroached on for building, amenities, etc. and even remote sites may be used in time for dumping human waste or, for example, the siting of national military or nuclear installations. It may be difficult to assess the short-term requirements for land use, but checks for development plans for potential sites should be routinely undertaken. It is, however, impossible to assess the long-term human requirements in one or two hundred years' time. Legislation ensuring that once conservation sites are designated they are maintained and not developed for other uses may assist with the security of the site, but experience has shown that legislation can be circumvented if the political will is sufficiently strong. Evidence of this is provided, for example, by the destruction of Sites of Special Scientific Interest (SSSIs) in Britain by road-building programmes. On the whole, legislation to protect individual habitats and ecosystems is not common. This does beg the question: are any reserves sustainable over a long period?

An answer might be to strengthen the legislation protecting reserve sites, but this will only be an answer if there is the equally long-term public and political will to support the legislation. In the short term, one approach is to designate multiple reserve sites. If multiple reserves are established for each target taxon, then the destruction of any one reserve will obviously have less overall impact. Moreover, if a species is extremely rare and restricted, *ex situ* techniques must assume greater importance. In fact *ex situ* approaches are absolutely essential if the population size of the species has become so low that survival *in situ* cannot be guaranteed or when ecosystems in which the species occurs are so degraded that survival of the target species is doubtful.

22.6.5 Formulation of the management plan

Allied to the selection of a reserve and its design, a practical management plan must be formulated to ensure that the target taxa remain sufficiently abundant at the site. The site will have been selected because it contains abundant and hopefully genetically diverse populations of the target taxon. Therefore, the first step in formulating the management plan is to observe the biotic and abiotic dynamics of the site. It should be surveyed so that the species present in the ecosystem are known; the ecological interactions within the reserve should be understood; a clear conservation goal should be decided and a means of implementation agreed. A detailed discussion of the formulation of a management plan is provided in Chapter 9.

The actual content or style of a management plan will vary depending on the location, target species, organization, staff, etc. that are involved. There is no standard format, but items generally included are: conservation objectives; site biotic and abiotic description; site history; public interest; factors influencing management; management prescription (what work needs to be carried out and precisely how and when to do it); ecological and genetic survey and monitoring schedule; budget and manpower. As the specific focus of establishing the genetic reserve will be to conserve a specific target taxon, the management plan will require details associated with the target taxon being conserved, both at the general level describing the taxon (taxonomy, phenology, habitat preference, breeding system, minimum population size, etc.) and description of the target populations at the site (e.g. mapping of populations and density within the site, autecology within the reserve, synecology with associated fauna and flora). The formulation of an appropriate management plan is critical to the safe conservation of the target taxon and will detail the appropriate level of intervention through grazing, mowing, burning, planting and harvesting regimes, as well as appropriate levels of biotic competition within the habitat.

Changes in population levels and density are a natural component of community dynamics. The management plan must allow for natural fluctuations due to stochastic (severe weather, floods, fire and epidemics) as well as cyclical and successional changes as long as they do not threaten the long-term viability of the target taxon. Stochastic and cyclical changes in the short term may be quite dramatic, but will rarely lead to species extinctions (Hellawell, 1991), although they are likely to lead to genetic drift (Chapter 7). The management plan should contain actual limits for the target taxon populations which take into account potential natural changes in population size, but beyond which management action is triggered. Having emphasized the natural changes seen in plant populations, humans undoubtedly have the most dramatic effect on communities, through incipient urbanization and pollution, or changes in agricultural and forestry practice, for example. Therefore the management plan must be flexible enough to accommodate superficial anthropogenic factors, but recognize those factors that could seriously threaten the levels of the target population.

Given (1994) stresses that it is important to realize that preserving communities is not necessarily the same as preserving genes. It is quite possible to preserve a community type and still lose genetic diversity, if not species. Therefore, it is vital that the reserve is designed and managed in an appropriate manner to maintain genetic diversity of the target taxon or taxa and that corrective action is automatically taken if this objective is threatened.

22.7 RESERVE MANAGEMENT AND MONITORING

22.7.1 Initiation of reserve management plan

The management of any genetic reserve will involve an element of experimentation, and it is unlikely that the ideal management regime will be known when initially establishing the reserve. For example, how can one accurately estimate the appropriate level of grazing in the reserve prior to initiating the management plan? A knowledge of the area and the current and historic grazing level will be important, but the actual level of grazing recommended once the reserve is established can only be known through scrupulous experimentation. Thus the initiation of the management plan will require careful introduction combined with evaluation, revision and refinement in the light of its practical application. Therefore, the initial level of management will be high, with intensive and extensive monitoring procedures, and the plan will need to be flexibly applied.

22.7.2 Reserve monitoring

Changes in the structure or size of populations of the target species within the reserve will obviously affect the conservation integrity of the reserve. The populations of the target species in the reserve will require regular monitoring to identify any actual or incipient change and, if detected, instigate appropriate management review and amendment. The monitoring process is likely to involve defining objectives, identifying key associated taxa and sample quadrat locations, selection of data for quadrat recording, determination of desirable frequency and timing of quadrat recording, accumulation of data sets, statistical analysis and production of recommendations on the management plan (Chapter 9 discusses these points in detail). The management plan should include a minimum and maximum number for the population size and the population monitoring process should act as a feedback mechanism indicating when these levels are reached. This would allow management changes to be made, thus ensuring the secure conservation of the target taxa.

It is impossible to record and monitor every species or individual plant within the reserve, so the conservationist is forced to take samples of data that, if effectively selected, will reflect the overall picture in the reserve as a whole. Key indicator taxa and sites within the reserve are selected for monitoring on a regular basis. This form of monitoring usually involves the establishment of both fixed and random quadrats or transects within the reserve. The key indicator taxa are likely to include the target taxon, but may also include the other plant and animal species, such as those without which the population of target taxa would decline – for example, primary herbivores or necessary pollinators.

There are numerous methods of assessing species abundance or diversity. The most commonly used is presence or absence linked to some estimate of density of a species (vegetative, flowering or fruiting) in a particular quadrat. Absolute measures of the quantity of a given taxon may be assessed in the form of number of individuals, demographic structure, reproductive fecundity (particularly for a long-lived perennial), proportion of ground cover, distribution pattern and biomass or yield (this has the advantage that it is accurate, but the drawback that it is destructive). There is no recommended size for the quadrats: they should be small enough to be searched easily and permit a larger number of quadrats to be taken overall, but large enough to lessen any quadrat edge effect and enable the quadrat to include large specimens or patches. Having begun monitoring, the quadrat size should be kept constant to permit easy comparison of data from subsequent surveys. The actual number of quadrats to be recorded will be a compromise between the resources required to record the quadrats and the information gained from recording extra quadrats. Monitoring is most likely to occur annually, once in a season's growth, though for a long-lived perennial the monitoring event may be less frequent. It is important to monitor at similar stages in the target species' life cycle on each occasion, to be able to record comparable results.

Increasingly, molecular methods of measuring genetic diversity will have an important role within conservation programmes whether *ex situ* or *in situ*. Within *in situ* programmes the monitoring of any change in the pattern of diversity within a reserve can be effectively undertaken using molecular markers, and cannot be approached at all if such markers are not employed. The technologies involved in the use of these markers and the acquisition of data derived from them can be criticized for being time consuming and expensive, but this is rapidly becoming much less so. Consequently – if not now, then soon – we can expect that not only will data be collected *vis à vis* species lists, and density and range of individual species, but also target populations within any reserve will be assessed in terms of their true genetic diversity. Whether genetic markers will prove practical for repeated monitoring of populations within reserves is currently unknown. Their immediate value lies in establishing accurately the levels of genetic diversity between and within the potential populations and aiding the selection of populations to be conserved in a genetic reserve. Following the establishment of the reserve, loss of diversity could be directly deduced if the extent to which population numbers are varying in sequential monitoring operations is known.

Having collected the data for a particular survey the conservationist will want to compare the population characteristics of the survey with previous surveys to draw conclusions about any significant changes that may have occurred in the target populations. Routine statistical analysis of the data sets will indicate whether there has been significant change in population density from the previous surveys and whether a trend is becoming apparent over the longer term. Care must be taken to distinguish between the natural ranges for population characteristics and those induced by management or other intervention.

22.7.3 Community interrelationships

Although the goal of genetic conservation may be explicitly the conservation of a single species, as opposed to ecological conservation (where the goal is often broadened to include conservation of whole communities), single target species do not exist in isolation. They each exist within a community and therefore will naturally interact with a range of abiotic and biotic factors in the reserve. These interactions must be understood if they are to be effectively incorporated into the management plan. For example, if the target taxon is a pasture species, it is by definition routinely grazed. If instigation of the management plan either increases or decreases the grazing frequency or density, it may alter the habitat sufficiently to have an adverse effect on population numbers of the target taxon.

Disturbance of the population equilibria within the reserve may result from abiotic as well as biotic factors. Periodic storms, floods, droughts or fires can open gaps in the canopy, encourage or discourage dispersal and seed production and may be a regular feature of the habitat. Reserves should be sufficiently large and the management regime sufficiently flexible to allow for this continued habitat disturbance and the resultant natural fluctuation in populations (Chapter 8). Owing to their nature, many abiotic factors associated with the weather will be impossible for the reserve manager to control, but their potential effect on the target taxon population should be considered as part of the management plan. The reserve manager can, however, exert some control over fire, which can be used as an effective management tool that encourages the sprouting and seed production of many species (Gill *et al.*, 1981).

Invasive plants can pose a serious threat to target populations (see the example for Mauritius discussed in Chapter 9); careless introduction of alien keystone species can quickly out-compete native species and totally alter the nature of the habitat and therefore its species composition. Frankel *et al.* (1995) list four ways in which alien introductions can radically alter the community structure: altering natural geomorphic processes, changing nutrient cycling and balances, changing fire regimes and altering seedling recruitment. It is difficult to predict in advance the result of any of these forms of alien introduction to a reserve; some may assimilate well, while others can totally disrupt the community. Therefore, the general rule is to try to exclude new alien species from the reserve. In contrast the beneficial effect of established keystone plant species may be essential to the target taxon (Chapter 3). Keystone species are usually the dominant species within a habitat which tend to define the general habitat in which the other species exist and without which the target taxon may not be able to survive. So the encouragement of the established keystone species may be essential to the health of the target taxon.

The introduction or invasion of exotic animals to a reserve can have a deleterious effect by directly altering the grazing regime or indirectly altering the carnivore/herbivore balance within the reserve. However, as discussed above, animals may also have a beneficial effect by not only grazing and maintaining a population at a pre-climax vegetational state, but also acting as pollinators, seed dispersal agents, etc. The effect of insects on plant conservation is discussed in Chapter 14. The spread of pathogens or pests to the reserve can directly harm the target taxon or, indirectly, the species that have a relationship with the target taxon and without which the target taxon may decline. For example, the decimation of elm forest by Dutch elm disease has undoubtedly led to a decline in species with an obligate relationship with the various *Ulmus* species.

It should be stressed that biotic interactions can have a beneficial as well as a deleterious effect on the target taxon. It may be necessary for the reserve manager to simulate desirable biotic interaction in the absence of the appropriate animal or plant by, for example, mowing, burning, selective extraction or felling, or even hand-pollination.

22.8 RESERVE UTILIZATION

The establishment and management of the reserve is not an end in itself. Throughout this text an explicit link has been made between conservation and utilization: humans conserve because they wish to utilize, either now or in the future. Therefore there is a need to have an appreciation and to discuss utilization of the material conserved in the reserve as part of the model for genetic reserve conservation. This point is highlighted in the Convention on Biological Diversity and in this context any utilization should be 'sustainable' and 'meet the needs and aspirations of present and future generations'.

22.8.1 Traditional, general and professional utilization

The utilization of the material conserved in the reserve may be divided among traditional, general and professional users and, as highlighted above, each user community should be surveyed and their requirements considered within the strategic management plan.

Reserves are very rarely established in an anthropogenic vacuum. There are likely to be local farmers, land owners and other members of the local population who utilize the proposed reserve site and who are likely to remain neighbouring communities. Their traditional use of the site is likely to be disrupted by the establishment of the reserve. These local people may have historically harvested or collected from, hunted over or may simply have enjoyed visiting the site on which the reserve has been established. It is unlikely that any *in situ* conservation project could succeed in the absence of local support and it will definitely fail if the local population opposes the establishment of the reserve. Therefore where traditional utilization is compatible with conservation objectives, sustainable exploitation of the buffer or transition zones by local, traditional user communities should be encouraged. However, to avoid negation of the conservation objectives their access and any harvesting, hunting, etc. may need to be regulated. There may be a need to compromise between traditional utilization and conservation objectives in order to ensure success of the reserve (Shands, 1991). The involvement of local people should not be considered as a distraction or disadvantage, since they may be able to assist through volunteer schemes in the routine management and monitoring of the reserve. The personal experience of the authors has indicated that local people are very proud to find that their environs contain 'important' plant species and are very willing to assist in their conservation. This applies to rural communities in both developing and developed countries. It facilitates the development of goodwill between the professional conservationist and local communities neighbouring the reserve. In some cases, though, there may be a need to provide other incentives to engender goodwill.

The second user group is the population at large, whether local, national or international: their support may be essential to the long-term political and financial viability of the reserve. As discussed in Chapter 3, the ethical and aesthetic justification for species conservation is of increasing importance to professional conservationists. Commonly the general public will ultimately finance the establishment and continuation of the reserve through taxation, so their use, even if only through ethics and aesthetics, should not be understated. It is surely not just a coincidence that there is a direct relationship between the growth of public awareness of conservation issues and the growth of professional conservation activities in recent years. As conservationists, we have the advantage that the majority of people feel an inherent love of nature and abhorrence of its careless eradication. Wilson (1984) has referred to this human trait as biophilia. We must ensure that this user community is supplied with an appropriate level and quantity of information to ensure their continued support.

Some members of the general public may wish to visit the reserve and this should be clearly encouraged as an educational exercise. If local people who traditionally utilized the site are opposed to the reserve, they may be converted to supporters if they can see direct financial benefits to their community resulting from eco-tourist, school and other groups of visitors to the reserve. The reserve design must take into account the needs of visitors, such as visitors centres, nature trails, lectures, etc. They are also likely to bring additional income to the reserve itself through guided tours and the sale of various reserve information packs.

Professional utilization of germplasm conserved in the genetic reserve will be on a similar basis to professional utilization of *ex situ* conserved germplasm. One of the main disadvantages of *in situ* as opposed to *ex situ* conservation is that it is more difficult for the plant breeder to gain access to the germplasm (Hawkes, 1991a). To avoid or lessen this problem those managing the reserve should ensure that they characterize, evaluate and publicize the germplasm held in the reserve. The onus is surely on reserve managers, just as it is on gene bank managers, to promote utilization of the material in their care. Just as gene banks and botanical gardens publish catalogues of their collections, so the reserve manager should regularly publish a catalogue and description of the germplasm held in the reserve to assist potential users. The level of documentation of passport, characterization and evaluation data recorded should be just as extensive for *in situ* as for *ex situ* conserved germplasm (Chapter 11). The quantity and level of documentation has a direct relationship with the potential of the germplasm for exploitation. Having made this point, it is unlikely that the plant breeder will ever have such easy access to material conserved in a genetic reserve as in a gene bank, especially if the gene bank is located in a plant breeding institute, which is often the case. Seasonality also limits access to germplasm in a genetic reserve, since the seed can only be collected during the fruiting season of the plant, whereas germplasm is available throughout the year from the *ex situ* collection.

In many cases, the work of professional users, the general public and local people can be linked through partnership within non-governmental organizations, especially those involved in sustainable rural development, conservation volunteers or use of resources in accordance with traditional cultural practices. All partners will therefore share the goals of sustainable use of biological resources, taking into account social, economic, environmental and scientific factors which form a cornerstone to the nations' proposals to implement Agenda 21.

22.8.2 Linkage to *ex situ* conservation, duplication, research and education

To provide a back-up to the conservation of the germplasm in the genetic reserve, the germplasm should be sampled and deposited in appropriate *ex situ* collections. Although both *ex situ* and *in situ* techniques have their advantages and disadvantages (Chapter 2), the point is re-emphasized here that they should not be seen as being alternatives or in opposition to one another. The two strategies are complementary: good gene bank managers will duplicate their collection in other gene banks, and reserve managers should also duplicate their collection in *ex situ* collections. The point is made in Chapter 2 that it is by definition not possible to duplicate material from one reserve to another without the material being taken *ex situ*, but it is worth repeating here that it would be foolish to focus your entire *in situ* conservation effort on a single reserve. Multiple reserves should be established, where possible, to duplicate effectively the conservation of the material *in situ*. In this context, if the germplasm user does not have a specific requirement for material from a reserve, the gene bank may be seen to act as a staging post for those wishing to utilize the germplasm originally conserved *in situ*. A clear protocol establishing where the material is to be duplicated should be included in the management plan of the reserve.

A well-founded reserve should act as a research platform. As has been shown in Chapters 15, 16 and 21, genetic reserves make ideal locations for field experimentation. There is a real need for a better understanding of intrareserve species dynamics to aid the sustainable management of the specific taxa included within the reserve, but also as a general tool for experimentation of ecological and genetic studies of *in situ* conserved species. Research activities based on the material conserved should be encouraged, as they provide another use for the material conserved and another justification for establishing the reserve. However, these activities should take place in the inner buffer zone, not the core reserve itself, so that they cannot threaten the populations of the target taxa.

Increasing public awareness, education and appreciation of the environment among non-professionals is crucial if any conservation activities are to succeed in the longer term. The major part of conservation is funded from the public purse. Thus if we, as conservationists, are to maintain or increase the levels of funding for the work we consider so vital, we must ensure that the public are supportive. To obtain this support local people must be environmentally aware. This awareness can be raised by various means, such as media publicity campaigns, use of educational and training materials, lecture programmes, nature teaching in schools, the establishment of school nature areas or wildlife groups, local conservation action initiatives, etc. (Maxted and Ford-Lloyd, 1996). Perhaps one of the most effective means of increasing public awareness is to promote public visits to reserves where germplasm is being actively conserved. It seems likely that the public will be more interested in visiting a reserve to see 'real' plants being conserved, than visiting a gene bank to see sealed tins or foil packets. Therefore, genetic reserves can play an important role in increasing public awareness of conservation issues. The support of an informed public is a valuable asset and should be promoted by the scientific conservation sector.

To aid those trying to practise *in situ* conservation in a genetic reserve, the various activities outlined in the model in Figure 22.2 and discussed above are summarized in Appendix 22A. It is worth noting here that the list of activities is extensive and it is unlikely that all of these will be considered in detail when considering and practising *in situ* conservation in a genetic reserve. If the model is applied uncritically, it may take years before it is possible to conserve the target gene pool safely. In many cases actual genetic erosion is occurring now, so we cannot afford to wait for a long time before we commence active conservation. Therefore the model is proposed as an outline of the activities that could be undertaken in an ideal situation, a summary of good practice, and not something that could or should be followed slavishly.

ACKNOWLEDGEMENTS

We thank Luigi Guarino and Mike Maunder for their contributions to this work.

APPENDIX 22A SUMMARY OF BASIC REQUIREMENTS FOR ESTAB-
LISHING *IN SITU* GENETIC RESERVES

1. Selection of target taxa

Decide which species need to be protected and included in *in situ* genetic
reserves. If possible include more than one chosen species in each
reserve.

2. Project commission

Formulate a clear, concise conservation statement establishing what species, why and in general terms where the species are to be conserved.

3. Ecogeographic survey/preliminary survey mission

Obtain basic information for the planning of effective conservation.
Survey the distribution of taxonomic and genetic diversity, ecological
requirements and the reproductive biology of the chosen species over its
entire geographical range. Where little ecogeographic data are available a
preliminary coarse grid survey mission to collate the necessary background biological data on the species may be required.

4. Conservation objectives

Formulate a clear, concise set of conservation objectives which state the
practical steps that must be taken to conserve the species, and specifically
where and how the species might be conserved.

5. Field exploration

Visit potential sites indicated as having high levels of species and genetic
diversity by the ecogeographic survey or preliminary coarse grid survey
mission to verify the predictions of species and genetic diversity.

6. Conservation strategy – *in situ* genetic reserve

This involves the designation, management and monitoring of the
reserve.

6.1 Reserve planning and establishment

6.1.1 Site assessments

Select the actual sites where genetic reserves will be established; where possible they should cover the range of morphological and genetic diversity, and the ecological amplitude exhibited by the chosen species. Several reserves (possibly dozens of reserves) spread over the geographical range and the ecological environments occupied by the species may be required to cover sufficiently large fractions of its gene pool. Ensure that each reserve represents the fullest possible ecological range (microniches), to help secure maximal genetic variation, and to buffer the protected population against environmental fluctuations, pests and pathogens, and man-made disturbances. As part of this evaluation prepare a vegetation map of the area, surveying in detail the plant communities (and habitats) in which the target species grows.

6.1.2 Assessment of local socio-economic and political factors

Constraints ranging from economic to scientific and organizational will affect the establishment of the reserve. The simplest way forward in economic and political terms is for countries to take action on establishing a series of national parks or heritage sites, as this is likely to be of some benefit to the people of the countries and will gain their support.

6.1.3 Site size, number, distribution and design

Sites should be large enough to contain at least (1000–)5000–10 000 individuals of each target species to prevent natural or anthropogenic catastrophes causing severe genetic drift or population inviability. Sites should be selected to maximize environmental heterogeneity. Each reserve site should be surrounded by a buffer zone of the same vegetation type, where experiments on management regimes might be conducted and visits by the public allowed, under supervision.

6.1.4 Taxon and reserve sustainability

Establishing and managing an *in situ* genetic reserve is resource expensive and therefore both the taxon and reserve must be sustainable over an extended period of time or the investment will be forfeit.

6.1.5 Formulation of the management plan

The reserve site will have been selected because it contains abundant and, it is hoped, genetically diverse populations of the target taxon. Therefore, the first step in formulating the management plan is to observe the biotic and abiotic qualities and interactions at the site. Once these ecological dynamics within the reserve are known and understood, a management plan that incorporates these points, at least as they relate to the target taxon, can be proposed.

6.2 *Reserve management and monitoring*

6.2.1 *Initiation of reserve management plan*

It is unlikely that any management plan will be wholly appropriate when first applied; it will require detailed monitoring of target and associated taxa and experimentation with the management plan before a more stable plan can be used. The plan may encompass several management regimes (a range of grazing practices, tree-felling, burning, etc.) within the reserve.

6.2.2 *Reserve monitoring*

Each site should be monitored systematically at a set time interval and the results fed back in an iterative manner to enhance the evolving management regimes. The monitoring will take the form of measures of taxon number, diversity and density as measured in permanent transects, quadrats, etc.

6.3 *Reserve utilization*

6.3.1 *Traditional, general and professional utilization*

Humans should conserve because they wish to utilize and it is necessary to make an explicit link between the material conserved and that currently or potentially utilized *ex situ* by humankind. There are three basic user communities: traditional, general and professional.

6.3.2 *Linkage to* ex situ *conservation, duplication, research and education*

There is a need to form links with *ex situ* conserved material to ensure utilization but also as a form of safety duplication. The reserve forms a natural platform for ecological and genetic research, as well as providing educational opportunities for school, higher educational and general public levels.

7. Conservation products

These will be primarily populations of live plants held in the reserve, voucher specimens and the passport data associated with the reserve and plant populations. Germplasm taken from the reserve may be held *ex situ* for safety duplication.

8. Conserved product deposition and dissemination

The main conserved product, the plant populations of the target taxon, are held in the reserve. However, there is a need for safety duplication and a sample of germplasm should also be deposited in an appropriate *ex situ* collection (gene bank, field gene bank, *in vitro* banks, botanical gardens or conservation laboratory) with the appropriate passport data.

9. Characterization/evaluation

The first stage of utilization will involve the recording of genetically controlled characteristics (characterization) and the material may be grown out under diverse environmental conditions to evaluate and screen for drought or other tolerance, or the experimental infection of the material with diseases or pests to screen for particular biotic resistance (evaluation).

10. Plant genetic resource utilization

The conserved material is likely to be used in breeding and biotechnology programmes, and to provide food, fuel, medicines and industrial products, as well as a source of recreation and education. Locally the materials held in the reserve may have traditionally been used in construction and technology (hunting, craft, adornment, transport), or been harvested, hunted or collected. This form of traditional utilization of the reserve should be encouraged, provided it is sustainable and not deleterious to the target taxon or taxa.

23

Towards the future

G.C. Hawtin and T. Hodgkin

23.1 INTRODUCTION

There is growing awareness of the importance of *in situ* conservation for plant genetic resources for food and agriculture (PGRFA). This recognition has grown out of several factors:

- **The increasing influence of the environmental movement**, which has resulted in greater public awareness of the importance of conservation. Although the primary interest of many in this movement has been with conserving ecosystems (areas of natural beauty) and preserving high-profile endangered species (pandas, whales, butterflies and orchids), it has generated a recognition of the importance of maintaining a wide range of different ecosystems and of conserving individual species *in situ*.
- **The increasing importance of wild species as a source of genes for crop improvement**. With modern genetic manipulation techniques, all organisms become potential sources of useful genes. However, species that are more closely related to crops contain, arguably, the most valuable reservoir, whether these are introduced to crop species by conventional interspecific hybridization methods or through genetic engineering.
- **The recognition that it is impractical, for cost and technical reasons, to consider conserving all potentially useful genes *ex situ***. This is the case, for example, for clonally propagated crop species such as sweet potato and yam, and for those species that produce seed that cannot be stored *ex situ* for any length of time – recalcitrant seeded species. Many tropical trees, such as mangoes and jackfruit, fall into this cate-

Plant Genetic Conservation.
Edited by N. Maxted, B.V. Ford-Lloyd and J.G. Hawkes.
Published in 1997 by Chapman & Hall. ISBN 0 412 63400 7 (Hb) and 0 412 63730 8 (Pb).

gory. In addition, many crop relatives produce only limited quantities of seed and are also extremely difficult to manage in *ex situ* gene banks. Maintaining populations of such species *ex situ*, in populations that are of sufficient size to capture the full range of intraspecific diversity, is often impossible in practice.

- **The recognition of the importance of allowing evolution to continue under natural or human selection.** *In situ* conservation allows the continued evolution of the adaptive genetic variation of species and their populations.
- **The appreciation that the continued maintenance and use of a wide range of inter- and intraspecific diversity can constitute an important element of the production strategies of farmers and communities.** In this way the continuing maintenance of traditional cultivars and types can effectively contribute to sustainable development.
- **An increased recognition by many countries of the political, social and economic importance of retaining the useful plant genetic diversity that occurs within their borders.** *In situ* conservation is regarded as an effective component of strategies to realize the value of what are increasingly seen as important national resources.

These factors are likely to continue to influence conservation thinking in agricultural circles in the future as ever more diversity is lost – backed up by greater amounts of data on the true extent of genetic erosion. Over time, it will be possible to evaluate more fully the extent to which *in situ* conservation provides an effective mechanism for maintaining adaptive gene complexes, for allowing evolution to continue and for maintaining more diversity than can be maintained *ex situ*.

The benefits of *in situ* conservation thus derive both from the particular contribution it can make to the maintenance of useful (and usable) inter- and intraspecific diversity, and from its wider social and economic contribution. The arguments that support the conservation of ecosystems and the conservation of biodiversity in general are extended by the particular contribution that *in situ* conservation of PGRFA can make in a wider social context and to improved agricultural production.

As scientific understanding (biological, ecological and social) of *in situ* conservation grows and new techniques become available for both *in situ* and *ex situ* conservation, so it will become increasingly possible to devise conservation strategies that are most appropriate for given situations – taking into account such factors as new scientific and technical knowledge, and institutional, human and financial resource limitations. As noted in earlier chapters, optimal strategies are likely to include both *in situ* and *ex situ* elements as components of complementary strategies for the target gene pools.

The growing recognition of the importance of conserving genetic resources *in situ* is exemplified by, *inter alia*:

- the Convention on Biological Diversity and Agenda 21;
- FAO's Global Plan of Action for Plant Genetic Resources, in which four out of 20 sections are specifically devoted to *in situ* (including on-farm) conservation;
- the growing number of national, regional and international programmes that are devoted to *in situ* conservation;
- the growing scientific literature (including this book) and conferences on the topic.

In spite of this trend, the science of *in situ* genetic resources conservation is relatively young and investigations necessarily have to go hand in hand with the application of management strategies which, in many cases, have yet to be proven. In some cases, as in forestry, where *in situ* conservation is the primary method of conservation, considerable information has accumulated on the management of resources in ways that ensure the maintenance of the desired ecosystem or species. However, even in forestry, there is relatively little experience of conserving intraspecific diversity and this is certainly so in the case of conserving the genetic resources of annual crop relatives in genetic reserves and of the on-farm conservation of traditional cultivars.

In developing *in situ* conservation systems for the future, a wide range of issues needs to be addressed. These include both the selection of target taxa and scientific and technical aspects of conservation. Other factors also have a substantial influence on the effectiveness of *in situ* conservation measures, including the policy environment, institutional roles and responsibilities, the availability of funding, the existence of effective information exchange systems, and broad public awareness and support. These are each considered briefly in the following sections.

23.2 SCIENTIFIC AND TECHNICAL CONSIDERATIONS

This book has concentrated to a large extent on scientific and technical issues as they concern the conservation of PGRFA. Particular attention has been paid to conserving wild crop relatives and traditional cultivars, with a focus on specific taxa or gene pools. In practice, the situation is much more complex. Wild relatives are conserved within ecosystems that have to be maintained as part of an overall environmental management strategy. Similarly, individual crops or cultivars have to be conserved within the framework of a range of different production systems and farm management strategies. Within the former there may be many different types of reserve or other management units, including managed woodlands, pastures and other semi-natural ecosystems. Within the latter there will be hedgerows, field margins and other areas managed by farmers or communities for a range of different purposes. In contrast to the situation in nature reserves, where the conservationist may be able to play a substantial role in management, in the farm situation the farmer is the major decision maker. However, the conservationist may also play an active role in on-farm conservation, whether directly (for example, through reintroducing land races) or indirectly (for example, through promoting supportive policies in areas such as subsidies, plant breeders' rights, seed legislation and grazing regulations, or through promoting markets for 'diversity-rich' products). The development of scientifically based technologies and strategies for *in situ* conservation must take into account this enormous complexity of approaches and possibilities.

Throughout this book there are numerous examples of areas where further research is needed. These relate to such biological and physical areas as taxonomy, the extent and distribution of genetic diversity, population size, reproductive biology, phenology, plant geography and ecology (relationships with both the biotic environment, such as competition and species interactions, and with the abiotic environment), climatic cycles and other variables. There are also social, cultural, policy and economic areas such as understanding farmer and community needs and decision-making processes, indigenous knowledge, participatory breeding methods, land and plant ownership rights, socio-economic policy alternatives and new marketing possibilities.

In general, there are three broad needs: new research methods, new conservation techniques and practical applications.

23.2.1 New research methods

There is a need for new research methods that can contribute to locating and monitoring the extent and distribution of genetic diversity. A major concern is to bring together information on the environment with that on the distribution of taxa and genetic diversity within taxa. Geographic Information Systems (GIS) are expected to contribute increasingly to this in the future. While a wide range of techniques already exists, there is a need to be able to integrate them in order to be able to create a coherent understanding of the spatial distribution of diversity and the factors that affect it. The distribution of neutral genetic markers, such as isozyme or molecular markers, as well as genes determining adaptive traits needs to be further explored. Environmental factors which need to be studied include not only physical environmental factors such as altitude, soil type or rainfall levels, but also human factors such as population densities and movement that influence decisions on crop and land use. Finally, socio-economic and anthropological research is needed that will permit an understanding of decision making at a local level so that a knowledge of the concerns of farmers and communities, and the ways in which they meet their needs, can inform the conservation process and make it realistic and appropriate.

Of equal importance, though less well understood, are factors that affect the temporal distribution of diversity. As *in situ* conservation gains momentum, the effectiveness of strategies in maintaining diversity over substantial periods will become an essential measure of their effectiveness in terms of biodiversity conservation. We need to know not only what changes are occurring but also the significance of these changes in terms of population viability and survival. Wherever relevant, therefore, research work should involve the development of protocols which will allow meaningful sampling at repeated intervals to monitor any changes in the target populations.

23.2.2 New conservation techniques

There is a need for new conservation techniques – for example, for managing reserves, for controlling practices in public, community or privately owned lands, or for promoting the conservation of land races on-farm. Each situation will make unique management demands and raise unique management questions. None the less there are general issues that also need to be explored, depending on the resource and situation. In many forestry situations, for example, there is a need to explore sustainable use practices for those species that are the most important to the economies of the local people or national industries, and to determine what degree of use and extraction is compatible with the maintenance of sufficient levels of diversity for species and population maintenance. For the wild relatives of crops, a major concern may be with the way in which the survival of individual species can be supported in the context of other aspects of ecosystem maintenance or use. Thus, many crop wild relatives occur as part of pasture ecosystems and their survival may depend on appropriate grazing intensities or burning frequencies. Traditional crops are often maintained because they meet farmers' needs and better alternatives do not exist. The factors responsible for this need to be determined, such as the role of agricultural policies, price and cost regimes and local community and farmer needs.

23.2.3 Practical applications

There is a need for the practical adaptation and application of suitable methodologies and techniques in a wide range of real situations – different gene pools, in different geographical, ecological, policy and social environments – thus building up a broad body of knowledge relating to a wide range of circumstances. From such empirical approaches, using the most appropriate techniques and methods, it should be possible, over time, to develop widely applicable conservation models and decision-making strategies. The experiences described in earlier chapters make an important contribution in this area and need to be repeated and extended to different situations and circumstances.

23.2.4 Action

While such an agenda may seem very large, almost daunting, there are many actions that can be undertaken with limited resources to provide a basis on which to plan more extensive or substantial investigations.

With respect to wild relatives of crops, the collection of information on the distribution and status of different crop relatives within a country, and the extent to which they can be found in protected areas, provides a basis for planning and locating target populations. The work described in Chapter 3 for *Solanum* and *Vicia* spp. shows one way in which a relatively simple collation of existing knowledge, based on herbaria and gene bank passport data, can be used in the development of complementary conservation strategies. For many important gene pools, much information already exists and the work is largely one of bringing together and analysing existing data which may be more or less widely scattered. Given adequate resources, such analyses can then provide the basis on which to plan the location of genetic reserves and begin conservation activities. Chapter 16 gives an example of how an extensive knowledge of species distribution in Turkey was used to develop an overall plan for conserving the target taxa selected in the project.

The on-farm conservation of traditional crop cultivars can also be based on fairly simple initial activities involving the development of collaborative links to communities who use and maintain such materials. Participatory discussions and rapid rural appraisal techniques can provide a good understanding of the interests and needs of local communities and aid in the identification of opportunities for promoting the continued maintenance of traditional cultivars. Such activities can create openings for a more systematic exploration of various options, including how conservation can be linked to farmers' development needs.

23.3 THE SELECTION OF TARGET TAXA

Selecting taxa for *in situ* conservation is complex and many factors need to be taken into account when choosing which species to conserve and where and how to employ resources. In the case of wild species, considerations will include the closeness of their relationship to the cultivated species and the ease with which they can be used as donors of useful traits. The presence of known useful variation will also be important, as will the status and distribution of the species and the degree to which they are threatened in particular habitats. For some extremely important wild relatives, such as *Triticum dicoccoides* or teosinte, it has been considered worth while to set up specific reserves to ensure their maintenance. However, most taxa are likely to be conserved as components of ecosystems which are being conserved for other reasons.

Decisions on where to concentrate resources will therefore be based on general criteria (for example, those developed by organizations such as IUCN) that are relevant to all species, together with specific criteria concerning potential usefulness with respect to present or future crop production. One feature particular to wild relatives of crop species is the potential value of conserving populations adapted to marginal or extreme environments and that are likely to have adaptive characters such as drought, cold or salinity tolerance.

Selecting taxa for on-farm conservation raises equally difficult problems. Since the species concerned constitute part of functional agricultural production systems, a focus on specific species may not always be appropriate, or relevant to the farmer. National programmes may wish to ensure that species of greatest importance for national production receive the most attention, such as potato in the Andes or rice in south-east Asia. On the other hand, it has been argued that for such major species there is already a substantial *ex situ* conservation effort and that an emphasis on underutilized or neglected crops is more appropriate. Since farmers are the primary decision makers, external priorities may be somewhat irrelevant. However, resources and facilities available for *ex situ* conservation may be very limited for minor or neglected species such as buckwheat, *Musa ensete*, fonio (*Digitaria exilis*), sapote and bambara groundnut, and on-farm conservation may represent the only viable strategy for maintaining their genetic diversity.

The setting of priorities among species has both biological and social dimensions. Where conservation of a species depends on maintaining a particular ecosystem, due regard must also be paid to those other species that are essential components of the ecosystem, even when these are not the direct target of a conservation plan. This is particularly the case with the conservation of forest species.

23.4 THE POLICY ENVIRONMENT

The effectiveness of future *in situ* genetic resources conservation programmes will depend to a very great extent on the national and international policy environment.

National policies to promote *in situ* conservation are urgently needed in many countries. They must take into account factors such as the establishment of reserves, regulations to ensure their effective and sustainable management, and the creation of long-term incentives for on-farm conservation.

While the Convention on Biological Diversity (CBD) emphasizes the importance of *in situ* conservation, the driving force behind its formulation and adoption was the broader imperative of conserving biodiversity generally rather than the narrower – but arguably more economically important – objective of conserving PGRFA. The ratification of the CBD implies a recognition by the governments concerned of their responsibility for conserving biodiversity.

Policies adopted by national governments to promote the objectives of the CBD will have a profound influence on the extent and effectiveness of *in situ* conservation efforts. Where national policies emphasize the conservation and sustainable use of biodiversity, it will be possible to undertake conservation in a policy environment that recognizes and supports the long-term nature of *in situ* conservation. Given the importance of land ownership and access rights, policy support in areas such as appropriate land tenure and forest extraction practices will be crucial to the effective implementation of *in situ* conservation plans. The need to ensure sustainability must be a key determinant of policies regarding conservation. This will require the adoption of flexible approaches that spread risks widely. Populations in any given gene management zone may always be at risk from new developments such as the discovery of oil or a decision to build a dam. It is therefore important that policies promote such safety measures as the development of duplicate or back-up reserves.

Similarly, national agricultural policies can greatly influence the success of on-farm conservation. For example, certain agricultural policies may have a pronounced negative effect on the extent to which farmers find it in their interest to maintain traditional cultivars. These might include the linking of access to fertilizer, or credit, with the adoption of specific cultivars, or the implementation of seed regulations that hamper the continuation of informal seed exchange practices. Policies which recognize the value to farmers of local traditional cultivars, and which promote their continued use in certain production environments, could substantially increase the effectiveness of on-farm conservation. An example is the European Union's support of local named products.

From a utilitarian perspective, conservation cannot be regarded as an objective in itself. The CBD is concerned with not only conservation but also the sustainable use of biodiversity for development. Thus national and international policies are needed not only to promote conservation, but also to promote access to it and its sustainable use.

There is currently considerable debate and controversy on the subject of access to genetic resources. The prevailing, though not universally accepted, philosophy of the 1970s and early 1980s was to regard PGRFA as the 'common heritage of mankind', as was enshrined in the International Undertaking on Plant Genetic Resources. Access was promoted on a free and open basis. Although the concept has been progressively modified and the CBD recognizes genetic resources as a resource over which nations have sovereign control, it nevertheless obliges signatory governments to promote access for environmentally sound purposes. The focus of the debate as regards PGRFA is currently centred in FAO. Regimes for promoting access internationally, while respecting the sovereign rights of nations to control the resources and receive an equitable sharing of benefits, are under negotiation in the Commission on Genetic Resources. These negotiations are taking place in the context of revising the International Undertaking to bring it in line with the CBD. Also under discussion are ways of giving practical recognition to the concept of Farmers' Rights, defined as the rights of farmers arising from their past, present and future contribution to the conservation and development of PGRFA.

The CBD gives signatory countries the right to demand their Prior Informed Consent (PIC) to the granting of access by others to the genetic diversity within their borders. In many countries, the concept of PIC is also being applied to the local level. Legislation is being enacted in several countries (e.g. Philippines) requiring germplasm collectors to obtain the consent of local and indigenous communities prior to carrying out any collecting activities.

Whatever the outcome of these and related negotiations, the next few years are likely to see the creation of new international systems for conserving and exchanging PGRFA and for ensuring that the benefits arising from their exploitation are equitably shared between conservers/providers and those that use them. Such systems may operate multilaterally, as has been predominantly the case for PGRFA in the past, or may come to depend increasingly on bilaterally negotiated agreements. Such developments will inevitably impact on the priority afforded by governments to conservation – as PGRFA increasingly come to be seen as having a direct 'value' for those that conserve them and make them available. There is a danger that, largely as a result of developments in 'bioprospecting' by the pharmaceutical industry, expectations of monetary benefits to be gained from PGRFA are unrealistically high. While such expectations are likely to lead to a greater awareness of the importance of conservation, they might also result in the imposition of greater restrictions on access, to the detriment of the flow of germplasm upon which plant breeding depends. Policy developments are also likely to impact on such areas as institutional responsibilities and on information systems set up to inventory genetic resources and to monitor and regulate their international movement and use.

23.5 INSTITUTIONAL DEVELOPMENT

A wide range of institutions are directly or indirectly concerned with *in situ* conservation at the local and national levels. While each country differs in its arrangements, some of the key institutions involved typically include the following:

- **National parks and wildlife services**, which are the managers of nature reserves or other protected areas. Much of the work of these organizations has traditionally concentrated on the macro scale, involving the management of large areas or emphasizing high profile species such as elephants or pandas. To contribute more effectively to the conservation of plant genetic resources, such institutions generally need to take more account of the intraspecific as well as the interspecific diversity in the areas they manage.
- **Forestry departments and institutions**, which are also often responsible for managing very large areas. They have a range of approaches to conservation of the species that occur within areas under their control, but again have traditionally given relatively little attention to intraspecific diversity or to the wild relatives of crop species.
- **Agricultural research institutions**, especially gene banks and plant genetic resources programmes, which generally have the major national responsibility for the *ex situ* conservation of PGRFA. Such centres can often provide expertise on the species of concern, but frequently lack specific expertise on *in situ* conservation.
- **University biology, geography, botany, agriculture, genetics, anthropology and economics departments**, which can often play a key role in supporting the research needed to make *in situ* conservation effective. Given the wide range of disciplines needed in an effective *in situ* conservation programme, it is important that there be adequate mechanisms for encouraging multidisciplinary research approaches.
- **National herbaria and botanical gardens**, which can support the *ex situ* conservation of reserve collections and provide essential information on species identity and distribution.
- **NGOs, farmers' organizations, local cooperatives and local government institutions**, which are the keys to ensuring the support of local communities. Without such support, *in situ* conservation programmes are unlikely to be sustainable.
- **Agricultural support services**, such as extension and advisory services, whose collaboration is likely to prove invaluable for promoting and facilitating the effective maintenance of traditional cultivars by farmers and local communities.
- **National gene banks** – or their equivalent – are normally the only agencies specifically charged with the responsibility of carrying out PGRFA conservation. As such, in many countries they are the logical

agency to promote *in situ* conservation of PGRFA, in spite of normally having mainly, or exclusively, *ex situ* responsibilities.

Given the large number of organizations within most countries that have an actual or potential function relevant to *in situ* conservation, there is normally a need for a national mechanism for coordination – a national committee, commission, lead agency or other such arrangement. There is also the need to develop national plans for *in situ* conservation, identifying priorities (for gene pools, ecoregions) and appropriate strategies. The roles and responsibilities of the different institutions need to be clearly specified.

At the international level, there are several institutions and programmes active in the *in situ* conservation of PGRFA. Over the coming years it is expected that many will expand their activities in this area and that new institutions or programmes will be created, or take on new roles. The United Nations Environment Programme (UNEP) and the United Nations Economic, Social and Cultural Organization (UNESCO), through its Man and Biosphere Programme, are both concerned with environmental conservation in a wide sense and with the development and management of genetic reserves. In both organizations there is increasing interest in the development of programmes which involve local people in the management and use of protected areas. FAO is involved in the conservation of forest genetic resources (which is largely carried out *in situ*) and with the conservation of crop and forage species and their wild relatives. Its Commission on Plant Genetic Resources is a key intergovernmental decision-making body concerned with the policy aspects of PGRFA conservation.

The Parties to the Convention on Biological Diversity (COP) have established an advisory body, the Subsidiary Body on Scientific, Technical and Technological Advice (SBSTTA), that has among its responsibilities the overseeing of conservation policies and action in the context of the Convention. This body has a particular concern with the identification and clarification of issues to be addressed by the COP and maintains an oversight of areas requiring research and development.

Several international NGOs are concerned with conservation, particularly the World Wide Fund for Nature (WWF) and the World Conservation Union (IUCN). Both have been traditionally linked with the conservation of high profile species, but have an increasing interest in a much wider approach to biodiversity conservation. WWF has collaborated in the recent People and Plants Programme which seeks to involve communities in *in situ* conservation and IUCN has collaborated with the International Plant Genetic Resources Institute (IPGRI), UNEP and FAO in the production of technical guidelines for collecting PGRFA. A number of other NGOs have supported or are supporting substantial *in situ* conservation work relevant to PGRFA, including the Community Biodiversity Development and Conservation (CBDC) Programme (an international collaborative programme involving NGOs and national programmes) and Seeds of Survival, which supports the maintenance and improvement of land races in Ethiopia.

The centres of the Consultative Group on International Agricultural Research (CGIAR) are also expanding their involvement with *in situ* conservation through their research and training programmes. The International Plant Genetic Research Institute (IPGRI), for example, is initiating a substantial programme of investigation on the scientific basis of on-farm conservation, and collaborates closely with the Centre for International Forestry Research (CIFOR) in research on the conservation of forest genetic resources. The International Rice Research Institute (IRRI), the International Potato Centre (CIP) and the International Maize and Wheat Improvement Centre (CIMMYT) are all involved in research programmes concerned with on-farm conservation, and the International Centre for Agricultural Research in the Dry Areas (ICARDA) is participating in work on the *in situ* conservation of wild wheats.

For institutions to be effective in promoting and supporting the conservation of PGRFA *in situ*, especially at the national level, there is a need for a greatly strengthened human resource base. People with relevant skills are needed across a wide front, from nature reserve and wildlife management officers, agricultural extension workers and foresters, to information specialists and biological, environmental and social scientists and technicians.

National *in situ* conservation strategies should include human resource plans and appropriate education and training strategies. Training will be needed at the local and national levels for local conservation practitioners, while scientific training, if not available nationally, may have to be carried out in universities and other training and educational institutions abroad. Globally, there is a need to expand training opportunities, especially in relevant scientific disciplines including such areas as taxonomy, genetic diversity assessment, population genetics, reproductive biology, Geographic Information Systems, ecology, environmental sciences, anthropology, ethnobotany, and information sciences. Such training will often be most effective if carried out in the context of active research and conservation programmes.

23.6 FINANCIAL RESOURCES

Conservation of PGRFA, whether *in situ* or *ex situ*, is expensive. It is also a long-term venture. Benefits arising from its use accrue over time frames that make conservation *per se* unattractive for private investment. This market failure implies that conservation costs must, to a very great extent, be borne or at least controlled by the public sector, whether the funding is derived largely from the immediate users (e.g. a tax on the seed trade, the selling of logging concessions or revenues from entry fees to national parks) or less directly from the public purse.

Conserved genetic resources are a local, national and international public good and funding is needed to support activities at all these levels. Funds are needed, for example, to support the efforts of local communities in conserving the resources under their direct control – for example, through the creation of community gene banks, support for participatory breeding efforts in collaboration with the formal sector, and the promotion of 'minor' crops and 'diversity-rich' products. The practical recognition of Farmers' Rights through such means as the creation of National Community Gene Funds, as proposed by Swaminathan, represents one possibility for securing the funding needed to support local and community conservation activities. At the national level, funding is needed for such activities as the establishment and ongoing management of reserves, and for developing appropriate institutions to support the effort.

At the international level, and particularly in the wake of UNCED, funding for the *in situ* conservation of PGRFA is being made available both bilaterally to developing countries by donor agencies in the industrialized countries, and multilaterally through various UN agencies, through the CGIAR system, and through various international NGOs and others. In addition, the Global Environment Facility (GEF) represents a significant source of new funding for *in situ* conservation (Chapters 16 and 18). Other funding agencies, e.g. the World Bank, are also considering expanding their support to *in situ* conservation. Given the global public good nature of conserving PGRFA, it is arguable that international funding should come not only through development assistance channels but, more importantly, in ways that recognize the ultimate benefit to all people in all countries.

23.7 INFORMATION SYSTEMS

Chapter 11 pointed out the need for appropriate and effective information systems. Information is critical for all aspects of *in situ* conservation, from the determination of priority species, the estimation of genetic erosion, the location of *in situ* conservation areas, the management of reserves and the monitoring of gene conservation on-farm, to the information needed to promote the use of the genetic resources such as passport data, information on environmental variables, the distribution of diversity of useful traits, and indigenous and local knowledge about specific characteristics.

With the growing application of scientific techniques to *in situ* conservation, the volume of data and information generated will expand dramatically. Thus demands for effective information management systems will inevitably increase considerably in the coming years. The design and implementation of effective *in situ* conservation systems requires an ability to access and manipulate very large and complex data sets, and to be able to draw conclusions from very diverse sources of information. Computerized data management systems need to be developed further and there would appear to be very good opportunities for the application of Expert Systems as an aid to decision making.

Information systems for handling local knowledge require special attention, not only because of the nature of the data – being very difficult to structure formally – but also because of the very complicated and sensitive issues surrounding ownership rights to such information. Mechanisms need to be developed to ensure that the role of the providers of the information is fully recognized and their rights protected, while at the same time facilitating access to relevant information by those who need it.

While electronic access to large amounts of information and data is becoming increasingly possible through such means as the Internet, for conservationists in many parts of the world this is still a distant hope. There is an urgent need for international support for the expansion of information and data sharing systems globally, and for the development of mechanisms, such as CD-Rom systems, to enable relevant data and information to be accessed even by those who lack direct access to an electronic international information network.

23.8 PUBLIC AWARENESS

Ensuring the sustainability of efforts to conserve genetic resources requires a continuing commitment by governments to support these activities at local, national, regional and international levels. Many scientists and environmentalists around the world now understand the importance of conserving PGRFA *in situ*, but this does not necessarily confer the political will needed to translate objectives into reality. Even the requirement that signatory countries to the Convention on Biological Diversity develop *in situ* conservation strategies and programmes does not guarantee that sufficient financial and other resources will be available to allow the programmes to be viable in the long term. In addition, many agriculturists still fail to appreciate the need for *in situ* conservation, believing that the genes required for future plant breeding can best – or even only – be conserved *ex situ*. The role of farmers in the development and management of crop diversity has only recently begun to be appreciated by formal-sector scientists to any significant degree, and many of them regard the continued role of farmers in this respect as a relic.

The establishment of *in situ* conservation of PGRFA as a permanent item on global and national policy agendas is the only means of ensuring an ongoing commitment to *in situ* programmes. There is an opportunity cost to investing in conservation. To ensure continuing support for national *in situ* conservation efforts, it will be necessary to convince decision makers that such an investment offers a greater contribution than other options to a country's short-term as well as long-term development. At the same time, it will also be necessary to foster a broader constituency of support for *in situ* conservation among scientists and agriculturists.

Public awareness provides the key to mobilizing opinion and to fostering sustained political action nationally and globally. Well-focused and targeted public awareness activities aimed at policy makers and those who influence them (such as NGOs and the media) can increase the support and commitment of those who are in a position to set policies and make decisions on resource allocation. Public awareness targeted at scientists, technicians and even the conservation community at large can increase awareness of the importance of conserving PGRFA *in situ*. In addition, such efforts can be expected to facilitate and even create a demand from communities and local organizations to become involved in *in situ* conservation activities, thus helping to ensure a broader base for conservation.

23.9 CONCLUSIONS

Although a good start has been made in understanding the scientific underpinnings of *in situ* conservation, much remains to be done. As this book has clearly shown, there are many areas where further research is needed and where the development of new techniques will contribute substantially to conservation theory and practice.

There is an urgent need to begin more systematically to put theory into practice and to establish conservation programmes that will provide opportunities for empirical learning by the conservation community. There is currently a dearth of practical examples in the field – situations in which the scientific community can develop and test new hypotheses and try out new techniques.

While science will inevitably make substantial advances over the coming years, future *in situ* conservation efforts will undoubtedly be shaped as much by policy developments as by the purely technical. The outcome of the current debates and negotiations, covering a wide range of sensitive policy issues at both national and international levels, will influence everything from the number, location and management of reserves, and the development of support systems for on-farm conservation, to funding mechanisms and procedures for gaining access to *in situ* genetic resources and the information about them.

It is important that all involved in the effort – whether scientists or practitioners – play their part not only in advancing the science or practice of conservation, but also in helping to make relevant decision makers aware of the potential contribution of PGRFA to the development of sustainable and productive agricultural systems, and the role that *in situ* strategies can play within the overall context of conserving these invaluable resources for the future.

References

Abele, L.G. and Conner, E.F. (1979) Application of island biogeography theory to refuge design: making the right decision for the wrong reasons, in *Proceedings of the First Conference on Scientific Research in the National Parks* (ed. M. Linn), National Parks Service, Washington, DC.

Adams, W.T. (1992) Gene dispersal within forest tree populations. *New Forests*, **6**, 217–240.

Adams, W.T., Strauss, S.H., Copes, D.L. and Griffin, A.R. (eds) (1992) *Population Genetics of Forest Trees*, Kluwer Forestry Sciences 42, 420 pp.

Adey, M.E., Allkin, R., Bisby, F.A. and White R.J. (1984) The Vicieae Database: an experimental taxonomic monograph, in *Databases in Systematics* (eds R. Allkin and F.A. Bisby), Academic Press, London and Orlando, pp. 175–188.

Ahmadi, N., Glaszmann, J.C. and Rabary, E. (1991) Traditional highland rices originating from intersubspecific recombination in Madagascar, in *Rice Genetics II*, Proceedings of the Second International Rice Genetics Symposium, held at IRRI, Los Baños, Philippines, pp. 67–79.

Alcorn, J.B. (1984) *Huastec Mayan Ethnobotany*, University of Texas Press, Austin, Texas.

Alefeld, F. (1860) *Hypechusa*, nov. gen. Viciearum. *Öst. Bot. Zeitschr.*, **19**, 165–166.

Alexander, H.M. (1989) Spatial heterogeneity and disease in natural populations, in *Spatial Components of Plant Disease Epidemics* (ed. M.J. Jeger), Prentice-Hall, New Jersey, pp. 144–164.

Alexander, H.M. (1992a) Evolution of disease resistance in natural plant populations, in *Plant Resistance to Herbivores and Pathogens: Ecology, Evolution and Genetics* (eds R.S. Fritz and E.L. Simms), University of Chicago Press, Chicago, pp. 326–344.

Alexander, H.M. (1992b) Fungal pathogens and the structure of plant populations and communities, in The *Fungal Community: Its Organisation and Role in the Ecosystem* (eds G.C. Caroll and D.T. Wicklow), Marcel Dekker, New York, pp. 482–496.

Allard, R.W. (1970) Population structure and sampling methods, in *Genetic Resources in Plants – their Exploration and Conservation* (eds O.H. Frankel and E. Bennett), IBP Handbook, No. 11, Blackwell Scientific Publications, Oxford, pp. 97–107.

Allard, R.W. (1988) Genetic changes associated with the evolution of adaptedness in cultivated plants and their wild progenitors. *Journal of Heredity*, **79**, 225–238.

Allard, R.W. (1990) The genetics of host–pathogen coevolution. Implication for genetic resources conservation. *Journal of Heredity*, **81**, 1–6.

Allegretti, M.H. (1990) Extractive reserves: an alternative for reconciling development and environmental conservation in Amazonia, in *Alternatives to Deforestation* (ed. A.B. Anderson), Columbia University Press, New York, pp. 252–264.

Allegretti, M.H. and Scwartzman, S. (1986) *Extractive reserves; a sustainable development alternative for Amazonia.* Unpublished report to WWF-US, Washington, DC.

Allkin, R., Goyder, D.J., Bisby, F.A. and White, R.J. (1986) *Names and Synonyms of Species and Subspecies in the Vicieae: Issue 3*, Vicieae Database Project, Southampton, 46 pp.

Altieri, M.A. (1994) *Biodiversity and Pest Management in Agro-ecosystems*, Food Products Press, New York.

Altieri, M.A. and Merrick, L.C. (1987) *In situ* conservation of crop genetic resources through maintenance of traditional farming systems. *Economic Botany*, **41**, 86–96.

Ambrose, J.D. (1995) *Status report on the Red Mulberry (*Morus rubra*) in Canada*, Committee on the Status of Endangered Wildlife in Canada, Ottawa, 17 pp.

Anderson, E. (1952) *Plants, Man and Life*, University of California Press, Berkeley, CA.

Andow, D.A. (1991) Vegetational diversity and arthropod population response. *Annual Review of Entomology*, **36**, 561–586.

Anikster, Y. and Noy-Meir, I. (1991) The wild-wheat field laboratory at Ammiad. *Israel Journal of Botany*, **40**, 351–362.

Anikster, Y., Eshel, A., Ezrati, S. and Horovitz, A. (1991) Patterns of phenotypic variation in wild tetraploid wheat at Ammiad. *Israel Journal of Botany*, **40**, 397–418.

Anikster, Y., Waldman, M., Ashri, A. and Noy-Meir, I. (1988) Population dynamics research for *in situ* conservation: wild wheat in Israel. *Plant Genetic Resources Newsletter*, **75–76**, 9–11.

Anon. (1990) *Natural Reserves of Central Asia and Kazakhstan*, Mysl' Publishers, Moscow, 399 pp.

Anon. (1995) Royal fungus saved. *New Scientist*, **148** (October 14), 11.

Arbez, M. (1994) Fondement et organisation des réseaux européens de conservation des ressources génétiques forestières. *Genet. Sel. Evol.*, **26**, Suppl. 1, 301s–314s.

Arnold, J.E.M. (1991) *Community forestry: ten years in review*, FAO Community Forestry Note 7, FAO, Rome, 31 pp.

Arnold, M.H., Astley, D., Bell, E.A. *et al.* (1986) Plant gene conservation. *Nature*, **319**, 615.

Ataro, A. and Bayush, T. (1994) *Survey on relative performance of land races for the 1993/94 cropping season. Consultancy Report*, Seeds of Survival, Addis Ababa, Ethiopia.

Auricht, G.C., Reid, R. and Guarino, L. (1995) Published information on the natural and human environment, in *Collecting Plant Genetic Diveristy: Technical Guidelines* (eds L. Guarino, V. Ramanatha Rao and R. Reid), CAB International, Wallingford, pp. 131–152.

Baker, W.L. (1989) Landscape ecology and nature reserve design in the Boundary Waters Canoe Area, Minnesota. *Ecology*, **70**, 23–35.

Balée, W. (1994) *Footprints of the Forest: Ka'apor Ethnobotany. The Historical Ecology of Plant Utilisation by an Amazonian People*, Columbia University Press, New York, 396 pp.

Barker, A.M., Wratten, S.D. and Edwards, P.J. (1995) Wound-induced changes in tomato leaves and their effect on the feeding patterns of larval Lepidoptera. *Oecologia*, **101**, 251–257.

Batisse, M. (1986) Developing and focusing the biosphere reserve concept. *Nature and Resources*, **22**, 1–10.

Baudry, J. and Merriam, H.G. (1988) Connectivity and connectedness: functional versus structural patterns in landscapes, in *Connectivity in Landscape Ecology: Proceedings of the 2nd International Association for Landscape Ecology Seminar* (ed. K.F. Schreiber), *Münstersche Geographische Arbeitung*, **29**, 23–28.

Bawa, K.S. (1994) Effects of deforestation and forest fragmentation on genetic diversity in tropical tree populations, in *Proceedings of International Symposium on Genetic Conservation and Production of Tropical Forest Tree Seed* (eds R.M. Drysdale S.E.T. John and A.C. Yappa), ASEAN–Canada Forest Tree Seed Project, Muak-Lek, Thailand, pp. 10–16.

Bawa, K.S. and Ashton, P.S. (1991) Conservation of rare trees in tropical rainforests: a genetic perspective, in *Genetics and Conservation of Rare and Endangered Plants* (eds D.A. Falk and K.E. Holsinger), Oxford University Press, Oxford, pp. 62–71.

Bawa, K.S. and Krugman, S.L. (1991) Reproductive biology and genetics of tropical trees in relation to conservation and management, in *Rainforest Regeneration and Management* (eds A. Gómez-Pompa, T.C. Whitmore and M. Hadley), MAB Series 6, UNESCO and Parthenon Publishing Group, Paris, pp. 119–136.

Beerling, D.J. (1993) The impact of temperature on the northern distribution limits of the introduced species *Fallopia japonica* and *Impatiens glandulifera*. *Journal of Biogeography*, **20**, 45–53.

Bellon, M.R. (1991) The ethnoecology of maize variety management: A case study from Mexico. *Human Ecology*, **19**, 389–418.

Bellon, M.R. and Brush, S.B. (1994) Keepers of maize in Chiapas, Mexico. *Economic Botany*, **48**, 196–209.

Bellon, M.R. and Taylor, J.E. (1993) Farmer soil taxonomy and technology adoption. *Economic Development and Cultural Change*, **41**, 764–786.

Belsky, A.J. (1985) Does herbivory benefit plants? A review of the evidence. *American Naturalist*, **127**(6), 870–892.

Bennet, M.D. and Smith, J.B. (1991) Nuclear DNA amounts in angiosperms. *Philosophical Transactions of the Royal Society (London), Series B*, **334**, 309–345.

Bennett, E. (1970) Tactics in plant exploration, in *Genetic Resources in Plants – their Exploration and Conservation* (eds O.H. Frankel and E. Bennett), IBP Handbook No. 11, Blackwell, Oxford, pp. 157–179.

Bently, S. and Whittaker, J.B. (1979) Effects of grazing by a chrysomelid beetle (*Gastrophysa viridulatus*) on competition between *Rumex obtusifolius* and *Rumex criopus*. *Journal of Ecology*, **67**, 79–90.

Berlin, B., Breedlove, D.E. and Raven, R.H. (1974) *Principles of Tzeltal Plant Classification: an Introduction to Botanical Ethnography of a Mayan-speaking Community in Highland Chiapas*, Academic Press, New York.

Bierregard Jr, R.O., Lovejoy, T.E., Kapos, V. *et al.* (1992) The biological dynamics of tropical rainforest fragments. *BioScience*, **42**, 859–866.

Bisby, F.A. (1994) Global master species databases and biodiversity. *Biology International*, **29**, 33–40.

Blanco, O. (1988) *Agricultura de Subsistencia y Agricultura Comercial en los Andes*. *Cuadernos Informativos*, Revista de la CCTA, Lima, Peru.

Blanco O. (1993) Los Recursos Genéticos en los Sistemas Productivos Andinos – Conservación *In-situ*, in *Biotecnologia, Recursos Fitogenéticos y Agricultura en los Andes*, CCTA, Lima, Peru.

Boecklen, W.J. (1986) Optimal design of nature reserves: consequences of genetic drift. *Biological Conservation*, **38**, 323–338.

Boecklen, W.J. and Bell, G.W. (1987) Consequences of faunal collapse and genetic drift to the design of nature reserves, in *Nature Conservation: the Role of Remnants of Native Vegetation* (eds D.A. Saunders, G.A. Arnold, A.A. Burbidge and A.J.M. Hopkins), Surrey Beatty/CSIRO/CALM, WA, Australia, pp. 141–149.

Boffey, S.A. (1991) in *Methods in Plant Biochemistry* (eds P.M. Dey and J.B. Harborne), Academic Press, London and New York, pp. 147–169.

Boissier, E. (1872) *Flora Orientalis, Vol. 2*. H. Georg, Geneva and Basle.

Bon, M.C., Alih, B.B. and Joly, H.I. (1995) Genetic diversity of two rattan species: application of isozyme markers, in *Measuring and Monitoring Biodiversity in Tropical and Temperate Forests: Proceedings, IUFRO Meeting, Chiangmai, Thailand, 28 August–2 September, 1994*, ASEAN–Canada Forest Tree Seed Project, Muak-Lek, Thailand (in press).

Borgel, A. and Second, G. (1978) *Prospections des variétés traditionnelles et des espèces sauvages de riz au Tchad et au Cameroun*. Rapport de mission du 19 Octobre au 23 Décembre 1977. Rapport multigr. ORSTOM, Abidjan.

Boshier, D.H., Chase, M.R. and Bawa, K.S. (1995a) Population genetics of *Cordia alliodora* (Boraginaceae), a neotropical tree. 3. Gene flow, neighborhood, and population substructure. *American Journal of Botany*, **82**, 484–490.

Boshier, D.H., Chase, M.R. and Bawa, K.S. (1995b) Population genetics of *Cordia alliodora* (Boraginaceae), a neotropical tree. 2. Mating system. *American Journal of Botany*, **82**, 476–483.

Boster, J. (1983) A comparison of the diversity of Jivaroan gardens with that of the tropical forest. *Human Ecology*, **11**, 47–68.

Bothmer, von R. and Seberg, H. (1995) Collecting strategies: wild species, in *Collecting Plant Genetic Diversity* (eds L. Guarino, V. Ramanatha Rao and R. Reid), CABI, Wallingford, pp. 93–111.

Bowman, I. (1980) *Traducción de Carlos Nicholson – Los Andes del Sur del Peru*. Primera Edición Editorial Universo SA, Lima, Peru.

Boyce, M.S. (1992) Population viability analysis. *Annual Review of Ecology and Systematics*, **23**, 481–506.

Breasted, J.H. (1938) *The Conquest of Civilization* (ed. E. Williams Ware), Literary Guild of America, New York.

Brezhnev, D.D. and Korovina, O.N. (1981) Wild relatives of cultivated plants, in *Flora of the USSR*, Kolos Publishers, Leningrad.

Briscoe, D.A., Malpica, J.M., Robertson, A. *et al.* (1992) Rapid loss of genetic variation in large captive populations of *Drosophila* flies: implications for the genetic management of captive populations. *Conservation Biology*, **6**, 416–425.

Brown, A.H.D. (1978) Isozymes, plant population genetic structure and genetic conservation. *Theoretical and Applied Genetics*, **52**, 145–157.

Brown, A.H.D. (1991) Population divergence in wild crop relatives. *Israel Journal of Botany*, **40**, 512.

Brown, A.H.D. and Allard, R.W. (1970) Estimation of the mating system in open-pollinated maize populations using isozyme polymorphisms. *Genetics*, **66**, 133–145.

Brown, A.H.D. and Moran, G.F. (1981) Isozymes and genetic resources of forest trees, in *Isozymes of North American Forest Trees and Forest Insects* (ed. M.T. Conkle), USDA, Berkeley, CA, pp. 1–10.

Brown, A.H.D., Frankel, O.H., Marshall, D.R. and Williams, J.T. (eds) (1989) *The Use of Plant Genetic Resources*, Cambridge University Press, Cambridge, 382 pp.

Brown, V.K. (1991) The effects of changes in habitat structure during succession in terrestrial communities, in *Habitat Structure: the Physical Arrangement of Objects in Space* (eds S.S. Bell, E.D McCoy and H.R. Mushohsky), Chapman & Hall, London, pp. 141–168.

Browning, J.A. (1991) *In situ* conservation: the only way of conserving genetic diversity and assuring continuing host–pathogen–hyperparasite evolution. *Israel Journal of Botany*, **40**, 515.

Browning, J.A., Manisterski, J., Segal, A. *et al.* (1982) Extrapolation of genetic and epidemiologic concepts from indigenous ecosystems to agroecosystems, in *Resistance to Diseases and Pests in Forest Trees* (eds H.M. Heybroek, B.R. Stephan and K. von Weissenberg), Pudoc, Wageningen, pp. 371–380.

Brummitt, R.K. and Powell, C.E. (1992) *Authors of Plant Names*, Royal Botanic Gardens, Kew.

Brush, S.B. (1991a) A farmer-based approach to conserving crop germplasm. *Economic Botany*, **45**, 153–165.

Brush, S.B. (1991b) Farmer conservation of New World crops. The case of Andean Potatoes. *Diversity*, **7**, Nos. 1 and 2, 82–86.

Brush, S.B. (1992) Ethnoecology, biodiversity and modernization in Andean potato agriculture. *Journal of Ethnobiology*, **12**, 161–185.

Brush, S.B. (1995) *In situ* conservation of landraces in centers of crop diversity. *Crop Science*, **35**, 346–354.

Brush, S.B., Carney, H.J. and Huamán, Z. (1981) Dynamics of Andean potato agriculture. *Economic Botany*, **35**, 70–88.

Brush, S., Kressell, R., Ortega, R. *et al.* (1995) Potato diversity in the Andean centre of crop domestication. *Conservation Biology*, **9**, 1189–1198.

Brush, S.B., Taylor, J.E. and Bellon, M.R. (1992) Biological diversity and technology adoption in Andean potato agriculture. *Journal of Development Economics*, **39**, 365–387.

Burdon, J.J. (1987a) Phenotypic and genetic patterns of resistance to the pathogen *Phakospora pachyrhizi* in populations of *Glycine canescens. Oecologia*, **73**, 257–267.

Burdon, J.J. (1987b) *Diseases and Plant Population Biology*, Cambridge University Press, Cambridge.

Burdon, J.J. and Jarosz, A.M. (1988) The ecological genetics of plant–pathogen interactions in natural communities. *Philosophical Transactions of the Royal Society of London, Series B*, **321**, 349–363.

Burdon, J.J. and Jarosz, A.M. (1989) Wild relatives as sources of disease resistance, in *The Use of Plant Genetic Resources* (eds A.H.D. Brown, D.R. Marshall, O.H. Frankel and J.T. Williams), Cambridge University Press, Cambridge, pp. 280–296.

Burdon, J.J. and Jarosz, A.M. (1991) Host–pathogen interactions in natural populations of *Linum marginale* and *Melampsora lini*: I. Patterns of resistance and racial variation in a large host population. *Evolution*, **45**, 205–217.

Burdon, J.J. and Marshall, D.R. (1985) Host–pathogen relationships: struggle of the genes, in *Pests and Parasites as Migrants* (eds A. Gibbs and R. Meischke), Ruskin Press, Victoria, Australia, pp. 104–107.

Burdon, J.J. and Muller, W.J. (1987) Measuring the cost of resistance to *Puccinia coronata* Cda, in *Avena fatua* L. *Journal of Applied Ecology*, **24**, 191–200.

Burdon, J.J., Jarosz, A.M. and Kirby, G.C. (1989) Pattern and patchiness in plant–pathogen interactions – causes and consequences. *Annual Review of Ecology and Systematics*, **20**, 119–136.

Burel, F. and Baudry, J. (1995) Farming landscapes and insects, in *Ecology and Integrated Farming Systems* (eds D.M. Glen, M.P. Greaves and H.M. Anderson), Wiley, Chichester, pp. 203–220.

Burley, J. (1993) Balance between development and conservation, in *Proceedings of the International Symposium and Genetic Conservation and Production of Tropical Forest Tree Seed* (eds R.M. Drysdale, S.E.T. John and A.C. Yapa), ASEAN–CFTSC, CIRAD-Forêt, IDRC, Bangkok, Thailand, pp. 17–25.

Burley, J., Hughes, C.E. and Styles, B.T. (1986) Genetic systems of tree species for arid and semi-arid lands. *Forest Ecology and Management*, **16**, 317–344.

Burrough, P.A. (1986) *Principles of Geographic Information Systems for Land Resources Assessment*, Clarendon Press, Oxford.

Byerlee, D. and Moya, P. (1993) Impacts of international wheat breeding research in the developing world, 1966–90. CIMMYT, Mexico, D.F.

Byron, R.N. and Waugh, G. (1988) Forestry and fisheries in the Asian-Pacific region: issues in natural resource management. *Asian-Pacific Economic Literature*, **2**, 46–80.

Caballero, A. (1994) Developments in the prediction of effective population size. *Heredity*, **73**, 657–679.

Caetano-Anolles, G., Bassam, B.J. and Gresshof, P.M. (1991) High resolution DNA amplification fingerprinting using very short arbitrary oligonucleotide primers. *Bio-Technology*, **9**, 553–557.

Caetano-Anolles, G., Bassam, B.J. and Gresshof, P.M. (1992) Primer–template interactions during DNA amplification fingerprinting with single arbitrary oligonucleotides. *Molecular Genetics*, **235**, 157–165.

Caetano-Anolles, G., Bassam, B.J. and Gresshof, P.M. (1993) DNA amplification fingerprinting with very short primers, in *Application of RAPD Technology to Plant Breeding* (ed. M. Neff), ASHS Publishers, St Paul, MN, pp. 18–25.

Callaway, R.M. and Davis, F.W. (1993) Vegetation dynamics, fire and the physical environment in coastal central California. *Ecology*, **74**, 1567–1578.

Campbell, B. (1983) *Human Ecology: the Story of our Place in Nature from Prehistory to the Present*, Heinemann Educational Books, London.

Campbell, D.G. and Hammond, H.D. (eds) (1989) *Floristic Inventory of Tropical Countries*, New York Botanical Garden, 545 pp.

Carter, R.N. and Prince, S.D. (1981) Epidemic models used to explain biogeographical distribution limits. *Nature*, **293**, 644–645.

Carter, R.N. and Prince, S.D. (1988) Distribution limits from a demographic viewpoint, in *Plant Population Ecology* (eds A.J.Davy, M.J. Hutchings and A.R. Watkinson), Blackwell, Oxford, pp. 165–184.

Cassells, D.S. (1992) Forested watershed controls in N-E Australia as an interim model for other humid tropical forest environments. *ITTO Tropical Forest Management Update* **2**(2), 6–8.

Caswell, H. (1989) *Matrix Population Models*, Sinauer Associates, Sunderland, MA.

Catibog-Sinha, C. (1993) Implications of the NIPAS law for the conservation of genetic resources in the Philippines, in *Proceedings International Symposium on Genetic Conservation and Production of Tropical Forest Tree Seed* (eds R.M. Drysdale, S.E.T. John and A.C. Yapa), ASEAN–CFTSC, CIRAD-Forêt, IDRC, Bangkok, Thailand, pp. 214–220.

Causse, M.A., Fulton, T.M., Cho, Y.G. *et al.* (1994) Saturated molecular map of the rice genome based on an interspecific backcross population. *Genetics*, **138**, 1251–1274.

CGIAR (1994) *Keeping Faith with the Future: Forests and their Genetic Resources*, IPGRI, Rome, 14 pp.

Chalmers, K.J., Waugh, R., Sprent, R.I. *et al.* (1992) Detection of genetic variation between and within populations of *Gliricidia sepium* and *G. maculata* using RAPD markers. *Heredity*, **69**, 465–472.

Chang, T.T. (1976) The origin, evolution, cultivation, dissemination and diversification of Asian and African rices. *Euphytica*, **25**, 425–441.

Chang, T.T. (1985) Crop history and genetic conservation: rice – a case study. *Iowa State Journal of Research*, **59**, 425–455.

Chang, T.T. (1994) Plant genetic resources conservation and utilization. *Encyclopedia Agricultural Science*, **3**, 295–304.

Chang, T.T. (1995) Rice, in *Evolution of Crop Plants* (eds J. Smartt and N.W. Simmonds), Longman, London, pp. 147–155.

Chapman, C.G.D. (1986) The role of genetic resources in wheat breeding. *Plant Genetic Resources Newsletter*, **65**, 2–5.

Chapman, C.G.D. (1989) Collection strategies for the wild relatives of field crops, in *The Use of Plant Genetic Resources* (eds A.H.D. Brown, D.R. Marshall, O.H. Frankel and J.T. Williams) Cambridge University Press, Cambridge, pp. 263–279.

Chase, M.R., Boshier, D.H. and Bawa, K.S. (1995) Population genetics of *Cordia alliodora* (Boraginaceae), a neotropical tree. 1. Genetic variation in natural populations. *American Journal of Botany*, **82**, 468–475.

Christanty, L., Abdoellah, O.S., Marten G.G. and Iskandar, J. (1986) Traditional agroforestry in West Java: The pekarangan (homegarden) and kebun-talun (annual–perennial rotation) cropping systems, in *Traditional Agriculture in Southeast Asia; a Human Ecology Perspective* (ed. G. Marten), Westview Press, Boulder, CO, pp. 132–158.

Clark, D.B. (1991) The role of disturbance in the regeneration of neotropical moist forests, in *Reproduction Ecology of Tropical Forest Plants: Man and the Biosphere, Vol. 7*, Parthenon Publishing Group, Paris, pp. 291–315.

Clarke, R. (1986) *The Handbook of Ecological Monitoring*, Oxford University Press, Oxford.

Cleveland, D.A., Soleri, D. and Smith, S.E. (1994) Do folk crop varieties have a role in sustainable agriculture? *BioScience*, **44**, 740–751.

Coats, A. (1969) *The Quest for Plants: a History of the Horticultural Explorers*, Studio Vista, London.

Coffman, W.R. and Smith, M.E. (1991) Roles of public, industry, and international research center breeding programs in developing germplasm for sustainable agriculture, in *Plant Breeding and Sustainable Agriculture: Considerations for Objectives and Methods* (eds D.A. Sleper, T.C. Barker and P. Bramel-Cox), Crop Science Society of America Special Publication No. 18, Madison, WI, pp. 1–9.

Cohen, J.I., Alcorn, J.B. and Potter, C.S. (1991a) Utilization and conservation of genetic resources: International projects for sustainable agriculture. *Economic Botany*, **45**, 190–199.

Cohen, J.I., Williams, J.T., Plucknett, D.L. and Shands, H. (1991b) *Ex situ* conservation of plant genetic resources: global development and environmental concerns. *Science*, **253**, 866–872.

Colchester, M. and Lohmann, L. (eds) (1993) *The Struggle for Land and the Fate of the Forests*, Zed Books, 389 pp.

CONDESAN (1993) *Planificación y Priorización de Actividades de Investigación y Acción para el Desarrollo sostenible de la Ecoregión Andina*, 6–14. Centro Internacional de la Papa, La Molina, Lima, Peru.

Conklin, H.C. (1957) *Hanunóo agriculture: a report on an integral system of shifting cultivation in the Philippines*, FAO Forestry Development Paper No. 12, FAO, Rome.

Conklin, H.C. (1986) Des orientements, de vents, de riz ... pour une étude lexicologique des savoirs traditionnels. *Journal d'Agriculture Traditionnelle et de Botanique Appliquée*, **33**, 3–10.

Connell, J.H. (1978) Diversity in tropical rain forests and coral reefs. *Science*, **199**, 1302–1310.

Cooke, R.J. (1984) The characterization and identification of crop cultivars by electrophoresis. *Electrophoresis*, **5**, 59–72.

Cooper, D., Vellvé, R. and Hobbelink, H. (1992) *Growing Diversity: Genetic Resources and Local Food Security*, Intermediate Technology Publication, London, 165 pp.

Cox, G.W. (1993) *Conservation Ecology*, W.C. Brown, Dubugue, IA.

Crawford, T.J. (1984) What is a population? in *Evolutionary Ecology* (ed. B. Shorrocks), Blackwell Scientific Publications, Oxford, pp. 135–173.

Crawley, M.J. (ed.) (1982) *Herbivory: Studies in Ecology*, Blackwell Scientific Publications, Oxford.

Crawley, M.J. (1982) Herbivory: the dynamics of animal–plant interactions, in *Herbivory: Studies in Ecology* (ed. M.J. Crawley), Blackwell Scientific Publications, Oxford, pp. 211–287.

Crawley, M.J. (1985) Reduction in oak fecundity by low-density herbivore populations. *Nature*, **314**, 163–164.

Crawley, M.J. (1989) Insect herbivores and plant population dynamics. *Annual Review of Entomology*, **34**, 531–564.

Cregan, P.B., Akkaya, M.S., Bhagwat, A.A. *et al.* (1994) Length polymorphisms of simple sequence repeat (SSR) DNA as molecular markers in plants, in *Plant Genome Analysis* (ed. P.M. Gresshof), CRC Press, London.

Cromwell, E. (1990) *Seed diffusion mechanisms in small farmer communities: lessons from Asia, Africa and Latin America*, Network Paper 21, Agricultural Administration (Research and Extension) Network, Overseas Development Institute, London.

Cromwell, E., Brodie, A. and Southern, A. (1996) *Germplasm for multipurpose trees: access and utility for small farm communities*, ODI Research Study, Overseas Development Institute, London, 93 pp.

Cronk, Q.C.B. (1993) Extinction and conservation in the St Helena flora: the palaeobiological and ecological background. *Boletim do Museu Municipal do Funchal*, Sup. No. 2, 69–76.

Cropper, S.C. (1993) *Management of Endangered Plants*, CSIRO Publications, East Melbourne.

Crossa, J., DeLacey, I.H. and Taba, S. (1995) The use of multivariate methods in developing a core collection, in *Core Collections of Plant Genetic Resources* (eds T. Hogkin, A.H.D. Brown, Th.J.L. van Hintum and E.A.V. Morales), John Wiley, Chichester, England.

Crow, J.F. and Aoki, K. (1984) Group selection for a polygenic behavioural trait: estimating the degree of population subdivision. *Proceedings of the National Academy of Science, USA*, **81**, 6073–6077.

Crow, J.F. and Kimura, M. (1970) *An Introduction to Population Genetics Theory*, Harper and Row, New York.

Croy, E.J., Ikemura, T., Shirsat, A. and Croy, R.R.D. (1993) Plant nucleic acids, in *Plant Molecular Biology, LabFax* (ed. R.R.D. Croy), Bios Scientific Publishers, Blackwell, Oxford.

Cuevas-Perez, F.E., Guimaraes, E.P., Berrio, L.E. and Gonzalez, D.I. (1992) Genetic base of irrigated rice in Latin America and the Caribbean, 1971 to 1989. *Crop Science*, **32**, 1054–1059.

Cunningham, A.B. (1994) Integrating local plant resources and habitat management. *Biodiversity and Conservation*, **3**, 104–115.

Dallas, J.F. (1988) Detection of DNA 'fingerprints' of cultivated rice by hybridisation with human minisatellite probe. *Proceedings of the National Academy of*

Sciences, **85**, 6831–6835.

Dallas, J.F., McIntyre, C.L. and Gustafson, J.P. (1993) An RFLP species-specific DNA sequence for the A genome of rice. *Genome*, **36**, 445–448.

Damania, A.B. (1994) *In situ* conservation of biodiversity of wild progenitors of cereal crops in the Near East. *Biodiversity Letters*, **2**, 56–60.

David, C.C. and Otsuka, K. (1994) *Modern Rice Technologies and Income Distribution in Asia*, L. Rienner, IRRI, Boulder, CO.

David, J. (1992) Approche méthodologique d'une gestion dynamique des ressources génétiques chez le blé tendre (*Triticum aestivum*). Thesis, Institut National Agronomique Paris-Grignon.

Davis, M.B. (1989) Lags in vegetation response to greenhouse warming. *Climatic Change*, **15**, 75–82.

Davis, P.H. and Plitmann, U. (1970) *Vicia* L., in *Flora of Turkey, Vol. 3*, Edinburgh University Press, Edinburgh, pp. 274–325.

Davis, S.D., Droop, S.J.M., Gregerson, P. *et al.* (1986) *Plants in Danger: What do we Know?* International Union for Conservation of Nature and Natural Resources, Gland, Switzerland.

Dawson, I.K., Chalmers, K.J., Waugh, R. and Powell, W. (1993) Detection and analysis of genetic variation in *Hordeum spontaneum* populations from Israel using RAPD markers. *Molecular Ecology*, **2**, 151–159.

de Angelis, D.L. (1992) *Dynamics of Nutritional Cycling and Food Webs*, Chapman & Hall, London.

de Beer, J.H. and McDermott, M.J. (1989) *The Economic Value of Non-timber Forest Products in Southeast Asia*, Netherlands Committee for IUCN, Amsterdam, 175 pp.

Debouk, D.G., Toro, O., Paredes, O.M. *et al.* (1993) Genetic diversity and ecological distribution of *Phaseolus vulgaris* (Fabaceae) in Northwestern South America. *Economic Botany*, **47**, 408–423.

Dedkov, V.P. (1990) Ways of maintaining stability in ecosystems of the Repetek biospheric natural reserve, in *Natural Reserves of the USSR, their Present and Future*, Part II, Novgorod, pp. 44–46.

de Klemm, C. and Shine, C. (1993) *Biological Diversity Conservation and the Law. Legal Mechanisms for Conserving Species and Ecosystems*, International Union for Conservation of Nature and Natural Resources, Gland.

del Amo, S. (1992) Problems of forest conservation: a feasible mechanism for biodiversity conservation, in *Changing Tropical Forests: Historical Perspectives on Today's Challenges in Central and South America*, Forest History Society, USA, pp. 154–164.

Delfosse, E.S., Hasan, S., Cullen, J.M. *et al.* (1986) Beneficial use of an exotic phytopathogen, *Puccinia chondrillina*, as a biological control agent for skeleton weed, *Chondrilla juncea*, in Australia, in *Pests and Parasites as Migrants* (eds A. Gibbs and R. Meischke), Cambridge University Press, Cambridge, pp. 171–177.

Dennis, J.V. (1987) Farmer management of rice variety diversity in northern Thailand. Unpublished PhD dissertation, Cornell University. Michigan University Microfilms, Ann Arbor.

Denslow, J.S. (1985) Disturbance-mediated coexistence of species, in *The Ecology of Natural Disturbance and Patch Dynamics* (eds S.T.A. Pickett and P.S. White), Academic Press, New York, pp. 308–321.

Diamond, J.M. (1976) Island biogeography and conservation: strategy and limitations. *Science*, **193**, 1027–1029.

Diamond, J.M. (1984) Normal extinction of isolated populations, in *Extinctions* (ed. M.H. Nitecki), Chicago University Press, Chicago, pp. 191–246.

Dietrich, W., Katz, H., Lincoln, S.E. *et al.* (1992) A genetic map of the mouse suitable for typing intraspecific crosses. *Genetics,* **131,** 423–447.

Dinoor, A. (1962) The effect of *Quercus ithaburensis* on its herbaceous associates. *Bulletin of the Research Council of Israel,* **11,** 1–16.

Dinoor, A. (1970) Sources of oat crown rust resistance in hexaploid and tetraploid wild oats in Israel. *Canadian Journal of Botany,* **48,** 153–161.

Dinoor, A. (1975) Evaluation of sources for disease resistance, in *Crop Genetic Resources for Today and Tomorrow* (eds O.H. Frankel and J.G. Hawkes), Cambridge University Press, Cambridge, pp. 201–210.

Dinoor, A. (1981) Dynamic preservation of germ-plasm in nature reserves. *Kidma,* **24,** 22–25.

Dinoor, A. and Eshed, N. (1983) Pathogen problems in relation to the exploitation of alien resistance. *Proceedings of the 4th International Congress of Plant Pathology,* Melbourne, Australia.

Dinoor, A. and Eshed, N. (1984) The role and importance of pathogens in natural plant communities. *Annual Review of Phytopathoogy,* **22,** 443–466.

Dinoor, A. and Eshed, N. (1987) The analysis of host and pathogen populations in natural ecosystems, in *Populations of Plant Pathogens: their Dynamics and Genetics* (eds M.S. Wolfe, and C.E. Caten), Blackwell Scientific Publications, Oxford, pp. 75–88.

Dinoor, A. and Eshed, N. (1990) Plant diseases in natural populations of wild barley (*Hordeum spontaneum*), in *Pests, Pathogens and Plant Communities* (eds J.J. Burdon and S.R. Leather), Blackwell Scientific Publications, Oxford, pp. 169–186.

Dinoor, A. and Eshed, N. (1991) Dynamic host–parasite conservation from the stand point of a pathologist. *Israel Journal of Botany,* **40,** 516.

Dinoor, A. and Peleg, N. (1972) The identification of genes for resistance or virulence without genetic analyses, by the aid of the gene-for-gene hypothesis. *Proceedings of the European and Mediterranean Cereal Rusts Conferences,* Praha, CR, **11,** 115–119.

Dinoor, A., Eshed, N., Ecker, R. *et al.* (1991) Fungal diseases of wild tetraploid wheat in a natural stand in Northern Israel. *Israel Journal of Botany,* **40,** 481–500.

Draper, D. and Scott, R. (1988) The isolation of plant nucleic acids, in *Plant Genetic Transformation and Gene Expression; a Laboratory Manual* (eds J. Draper, R. Scott, P. Armitage and R. Walden), Blackwell Scientific Publications, Oxford, pp. 199–236.

Dudley, N. (1993) *Forests in Trouble: a Review of the Status of Temperate Forests Worldwide,* WWF International, Gland, 260 pp.

Dunn, C.P., Stearns, F., Guntonspergen, G.R. and Sharpe, D.M. (1993) Ecological benefits of the conservation reserve program. *Conservation Biology,* **7**(3), 132–139.

Ehrman, T.A.M. and Maxted, N. (1990) Ecogeographic survey and collection of Syrian Vicieae and Cicereae (Leguminosae). *Plant Genetic Resources Newsletter,* **77,** 1–8.

Eldridge, K., Davidson, J., Harwood, C. and van Wyk, G. (1993) *Eucalypt Domestication and Breeding,* Clarendon Press, Oxford, 288 pp.

Ellstrand, N.C. (1992a) Gene flow among seed plant populations. *New Forests,* **6,** 241–256.

Ellstrand, N.C. (1992b) Gene flow by pollen: implications for plant conservation genetics. *Oikos,* **63,** 77–87.

Ellstrand, N.C. and Elam, D.R. (1993) Population genetic consequences of small population size: implications for plant conservation. *Annual Review of Ecology and Systematics,* **24**, 217–242.

Emberson, R.M., Early, J.W., Marris, J.W.M. and Syrett, P. (1993) *Research into the status and distribution of Chatham Island endangered invertebrates.* Final Report for the Department of Conservation, Canterbury Conservancy, New Zealand.

Eriksson, G., Namkoong, G. and Roberts, H. (1993) Dynamic gene conservation for uncertain futures. *Forest Ecology and Management,* **62**, 15–37.

ERIN (1991) *ERIN Newsletter,* Environmental Research Information Network, Canberra, Australia.

Erlich, H.A. (1989) *PCR Technology: Principles and Applications for DNA Amplification,* Stockton Press, Stockton.

Erlich, P.R. and Raven, P.H. (1954) Butterflies and plants: a study in coevolution. *Evolution,* **18**, 588–608.

Erwin, T.L. (1982) Tropical forests: their richness in Coleoptera and other arthropod species. *Coleopterists Bulletin,* **35**, 74–75.

Erwin, T.L. (1988) The tropical forest canopy: the heart of biotic diversity, in *Biodiversity* (ed. E.O. Wilson), National Academy Press, Washington DC, pp. 123–129.

Eshed, A., Dinoor, A. and Litwin, Y. (1994) The physiological specialization of wheat powdery mildew in Israel and the search for mildew resistance in wild wheat *Triticum dicoccoides. Phytoparasitica,* **22**, 75.

Eshel, A., Anikster, Y. and Horovitz, A. (1989) Strategies of phenotypic variation in an annual self-pollinating plant. *Abstracts of the 2nd Congress of European Society for Social and Evolutionary Biology,* Rome, 25–29 September 1989, 24.

Esquivel, M. and Hammer, K. (1992) The Cuban homegarden 'conuco': A perspective environment for evolution and *in situ* conservation of plant genetic resources. *Genetic Resources and Crop Evolution,* **39**, 9–22.

Everitt, B. (1993) *Cluster Analysis,* Edward Arnold, London.

Ewens, W.J. (1963) The diffusion equation and pseudodistribution in genetics. *Journal of the Royal Statistical Society B,* **25**, 405–412.

Falk, D.A. and Holsinger, K.E. (1991) *Genetics and Conservation of Rare Plants,* Oxford University Press, Oxford.

Fano, H. and Benavides, M. (1992) Los Cultivos Andinos en Perspectiva – Produccion y Utilización en el Cusco – Centro de Estudios Regionales Andinos, in *Bartolome de las Casas,* Cusco-Centro Internacional de la Papa (CIP), Lima, Peru.

FAO (1991) FAO activities on *in situ* conservation of plant genetic resources. *Forest Genetic Resources Newsletter,* **19**, 2–8.

FAO (1993a) *Forest Resources Assessment, 1990. Tropical Countries,* FAO Forestry Paper 112, FAO, Rome, 61 pp. + appendices.

FAO (1993b) Ex Situ *Storage of Seeds, Pollen and* In Vitro *Cultures of Perennial Woody Plant Species,* FAO Forestry Paper 113, FAO, Rome, 83 pp.

Felsenburg, T., Levi, A.A., Galili, G. and Feldman, M. (1991) Polymorphism of high-molecular-weight glutenin in wild tetraploid wheat: spatial and temporal variation in a native site. *Israel Journal of Botany,* **40**, 451–473.

Fiedler, P.L. and Jain, S.K. (1992) *Conservation Biology: the Theory and Practice of Nature Conservation, Preservation and Management,* Chapman & Hall, London.

Finegan, B.G. (1992) The management potential of neotropical secondary lowland rainforest. *Forest Ecology and Management,* **47**, 295–321.

Fisher, R.A. (1930) *The Genetical Theory of Natural Selection,* 2nd edn (1958), Dover, New York.

Flavell, R.B. (1982) Chromosomal DNA sequences and their organisation, in *Nucleic Acids and Proteins in Plants* (eds B. Parthier and D. Boulter), Springer-Verlag, Berlin.

Flint, M. (1991) *Biological Diversity and Developing Countries – Issues and Options*, UK Overseas Development Agency, London, 50 pp.

Flor, H.H. (1942) Inheritance of pathogenicity of *Melampsora lini. Phytopathology,* **32**, 653–669.

Foley, P. (1994) Predicting extinction times from environmental stochasticity and carrying capacity. *Conservation Biology,* **8**, 124–137.

Ford-Lloyd, B.V. and Jackson, M.T. (1986) *Plant Genetic Resources: an Introduction to their Conservation and Use,* Edward Arnold, London.

Ford-Lloyd, B.V. and Maxted, N. (1993) Preserving diversity. *Nature,* **361**, 579.

Ford-Lloyd, B.V., Jackson, M.T. and Parry, M.L. (1990) Can genetic resources cope with global warming? in *Climatic Change and Plant Genetic Resources* (eds M. Jackson, B.V. Ford-Lloyd and M.L. Parry), Belhaven Press, London, pp. 179–182.

Forey, P.L., Humphries, C.J. and Vane-Wright, R.I. (1994) *Systematic and Conservation Evaluation,* Oxford University Press, Oxford.

Forman, R.T.T. (1987) Emerging directions in landscape ecology and applications in natural resourcesmanagement, in *Conference on Science in National Parks* (eds R. Herrmann and T.B. Craig), The George White Society and the National Parks Service, Washington, DC, pp. 59–88.

Fowler, C. and Mooney, P. (1990) *The Threatened Gene: Food, Politics and the Loss of Genetic Diversity,* Lutterworth Press, Cambridge.

Frampton, G.K., Cilgi, T., Fry, G. and Wratten, S.D. (1995) Effects of grassy banks on the dispersal of some carabid beetles (*Coleoptera*: Carabidae) on farmland. *Biological Conservation,* **17**, 347–355.

Francis, C.A. (1985) Variety development for multiple cropping systems. *Critical Review of Plant Science,* **3**, 133–168.

Francisco-Ortega, J., Jackson, M.T., Catty, J.P. and Ford-Lloyd, B.V. (1992) Genetic diversity in the *Chamaecytisus proliferus* complex (Fabaceae: Genisteae) in the Canary Islands in relation to *in situ* conservation. *Genetic Resources and Crop Evolution,* **39**, 149–158.

Francisco-Ortega, J., Newbury, H.J. and Ford-Lloyd, B.V. (1993) Numerical analyses of RAPD data highlight the origin of cultivated tagasaste (*Chamaecytisus proliferus* ssp. *palmensis*) in the Canary Islands. *Theoretical and Applied Genetics,* **87**, 264–270.

Frankel, O. (1970) Genetic conservation in perspective, in *Genetic Rersources in Plants – their Exploration and Conservation* (eds O.H. Frankel and E. Bennett), IBP Handbook No. 11. Blackwell, Oxford and Edinburgh.

Frankel, O.H. (1971) Variation; the essence of life. *Proceedings of the Linnean Society, New South Wales,* **95**, 158–169.

Frankel, O.H. (1973) *Survey of Crop Genetic Resources in their Centre of Diversity,* FAO/IBP, Rome.

Frankel, O.H. (1974) Genetic conservation: An evolutionary responsibility. *Proc. 13th International Genetics Congress, Genetics,* **78**, 53–65.

Frankel, O.H. and Bennett, E. (1970) *Genetic Resources in Plants – their Exploration and Conservation,* Blackwell Scientific Publications, Oxford.

Frankel, O.H. and Brown, A.H.D. (1984) Plant genetic resources today: a critical appraisal, in *Crop Genetic Resources: Conservation and Evaluation* (eds J.H.W. Holden and J.T. Williams), George Allen and Unwin, pp. 249–257.

Frankel, O.H. and Hawkes, J.G. (eds) (1975) *Crop Genetic Resources for Today and Tomorrow,* Cambridge University Press, Cambridge.

Frankel, O.H. and Soulé, M.E. (1981) *Conservation and Evolution,* Cambridge University Press, Cambridge.

Frankel, O.H., Brown, A.H.D. and Burdon, J.J. (1995) *The Conservation of Plant*

Biodiversity, Cambridge University Press, Cambridge.

Frankham, R. (1994) Conservation of genetic diversity for animal improvement. *Proceedings of the 5th World Congress on Genetics Applied to Livestock*, **21**, 385–392.

Franklin, I.R. (1980) Evolutionary changes in small populations, in *Conservation Biology. An Evolutionary–Ecological Perspective* (eds M.E. Soulé and B.A. Wilcox), Sinauer Associates, Sunderland, MA, pp. 134–139.

Franklin, J.F. (1988) Structural and functional diversity in temperate forests, in *Biodiversity* (ed. E.O. Wilson), National Academy Press, Washington, DC, pp. 166–175.

Franklin, J.F. (1993) Preserving biodiversity: species, ecosystems or landscapes? *Ecological Application*, **3**, 202–205.

Frese, L. and Doney, D.L. (eds) (1994) *International Beta Genetic Resources Network. A report on the 3rd International Beta Genetic Resources Workshop and World Beta Network Conference, Fargo, USA*, International Crop Network Series No. 11, International Plant Genetic Resources Institute, Rome.

Frodin, D.G. (1984) *Guide to the Standard Floras of the World*, Cambridge University Press, Cambridge.

Fry, G. (1995) Landscape ecology of insect movement in arable ecosystems, in *Ecology and Integrated Farming Systems* (eds D.M. Glen, M.P. Greaves, H.M. Anderson) Wiley, Chichester, pp. 177–202.

Fukuoka, S., Hosaka, K. and Kamijima, O. (1992) Use of random amplified polymorphic DNAs (RAPDs) for identification of rice accessions. *Japanese Journal of Genetics*, **67**, 243–252.

Fuller, R.M. (1987) The changing extent and conservation interest of lowland grasslands in England and Wales: a review of grassland surveys 1930–1984. *Biological Conservation*, **40**, 281–300.

Gale, J.S. (1990) *Theoretical Population Genetics*, Unwin Hyman, London.

Gale, J.S. and Lawrence, M.J. (1984) The decay of variability, in *Crop Genetic Resources: Conservation and Evaluation* (eds J.H.W. Holden and J.T. Williams), Allen and Unwin, London, pp. 77–101.

Gardner, C.O. (1961) An evaluation of the effects of mass selection and seed irradiation with thermal neutrons on yield of corn. *Crop Science*, **1**, 241–245.

Gaston, K.J. (1991a) How large is a species' geographic range? *Oikos*, **61**, 434–437.

Gaston, K.J. (1991b) The magnitude of global insect species richness. *Conservation Biology*, **5**, 283–296.

Gaston, K.J. (1992) Regional numbers of insect and plant species. *Functional Ecology*, **6**, 243–247.

Gebre-Mariam, H. (1993) *Availability and Use of Seed in Ethiopia*, Program Support Unit, Canadian International Development Agency, Addis-Ababa, 40 pp.

GEF (1993) *Eastern Anatolia Watershed Rehabilitation Project, Global Environment Facility Subproject: in situ conservation of genetic diversity*, unpublished technical annex 11295 – TU, The World Bank, Washington, DC.

Gentry, A.H. (1986) Endemism in tropical versus temperate plant communities, in *Conservation Biology: Science of Scarcity and Diversity* (ed. M.E. Soulé), Sinauer Associates, Sunderland, MA, pp. 153–181.

Gentry, A.H. (1992) Tropical forest biodiversity: distributional patterns and their conservational significance. *Oikos*, **63**, 19–28.

Ghesquière, A. (1988) Diversité de l'espèce sauvage de riz *Oryza longistaminata* A. Chev. and Roehr et dynamique des flux géniques au sein du groupe Sativa. Thèse de Doctorat ès Sciences, Univ. Paris Sud.

Ghesquière, A. and Miézan, K. (1982) *Etude de la structure génétique des variétés traditionnelles de riz en Afrique*. Réunion ORSTOM-IRAT, Septembre 1982.

Gil, E., Ortega, R. and Baca, B. (1995) *Ecología de la Agricultura Ese'eja – Evaluación y Posibilidades*, INANDES–UNSAAC, Cusco, Peru.

Gill, A.M., Grove, R.H. and Noble, I.R. (1981) *Fire and the Australian Biota*, Australian Academy of Sciences, Canberra.

Gillman, M.P. and Dodd, M. (in press) The variability of orchid populations. *Botanical Journal of the Linnean Society*.

Gillman, M.P. and Silvertown, J. (in press) Population extinction and the uncertainty of measurement. *Proceedings of the JNCC/BES Symposium on the Role of Genetics in Conserving Small Populations*.

Gilmour, D.A. and Fisher, R.J. (1991) *Villagers, Forests and Foresters*, Sahoyogi Press, Kathmandu, Nepal, 212 pp.

Given, D.R. (1993) *Principles and Practice of Plant Conservation*, Chapman & Hall, London.

Glacken, C.G. (1967) *Traces on the Rhodian Shore*, University of California Press, 763 pp.

Glaszmann, J.C. (1987) Isozymes and classification of Asian rice varieties. *Theoretical and Applied Genetics*, **74**, 21–30.

Glaszmann, J.C. (1988) Geographical pattern of variation among Asian rice native cultivars (*Oryza sativa*) based on fifteen isozyme loci. *Genome*, **30**, 782–792.

Glowka, L., Burhenne-Guilmin, F., Synge, H. *et al.* (1994) *A Guide to the Convention on Biological Diversity*, IUCN Environmental Law and Policy Paper No. 30, 161 pp.

Goldsmith, F.B. (1991a) Selection of protected areas, in *The Scientific Management of Temperate Communities for Conservation* (eds I.F. Spellerberg, F.B. Goldsmith and and M.G. Morris), British Ecological Society by Blackwell Scientific Publications, Oxford, pp. 273–291.

Goldsmith, F.B. (1991b) Vegetation monitoring, in *Monitoring for Conservation and Ecology* (ed. F.B. Goldsmith), Chapman & Hall, London, pp. 77–86.

Goldsmith, F.B., Harrison, C.M. and Morton, A.J. (1986) Description and analysis of vegetation, in *Methods in Plant Ecology* (eds P.D. Moore and S.B. Chapman), Blackwell Scientific Publications, Oxford, pp. 437–524.

Golenberg, E.M. (1991) Gene flow and the evolution of multilocus structures in wild wheat. *Israel Journal of Botany*, **40**, 513.

Gomez-Campo, C. *et al.* (1992) *Libro rojo de especies vegetales amenazadas de España peninsular e Islas Balneares*, Ministerio de Agricultura y Alimentacion, Madrid.

Gomez-Pompa, A. (ed) (1991) *Rain Forest Regeneration and Management*, Paris and Parthenon Publishing Group, UNESCO, Paris.

Goodman, D. (1987) The demography of chance extinction, in *Viable Populations for Conservation* (ed. M.E. Soulé), Cambridge University Press, Cambridge, pp. 11–34.

Goodrich, W.J. (1987) *Monitoring genetic erosion: detection and assessment*, unpublished consultancy document report, AGPG:IBPGR/86/99, IBPGR, Rome.

Gounin, D.D. and Neronov, V.M. (1986) Ecological principles of genetic resources protection and problems of monitoring the desertification in the Asian arid zone, in *Results and Perspectives of Nature Protection in the USSR*, Nauka Publishers, Moscow, pp. 172–192.

Gounin, P.D., Drozdova, Yu.V. and Neronov, V.M. (1990) Current status and perspectives of development of a regional network of natural resources in Central Asia and Kazakhstan, in *Natural Reserves of Central Asia and Kazakhstan*, Mysl' Publishers, Moscow, 399 pp.

Gray, A. (1991) The impact of biological conservation on indigenous peoples, in *Biodiversity. Social and Economic Perspectives* (ed. V. Shiva), Zed Books, New Jersey, pp. 59–76.

Grayson, A.J. (ed.) (1995) *The World's Forests: International Initiatives Since Rio*, Commonwealth Forestry Association, Oxford, 72 pp.

Greenslade, P.J. (1991) *Collembola* (springtails), in *The Insects of Australia*, 2nd edn, Vol. 1, Melbourne University Press, Melbourne, pp. 252–264.

Griffin, A.R. (1991) Effects of inbreeding on growth of forest trees and implications for management of seed supplies for plantation programmes, in *Reproductive Ecology of Tropical Forest Plants* (eds K.S. Bawa and M. Hadley), MAB Series 7, UNESCO, Paris, pp. 355–374.

Grobman, A., Salhuana, W. and Sevilla, R. (1961) *Races of Maize in Peru*, Publication No. 915, National Academy of Sciences, Washington, DC, 374 pp.

Groombridge, B. (1992) *Global Biodiversity: Status of the Earth's Living Resources*, Chapman & Hall, London.

Guarino, L. (1995) Assessing the threat of genetic erosion, in *Collecting Plant Genetic Diversity: Technical Guidelines* (eds L. Guarino, V. Ramanatha Rao and R. Reid), CAB International, Wallingford, pp. 67–74.

Guarino, L., Ramanatha Rao, V. and Reid, R. eds (1995) *Collecting Plant Genetic Diversity: Technical Guidelines*, CAB International, Wallingford.

Guldager, P. (1975) *Ex situ* conservation stands in the tropics, in *Methodology of Conservation of Forest Genetic Resources* (ed. L. Roche), FAO, Rome.

Gullen, P.J. and Cranston, P.S. (1994) Insects and plants, in *The Insects: an Outline of Entomology*, Chapman & Hall, London, pp. 247–274.

Hafernik, J.E. (1992) Threats to invertebrate biodiversity: implications for conservation strategies, in *Conservation Biology: the Theory and Practice of Nature Conservation, Preservation and Management* (eds P.L. Fiedler and S.K. Jain), Chapman & Hall, London, pp. 171–195.

Haggar, J.J. and Peel, S. (1993) Grassland managment and nature conservation, in *Proceedings of a Joint Meeting Between the British Grassland Society and the British Ecological Society*, Leeds University 27–29 September, Occasional Symposium No. 28, British Grassland Society Publications, Reading, UK.

Hamada, H., Seidman, M., Howard, B.H. and Gorman, C.M. (1984) Enhanced gene expression by the poly(dT-dG) poly(dC-dA) sequence. *Molecular and Cellular Biology*, **4**, 2622–2630.

Hames, R. (1983) Monoculture, polyculture and polyvariety in tropical forest swidden cultivation. *Human Ecology*, **11**, 16–34.

Hammond, P.M. (1992) Species inventory, in *Global Biodiversity: Status of the Earth's Living Resources* (ed. B. Groombridge), Chapman & Hall, London, pp. 17–39.

Hamrick, J.L. (1992) Distribution of genetic diversity in tropical tree populations: implications for the conservation of genetic resources, in *Proceedings of IUFRO S2.02–08 Conference, Breeding Tropical Trees*, Cali, Colombia, 9–18 October, 1992, pp. 74–82.

Hamrick, J.L. and Godt, M.J.W. (1990) Allozyme diversity in plant species, in *Plant Population Genetics, Breeding and Genetic Resources* (eds A.H.D. Brown M.T. Clegg, A.L. Kahler and B.S. Weir), Sinauer, Sunderland, MA, pp. 43–63.

Hamrick, J.L. and Murawski, D.A. (1990) The breeding structure of tropical tree populations. *Plant Species Biology*, **5**, 157–165.

Hamrick, J.L., Godt, M.J.W. and Sherman-Brolyes, S.L. (1992) Factors influencing levels of genetic diversity in woody plant species. *New Forests*, **6**, 95–124.

Hamrick, J.L., Murawski, D.A. and Nason, J.D. (1993) The influence of seed dispersal mechanisms on the genetic structure of tropical tree populations. *Vegetatio*, **107/108**, 281–297.

Hanski, I. and Gilpin, M. (1991) Metapopulation dynamics: brief history and conceptual domain. *Biological Journal of the Linnean Society*, **42**, 3–16.

Hanski, I. and Gyllenberg, M. (1993) Two general metapopulation models and the core-satellite species hypothesis. *American Naturalist*, **142**, 17–41.

Hansson, L., Fahrig, L. and Merriam, G. (1995) *Mosaic Landscapes and Ecological Processes*, Chapman & Hall, London.

Harlan, J.R. (1951) Anatomy of gene centers. *American Naturalist*, **85**, 97–103.

Harlan, J.R. (1975) Our vanishing genetic resources. *Science*, **188**, 618–621.

Harlan, J.R. (1977) Sources of genetic defense. *Annals of the New York Academy of Science*, **287**, 345–355.

Harlan, J.R. (1992) *Crops and Man*, American Society of Agronomy, Madison.

Harlan, J.R. and de Wet, J.M.J. (1971) Towards a rational classification of cultivated plants. *Taxon*, **20**, 509–517.

Harlan, J.R. and Zohary, D. (1966) Distribution of wild wheats and barleys. *Science*, **153**, 1074–1050.

Harper, J.L. (1977) *Population Biology of Plants*, Academic Press, London.

Harper, J.L. (1990) Pests, pathogens and plant communities: an introduction, in *Pests, Pathogens and Plant Communities* (eds J.J. Burdon and S.R. Leather), Blackwell, Oxford, pp. 3–14.

Harris, L.D. and Silva-Lopez, G. (1992) Forest fragmentation and the conservation of biological diversity, in *Conservation Biology: the Theory and Practice of Nature Conservation, Preservation and Management* (eds P.L. Fiedler and S.K. Jain) Chapman & Hall, London, pp. 197–237.

Harris, S.A., Fagg, C.W. and Barnes, R.D. (in preparation) Evaluation of genetic resources in *Faidherbia albida* (Del.) A.Chev. (Leguminosae; Mimosoideae).

Harrison, R.P. (1992) *Forests: the Shadow of Civilization*, University of Chicago Press, 287 pp.

Harrison, S. (1991) Local extinction in a metapopulation context: an empirical evaluation. *Biological Journal of the Linnean Society*, **42**, 73–88.

Harrison, S. (1994) Metapopulations and conservation, in *Large-scale Ecology and Conservation Biology* (eds P.J. Edwards, R.M. May and N.R. Webb), Blackwell, Oxford, pp. 111–128.

Hart, A. (1974) *Perspectives of Survival: the Role of the Farmer*, The Soil Association, London.

Hart, D.D. and Horwitz, R.J. (1991) Habitat diversity and the species–area relationship: alternative models and tests, in *Habitat Structure: the Physical Arrangement of Objects in Space* (eds S.S. Bell, E.D. McCoy, E.D. and H.R. Mushohsky) Chapman & Hall, London, pp. 47–68.

Harvey, P.H., Nee, S., Mooers, A.O. and Partridge, L. (1991) These hierarchial views of life: phylogenies and metapopulations, in *Genes in Ecology* (eds R.J. Perry, T.J. Crawford and G.M. Hewitt), British Ecological Society, pp. 123–137.

Harwood, R.W.J., Wratten, S.D. and Nowakowski, M. (1992) The effect of managed field margins on hoverfly (Diptera: Syriphidae) distribution and within-field abundance. *Proceedings of the 1992 Brighton Crop Protection Conference – Pests and Diseases 3*, pp. 1033–1037.

Hawkes, J.G. (1979) Evolution and polyploidy in potato species, in *The Biology and Taxonomy of the Solanaceae* (eds J.G. Hawkes, R.N. Lester and A.D. Skelding), Linnean Society Symposium Series Number 7, London, pp. 637–645.

Hawkes, J.G. (1980) *Crop Genetic Resources Field Collection Manual*, IBPGR/EUCARPIA, Rome, Italy.

Hawkes, J.G. (1983) *The Diversity of Crop Plants*, Harvard University Press, Cambridge, MA.

Hawkes, J.G. (1990) *The Potato: Evolution, Biodiversity and Genetic Resources*, Belhaven Press, London.

Hawkes, J.G. (1991a) International workshop on dynamic *in situ* conservation of wild relatives of major cultivated plants: summary of final discussion and recommendations. *Israel Journal of Botany*, **40**, 529–536.

Hawkes, J.G. (ed.) (1991b) *Genetic Conservation of World Crop Plants*, Academic Press, London.

Hawksworth, D.L. (1991) The fungal dimension of biodiversity: magnitude, significance and conservation. *Mycological Research*, **95**, 441–456.

Hawthorne, W.D. (1995) *FROGGIE Manual. IUCN Conservation Programme*, IUCN/ODA/Forest Department Ghana, 137 pp.

Hawthorne, W.D. and Abu-Juam, M. (1995) *Forest Protection in Ghana*, IUCN, Gland.

Helentjaris, T., Slocum M., Wright S., Schaefer A. and Nienhuis J. (1986) Construction of genetic linkage maps in maize and tomato using restriction fragment length polymorphisms. *Theoretical and Applied Genetics*, **72**, 761–769.

Hellawell, J.M. (1991) Development of a rationale for monitoring, in *Monitoring for Conservation and Ecology* (ed. F.B. Goldsmith), Chapman & Hall, London, pp. 1–14.

Hellin, J.J. and Hughes, C.E. (1993) Leucaena salvadorensis: *Conservation and Utilization in Central America*, Serie Miscelanea de CONSEFORH, No. 39–21/93, CONSEFORH, Honduras, 41 pp.

Henry, J.P., Pontis C., David J. and Gouyon, P.H. (1991) An experiment on dynamic conservation of genetic resources with metapopulations, in *Species Conservation: a Population–Biological Approach* (eds A. Seitz and V. Loeschcke), Birkhäuser Verlag, Basel, pp. 185–198.

Herben, T., Rydin, H. and Soderstrom, L. (1991) Spore establishment probability and the persistence of the fugitive invading moss, *Orthodontium lineare*: a spatial simulation model. *Oikos*, **60**, 215–221.

Herdt, R.W. and Capule C. (1983) *Adoption, Spread, and Production Impact of Modern Rice Varieties in Asia*, IRRI, Los Baños, Philippines.

Hernandez, E. (1985) Maize and the greater Southwest. *Economic Botany*, **39**, 416–430.

Heywood, C.A., Heywood, V.H. and Wyse-Jackson, P. (1991) *International Directory of Botanic Gardens*, 5th edn, Koeltz Scientific Books, Koenigstein.

Heywood, V.H. (1994) The measurement of biodiversity and the politics of implementation, in *Systematics and Conservation Evaluation* (eds P.L. Forey, C.J. Humphries and R.I. Vane-Wright), Systematic Association Special Volume 50, Oxford University Press, Oxford, pp. 15–22.

Heywood, V.H. and Stuart, S.N. (1992) Species extinctions in tropical forests, in *Tropical Deforestation and Species Extinction* (eds T.C. Whitmore and J.A. Sayer), Chapman & Hall, London, pp. 91–117.

Hill, M.O. (1979) *Twinspan, a Fortran Program for Arranging Multivariate Data in an Ordered Two-way Table by Classification of the Individuals and Attributes*, Section of Ecology and Systematics, Cornell University, New York.

Hirons, G., Goldsmith, F.B. and Thomas, G. (1995) Site management planning, in *Managing Habitats for Conservation* (eds W.J. Sutherland and D.A. Hill), Cambridge University Press, Cambridge, pp. 22–41.

Hobbs, R.J. and Norton, Q.A. (1996) A conceptual framework for restoration ecology. *Restoration Ecology* (in press).

Hodgkin, T., Brown, A.H.D., van Hintum, Th.J.L. and Morales, E.A.V. (eds) (1995) *Core Collections of Plant Genetic Resources*, Wiley, New York.

Hollis, S. and Brummitt, R.K. (1992) *World Geographical Scheme for Recording Plant Distributions (Plant Taxonomic Database Standards No. 2)*, Hunt Institute for Botanical Documentation, Pittsburgh.

Holmgren, P.K., Holmgren, N.H. and Barnett, A. (1990) *Index Herbariorium I: The Herbaria of the World*, edn 8, New York Botanical Garden, NY.

Horn, H.S. (1975) Markovian properties of forest succession, in *Ecology and Evolution of Communities* (eds M.L. Cody and J.M. Diamond), Harvard University Press, Cambridge, MA, pp. 196–211.

Horn, H.S. (1981) Succession, in *Theoretical Ecology* (ed. R.M. May), Blackwell, Oxford, pp. 253–271.

House, S. and Moritz, C. (1991) The impact of rainforest fragmentation on flora and fauna, in *Tropical Rainforest Research in Australia* (eds N. Goudberg, M. Bonell and D. Benzaken), Institute for Tropical Rainforest Studies, Townsville, 10 pp.

Hoyt, E. (1988) *Conserving the Wild Relatives of Crops*, IBPGR/IUCN/WWF, Rome, Italy.

Hubbell, S.P. (1979) Tree dispersion, abundance and diversity in a tropical deciduous forest. *Science*, **203**, 1299–1309.

Hughes, C.E. (1987) Biological considerations in designing a seed collection strategy for *Gliricidia sepium* (Jacq.) Walp. (Leguminosae). *Commonwealth Forestry Review*, **66**, 31–48.

Hughes, C.E. (1995) Risks of species introductions in tropical forestry. *Commonwealth Forestry Review*, **73**, 243–252.

Hughes, C.E., Sorenson, C., Bray, R. and Brewbaker, J. (1995) *Leucaena* germplasm collections, genetic conservation and seed increase, in *Proceedings, Leucanea Workshop*, Bogor, January 1994, ACIAR, Canberra.

Huke, R.E., Cordova, V. and Sardido, S. (1982) *San Bartolome: Beyond the Green Revolution*, IRRI Research Paper Series 74, International Rice Research Institute, Manila, The Philippines.

Humphreys, M.O. and Gale, J.S. (1974) Variation in wild populations of *Papaver dubium*. VIII. The mating system. *Heredity*, **33**, 33–42.

Hutchings, M.J. (1986) Plant population biology, in *Methods in Plant Ecology* (eds P.D. Moore and S.B. Chapman), Blackwell, Oxford, pp. 377–435.

Hutchings, M.J. (1987) The population biology of the early spider orchid, *Ophrys sphegodes* Mill. I. A demographic study from 1975–1984. *Journal of Ecology*, **75**, 711–727.

Hutchings, M.J. (1991) Monitoring plant populations: census as an aid to conservation., in *Monitoring for Conservation and Ecology* (ed. F.B.Goldsmith), Chapman & Hall, London, pp. 61–76.

IBPGR (1985a) *Forages for Mediterranean and Adjacent Arid/Semi-arid Areas: Report of a Working Group Meeting held at Limossal, Cyprus*, International Board for Plant Genetic Resources, Rome, Italy.

IBPGR (1985b) *Ecogeographic Surveying and In Situ Conservation of Crop Relatives: Report of an IBPGR Task Force Meeting held at Washington DC, USA*, International Board for Plant Genetic Resources, Rome, Italy.

IBPGR (1985) *Revised Descriptor List for Wheat (Triticum spp.)*, International Board for Plant Genetic Resources, Rome, Italy.

Iltis, H.H., Doebley, J.F., Guzman, M.R. and Pazy, B. (1979) *Zea diploperennis* (Gramineae): a new teosinte from Mexico. *Science*, **203**, 186–188.

Ingram, G.B. and Williams, J.T. (1984) *In situ* conservation of wild relatives of crops, in *Crop Genetic Resources: Conservation and Evaluation* (eds J.H.W. Holden and J.T. Williams), George Allen and Unwin, London, pp. 163–179.

Ingram, G.B. and Williams, J.T. (1993) Gap analysis for *in situ* conservation of crop gene pools: implications for the Convention on Biological Biversity. *Biodiversity Letters*, **1**, 141–148.

International Standards Organisation (1981) *Listing of ISO Codes. ISO 3166*, British Standards Institute, London.

IPGRI (1993) *Diversity for Development: the Strategy of the International Plant Genetic Resources Institute*, International Plant Genetic Resources Institute, Rome, Italy.

IPGRI (1994) *Descriptors for Barley (*Hordeum vulgare *L.)*, International Plant Genetic Resources Institute, Rome, Italy.

IPGRI (1995) *Descriptors for Capsicum (*Capsicum *spp.)*, International Plant Genetic Resources Institute, Rome, Italy.

IRRI–IBPGR (1980) *Descriptors for Rice (*Oryza sativa*)*, International Rice Research Institute, Manila, The Philippines.

IUCN (1993) *The Convention on Biological Diversity: an Explanatory Guide*, draft text, IUCN Environmental Law Centre, Bonn, Germany.

IUCN (1994a) *IUCN Red List Categories*, IUCN Species Survival Commission, International Union for Conservation of Nature and Natural Resources, Gland, 21 pp.

IUCN (1994b) *1993 United Nations List of National Parks and Protected Areas*, WCMC and CNPPA, International Union for Conservation of Nature and Natural Resources, Gland, 315 pp.

IUCN and WWF (1994) *Centres of Plant Diversity. A Guide and Strategy for their Conservation*, IUCN Publications Unit, Cambridge.

Jackson, M.T. and Ford-Lloyd, B.V. (1990) Plant genetic resources – a perspective, in *Climatic Change and Plant Genetic Resources* (eds M. Jackson, B.V. Ford-Lloyd and M.L. Parry), Belhaven Press, London, pp. 1–17.

Jain, S.K. (1975) Genetic reserves, in *Crop Genetic Resources for Today and Tomorrow* (eds O.H. Frankel and J.G. Hawkes), Cambridge University Press, Cambridge, pp. 379–396.

Jana, S. (1993) Utilization of biodiversity from *in situ* reserves, with special reference to wild wheat and barley, in *Biodiversity and Wheat Improvement* (ed. A.B. Damania), John Wiley, Chichester.

Jana, S. and Nevo, E. (1991) Variation in response to infection with *Erysiphe graminis hordei* and *Puccinia hordei* in some wild barley populations in a centre of diversity. *Euphytica*, **57**, 133–140.

Janzen, D.H. (1983) *Costa Rican Natural History*, University of Chicago Press, 816 pp.

Janzen, D.H. (1986) Blurry catastrophes. *Oikos*, **47**, 1–2.

Janzen, D.H. (1988) Management of habitat fragments in a tropical dry forest: growth. *Annals Missouri Botanical Garden*, **75**, 105–116.

Jarosz, A.M. and Burdon, J.J. (1988) The effect of small-scale environmental changes on disease incidence and severity in a natural plant–pathogen interaction. *Oecologia*, **75**, 278–281.

Jarosz, A.M., Burdon, J.J. and Muller, W.J. (1989) Long-term effects of disease epidemics. *Journal of Applied Ecology*, **26**, 725–733.

Jeffreys, A.J. and Pena, S.D.J (1993) Brief introduction to human fingerprinting, in *DNA Fingerprinting: State of the Science* (eds S.D.J. Pena, R. Chakraborty, J.T Epplen and A.J. Jeffreys), Birkhäuser Verlag, Basel, pp. 1–20.

Jeffreys, A.J., Wilson, V. and Thein, S.L. (1985) Individual-specific 'fingerprints' of human DNA. *Nature*, **314**, 76–79.

Johnson, S.P. (1993) *The Earth Summit: the United Nations Conference on Environment and Development (UNCED)*, Graham and Trotman/Martinus Nijhoff, London, 532 pp.

Jordan, W.R., Golph, M.E. and Aber, J.D. (1987) *Restoration Ecology: a Synthetic Approach to Ecological Research*, Cambridge University Press, Cambridge.

Juma, C. (1989) *The Gene Hunters: Biotechnology and the Scramble for Seeds*, Zed Books, London.

Kageyama, P. and Reis, A. (1993) Areas of secondary vegetation in the Itajai

Valley, Santa Catarina, Brazil: perspectives for management and conservation. *Forest Genetic Resources Information*, **21**, 37–39.

Kaplan, L. (1973) Early cultivated beans (*Phaseolus vulgaris*) from an intermontane Peruvian valley. *Science*, **179**, 76–77.

Karban, R. and Myers, J.H. (1989) Induced plant responses to herbivory. *Annual Review of Ecological Systematics*, **20**, 331–348.

Kearsey, M.J. and Kojima, K. (1967) The genetical architecture of body weight and egg hatchability in *Drosophila melanogaster. Genetics*, **56**, 23–37.

Kemp, R.H. (1993) *Conservation of Genetic Resources in Tropical Forest Management. Principles and Concepts*, FAO Forestry Paper 107, 105 pp.

Khush, G. (1993) Breaking the yield frontier of rice, in *Food Security in Asia*, Royal Society Meeting November, 1994, pp. 17–21.

Kimber, C.T. (1973) Spatial patterning in the dooryard gardens of Puerto Rico. *Geographical Review*, **63**, 6–26.

Ko, H.L., Cowan, D.C., Henry, R.J. *et al.* (1994) Random amplified polymorphic DNA analysis of Australian rice (*Oryza sativa* L.) varieties. *Euphytica*, **80**, 179–189.

Koski, F.T.V. (1991) Preservation of genetic resources of forest trees in Finland (mimeo), Finnish Forest Research Institute, Vantaa, Finland, 5 pp.

Kranz, J. (1990) Fungal diseases in multispecies plant communities. *New Phytologist*, **116**, 383–405.

Kremen, C., Merenlender, A.M. and Murphy, D.D. (1994) Ecological monitoring: a vital need for integrated conservation and development programs in the tropics. *Conservation Biology*, **8**, 388–397.

Krugman, S.L. (1984) Policies, strategies and means for genetic conservation in forestry, in *Plant Genetic Resources: a Conservation Perspective* (eds C.W. Yeatman, D. Kafton and G. Wilkes), AAAS Selected Symp, No. 87, Westview Press, Boulder, CO.

Krusche, D. (1982) Selection in host–parasite systems, in *Resistance to Diseases and Pests in Forest Trees* (eds H.M. Heybroek, B.R. Stephan and K. von Weissenberg), Pudoc, Wageningen, The Netherlands, pp. 312–317.

Krusche, D. and Geburek, Th. (1991) Conservation of forest gene resources as related to sample size. *Forest Ecology and Management*, **40**, 145–150.

Kupicha, F.K. (1976) The infrageneric structure of *Vicia. Notes from the Royal Botanic Garden, Edinburgh*, **34**, 287–326.

Kuusipalo, J. and Kangas, J. (1994) Managing biodiversity in a forestry environment. *Conservation Biology*, **8**, 450–460.

Laarman, J.G. and Sedjo, R.A. (1992) *Global Forests: Issues for Six Billion People*, McGraw-Hill, New York, 337 pp.

Lake, C.M. and Whaley, K.J. (1995) *Rangitikei Ecological Region Survey Report for the Protected Natural Areas Programme*, Department of Conservation, Wanganui.

Lambert, D.H. (1985) *Swamp Rice Farming: the Indigenous Pahang Malay Agricultural System*, Westview Press, Boulder and London.

Lamola, L.M. and Bertram, R.B. (1994) Experts gather in Mexico to seek new strategies in preserving agrobiodiversity. *Diversity*, **10**(3), 15–17.

Lande, R. (1988) Genetics and demography in biological conservation. *Science*, **241**, 1455–1460.

Lande, R. and Barrowclough, G.F. (1987) Effective population size, genetic variation, and their use in population management, in *Viable Populations for Conservation* (ed. M.E. Soulé), Cambridge University Press, Cambridge, pp. 87–123.

Lando, R.P. and Mak, S. (1994a) *Rainfed Lowland Rice in Cambodia: a Baseline Study*,

IRRI Research Paper Series 152, International Rice Research Institute, Manila, the Philippines.

Lando, R.P. and Mak, S. (1994b) *Deepwater Rice in Cambodia: a Baseline Survey*, IRRI Research Paper Series 153, International Rice Research Institute, Manila, the Philippines.

Lando, R.P. and Mak, S. (1994c) *Cambodian Farmers Decisionmaking in the Choice of Traditional Rainfed Lowland Rice Varieties*, IRRI Research Paper Series 154, International Rice Research Institute, Manila, the Philippines.

Lane, M.D. and Lawrence, M.J. (1995) The genetics of seed dormancy in *Papaver rhoeas*. *Heredity*, **75**, 84–91.

Latterich, M. and Croy, R.R.D. (1993) Plant Gene Index, in *Plant Molecular Biology Labfax* (ed. R.R.D. Croy), Blackwell, Oxford, pp. 49–120.

Lavin, M., Mathews, S. and Hughes, C. (1991) Chloroplast DNA variation in *Gliricidia sepium* (Leguminosae): interspecific phylogeny and tokogeny. *American Journal Botany*, **78**, 1576–1585.

Lawrence, M.J. (1984) The genetical analysis of ecological traits, in *Evolutionary Ecology* (ed. B. Shorrocks), Blackwell, Oxford, pp. 27–63.

Lawrence, M.J., Marshall, D.F. and Davies, P. (1995a) Genetics of genetic conservation. I. Sample size when collecting germplasm. *Euphytica*, **84**, 89–99.

Lawrence, M.J., Marshall, D.F. and Davies, P. (1995b) Genetics of genetic conservation. II. Sample size when collecting seed of cross-pollinating species and the information that can be obtained from the evaluation of material in gene banks. *Euphytica*, **84**, 101–107.

Lawrence, M.J., O'Donnell, S., Lane, M.D. and Marshall, D.F. (1994) The population genetics of the self-incompatibility polymorphism in *Papaver rhoeas*. VIII. Sampling effects as a possible cause of unequal allele frequencies. *Heredity*, **72**, 345–352.

Lawton, J.H. (1983) Plant architecture and the diversity of phytophagous insects. *Annual Review of Entomology*, **28**, 23–29.

Leach, M. and Fairhead, J. (1994) *The Forest Islands of Kissidougou*, Report to UK Overseas Development Administration ESCOR, London, 96 pp.

Le Boulc'h, V., David, J.L., Brabant P. and De Vallavieille-Pope, C. (1994) Dynamic conservation of variability: responses of wheat populations to different selective pressures including powdery mildew. *Genetics Selection Evolution* **26** (Suppl. 1), 221s–240s.

Ledig, F.T. (1986) Conservation strategies for forest gene resources. *Forest Ecology and Management*, **14**, 77–90.

Ledig, F.T. (1988) The conservation of diversity in forest trees. *BioScience*, **38**, 431–439.

Ledig, F.T. (1992) Human impacts on genetic diversity in forest ecosystems. *Oikos*, **63**, 87–108.

Ledig, F.T., Millar, C.I. and Riggs, L.A. (1990) Conservation of diversity in forest ecosystems. *Forest Ecology and Management*, **35**, 1–4.

Leith, H. and Lohmann, M. (eds) (1993) *Restoration of Tropical Forest Ecosystems. Tasks for Vegetation Sciences 30*, Kluwer Academic, Dordrecht.

Lenné, J.M. and Smithson, J.B. (1994) Varietal mixtures: a viable strategy for sustainable productivity in sustainable agriculture? *Aspects of Applied Biology*, **39**, 161–162.

Lenné, J.M. and Wood, D. (1991) Plant diseases and the use of wild germplasm. *Annual Review of Phytopathology*, **29**, 35–63.

León, J. (1964) *Plantas Alimenticias Andinas*, Boletín Técnico No. 6, IICA, Zona Andina, Lima, Peru.

León J. (1968) *Fundamentos Botánicos de los Cultivos Tropicales*, IICA, San Jose, Costa Rica.

Lesica, P. and Allendorf, F.W. (1992) Are small populations of plants worth preserving? *Conservation Biology*, **6**, 135–139.

Lesser, W.H. and Kratiger, A.F. (1994) The complexities of negotiating terms for gerplasm collection. *Diversity*, **10**(3), 6–10.

Levins, R. (1969) Some demographic and genetic consequences of environmental heterogeneity for biological control. *Bulletin of the Entomological Society of America*, **15**, 237–240.

Levins, R. (1970a) Extinction, in *Some Mathematical Problems in Biology* (ed. M. Gerstenhaber), Mathematical Society, Providence, pp. 77–107.

Levins, R. (1970b) Extinction. *Lectures on Mathematics in the Life Sciences*, **2**, 75–107.

Levy, A.A. and Feldman, M. (1988) Ecogeographical distribution of HMW glutenin alleles in populations of the wild tetraploid wheat *Triticum turgidum* var. *dicoccoides*. *Theoretical and Applied Geneics*, **75**, 651–658.

Lim, M.T. and Chin, T.Y. (1995) A practical approach to conservation of genetic diversity in Malaysia: genetic resource area, in *Measuring and Monitoring Biological Diversity in Tropical and Temperate Forests* (ed. T.J. Boyle), Proceedings of IUFRO Meeting, Chiangmai, Thailand, 28 August–2 September, 1994, ASEAN–Canada Forest Tree Seed project, Muak-Lek, Thailand (in press).

Linhart, Y.B. (1989) Interactions between genetic and ecological patchiness in forest trees and their dependent species, in *The Evolutionary Ecology of Plants* (eds J.H. Bock and Y.B. Linhart), Westview Press, Boulder, CO, pp. 393–430.

Litt, M. and Luty, J.A. (1989) A hypervariable microsatellite revealed by *in vitro* amplification of a dinucleotide repeat within the cardiac muscle actin gene. *American Journal of Human Genetics*, **44**, 397–401.

Lleras, E. (1991) Conservation of genetic resources *in situ*. *Diversity*, **7**(1,2), 72–74.

Loegering, W.Q., McIntosh, R.A. and Burton, C.H. (1971) Computer analysis of disease data to derive hypothetical genotypes for reaction of host varieties to pathogens. *Canadian Journal of Genetics and Cytology*, **13**, 742–748.

Longley, C. and Richards, P. (1993) Selection strategies of rice farmers in Sierra Leone, in *Cultivating Knowledge* (eds W. de Boef, K. Amanor, K. Wellard and A. Bebbington), Intermediate Technology Publications, London, pp. 51–57.

Louda, S.M. (1994) Experimental evidence for insect impact on populations of short-lived, perennial plants, and its application in restoration ecology, in *Restoration of Endangered Species* (eds M. Bowles and C.J. Whelan), Cambridge University Press, Cambridge, pp. 118–138.

Louette, D. (1994) Gestion traditionnelle de variétés de maïs dans la Réserve de la Biosphère Sierra de Manatlan (RBSM, états de Jalisco et Colima, Mexique) et conservation *in situ* des ressources génétiques de plantes cultivées. Thèse de l'Ecole Nationale Supérieure Agronomique de Montpellier.

Lovejoy, T.E., Bierregaard Jr, R.O., Rylands, A.B. *et al.* (1986) Edge and other effects of isolation on Amazonian forest fragments, in *Conservation Biology: the Science of Scarcity and Diversity* (ed. M.E. Soulé), Sinauer Associates, Sunderland, MA, pp. 257–285.

Lovejoy, T.E., Rankin, J.M., Bierregaard Jr *et al.* (1984) Ecosystem decay of Amazon rainforest remnants, in *Extinctions* (ed. M.H. Nitecki), Chicago University Press, Chicago, pp. 95–325.

Loveless, M.D. (1992) Isozyme variation in tropical trees: patterns of genetic organization. *New Forests*, **6**, 67–94.

Lucas, G. and Synge, H. (1978) *The IUCN Plant Red Data Book*, International Union for Conservation of Nature and Natural Resources, Gland.

Lugo, A.E. (1988) Estimating reductions in the diversity of tropical forest species, in *Biodiversity* (ed. E.O. Wison), National Academy Press, Washington, DC, pp. 58–70.

Lynch, M. (1991) Analysis of population genetic structure by DNA fingerprinting, in *DNA Fingerprinting: Approaches and Applications* (eds T. Burke, G. Dolf and A.J. Jeffreys), Birkhäuser Verlag, Basel, pp. 113–126.

Lynch, M. and Landé, R. (1993) Evolution and extinction in response to environmental change, in *Biotic Interactions and Global Change* (eds P.M. Kareiva, J.G. Kingsolver and R.B. Huey), Sinauer Associates, Sunderland, MA, pp. 234–250.

MacArthur, R.H. and Wison, E.O. (1967) *The Theory of Island Biogeography*, Princeton University Press, Princeton, New Jersey.

Mackay, I.J. (1980) Population genetics of *Papaver dubium*. PhD thesis, University of Birmingham.

Mackill, D.J. (1995) Classifying japonica rice cultivars with RAPD markers. *Crop Science*, **35**, 889–894.

Malecot, G. (1975) Heterozygosity and relationship in subdivided populations. *Theoretical Population Biology*, **8**, 212–241.

Mandelbrot, B.B. (1982) *The Fractal Geometry of Nature*, W.H. Freeman, San Francisco.

Manisterski, J., Treeful, L., Tomerlin, J.R. *et al.* (1986) Resistance of wild barley accessions from Israel to leaf rust collected in the USA and Israel. *Crop Science*, **26**, 727–730.

Margules, C.R., Higgs, A.J. and Rafe, R.W. (1982) Modern biogeographical theory: are there any lessons for nature reserve design? *Biological Conservation*, **24**, 115–128.

Margules, C.R. and Nicholls, A.O. (1988) Selecting networks of reserves to maximise biological diversity. *Biological Conservation*, **43**, 63–76.

Mariat, D. and Vergnaud, G. (1992) Detection of polymorphic loci in complex genomes with synthetic tandem repeats. *Genomics*, **12**, 454–458.

Marshall, D.R. (1990) Crop genetic resources: current and emerging issues, in *Plant Population Genetics, Breeding, and Genetic Resources* (eds A.H.D. Brown, M.T. Clegg, A.L. Kahler and B.S. Weir), Sinauer Associates, Sunderland, MA, pp. 367–388.

Marshall, D.R. and Brown, A.H.D. (1975) Optimum sampling strategies in genetic conservation, in *Crop Genetic Resources for Today and Tomorrow* (eds O.H. Frankel and J.G. Hawkes), Cambridge University Press, Cambridge, pp. 53–80.

Mather, K. (1960) Evolution in polygenic systems, in *Evoluzione e Genetica*, Academia Nazionale dei Lincei, Rome, pp. 131–152.

Mather, K. (1966) Variability and selection. *Proceedings Royal Society of London, Series B*, **164**, 328–340.

Mather, K. (1973) *Genetical Structure of Populations*, Chapman & Hall, London.

Mather, K. (1982) Response to selection, in *The Genetics and Biology of Drosophila*, Vol. 3c (eds M. Ashburner and J.N. Thompson, Jr), Academic Press, London, pp. 155–221.

Matsuyama T., Oyama, M. and Akiyama, T. (1993) DNA fingerprinting in *Citrus* cultivars, in *Techniques on Gene Diagnosis and Breeding in Fruit Trees* (eds T. Hayashi, M. Omura and N.S. Scott) Fruit Tree Research Station, Ibaraki, pp. 26–30.

Maunder, M. (1994a) Practical aspects of plant conservation in a botanic garden; the relationship between botanic gardens and the wild habitat. *Boissiera*, **47**, 155–165.

Maunder, M. (1994b) Botanical gardens: future challenges and responsibilities. *Biodiversity and Conservation*, **3**, 97–103.

Maunder, M. (1995) *Endemic plants: options for an integrated conservation strategy,* unpublished report for RBG Kew submitted to ODA and the Government of St Helena.

Maunder, M., Upson, T., Spooner, B. and Kendle, T. (1995) Saint Helena: sustainable development and conservation of a highly degraded island ecosystem, in *Islands: Biological Diversity and Ecosystem Function* (eds P. Vitousek and L. Loope), Springer Verlag, pp 205–215.

Mauremootoo, J.R., Wratten, S.D. and Worner, S.P. (1994) Permeability of hedgerows to predatory carabid beetles. *Agriculture, Ecosystems and Environment* (in press).

Maxted, N. (1991) New combinations and names in the genus *Vicia* (Leguminosae, Vicieae). *Kew Bulletin,* **47**(1), 129–130.

Maxted, N. (1993) A phenetic investigation of *Vicia* L. subgenus *Vicia* (Leguminosae, Vicieae). *Botanical Journal of the Linnean Society,* **111**, 155–182.

Maxted, N. (1995) *An Ecogeographic Study of* Vicia *subgenus* Vicia, Systematic and Ecogeographic Studies in Crop Genepools 8, International Plant Genetic Resources Institute, Rome, Italy, 184 pp.

Maxted, N. and Ford-Lloyd, B.V. (1996) Biodiversity conservation training. *Biological Conservation* (in press).

Maxted, N., Obari, H. and Tan, A. (1990) New and interesting Vicieae from the Eastern Mediterranean. *Plant Genetic Resources Newsletter,* **78**, 21–26.

Maxted, N., van Slageren, M.W. and Rihan, J. (1995) Ecogeographic surveys, in *Collecting Plant Genetic Diversity: Technical Guidelines* (eds L. Guarino, V. Ramanatha Rao and R. Reid), CAB International, Wallingford, pp. 255–286.

May, R.M. (1990) Taxonomy as destiny. *Nature,* **347**, 129–130.

McCouch, S.R., Kochert, G., Yu, Z.H. *et al.* (1988) Molecular mapping of rice chromosomes. *Theoretical and Applied Genetics,* **76**, 815–829.

McGowan, P., Gillman , M.P. and Dodd, M. (to be published) Assessment of changes in the distribution of pheasant and partridge species of south-east Asia. (Submitted ms.)

McNaughton, S.J. (1979a) Grazing as an optimization process: grass–ungulate relationships in the Serengeti. *American Naturalist,* **13**, 691–703.

McNaughton, S.J. (1979b) Grassland–herbivore dynamics, in *Serengeti: Studies of Ecosystem Dynamics in a Tropical Savanna* University of Chicago Press, Chicago, pp. 46–81.

McNaughton, S.J. (1983) Compensatory plant growth as a response to herbivory. *Oikos,* **40**, 329–336.

McNeely, J.A. (1988) *Economics and Biological Diversity; Developing and Using Economic Incentives to Conserve Biological Resources,* International Union for Conservation of Nature and Natural Resources, Gland.

McNeely, J.A. (1994) Social and economic values of genetic conservation, in *Proceedings of International Symposium on Genetic Conservation and Production of Tropical Forest Tree Seed* (eds R.M. Drysdale, S.E.T. John and A.C. Yappa), ASEAN–Canada Forest Tree Seed project, Muak-Lek, Thailand, pp. 26–33.

Meffe, G.K. and Carroll, C.R. (1994) *Principles of Conservation Biology,* Sinauer Associates, Sunderland, MA.

Meurk, C. (1996) Hang-ups and hopes for habitat restoration, in *Proceedings of a Workshop on Scientific Issues in Ecological Restoration,* Landcare Science Series Publications, Landcare Research, Christchurch.

Miézan, K. and Ghesquière, A. (1985) Genetic structure of African traditional rice cultivars, in *Rice Genetics,* Proceedings of the Rice Genetics Symposium, IRRI, Los Baños, Philippines, pp. 91–107.

Milligan, R.H. (1974) Insects damaging beech (*Nothofagus*) forests. *Proceedings of the New Zealand Ecological Society,* **21**, 32–40.

Milliken, W., Miller, R.P., Pollard, S.R. and Wandelli, E.V. (1992) *Ethnobotany of the Waimiri Atroari Indians of Brazil*, Royal Botanic Gardens, Kew.

Mills, L.S., Soulé, M.E. and Doak, D.F. (1993) The keystone species concept in ecology and conservation. *Bioscience*, **43**, 219–224.

Minka, N. (1987) *Tubérculos Andinos*, Edición Talpuy, Huancayo, Peru.

Monde, S.S. and Richards, P. (1994) Rice biodiversity conservation and plant improvement in Sierra Leone, in *Safeguarding the Genetic Basis of Africa's Traditional Crops* (ed. A. Putter), CTA, The Netherlands/IPGRI, Rome, pp. 83–100.

Moore, N.W. (1962) The heaths of Dorset and their conservation. *Journal of Ecology*, **50**, 369.

Moore, N.W. (1991) Observe extinction or conserve diversity, in *The Conservation of Insects and their Habitats, 15th Symposium of the Royal Entomological Society of London* (eds H.M. Collins and J.A. Thomas), Academic Press, London, pp. 1–8.

Moran, G.F. (1992) Patterns of genetic diversity in Australian tree species. *New Forests*, **6**, 49–66.

Moran, G.F. and Bell, J.C. (1987) The origin and genetic diversity of *Pinsu radiata* in Australia. *Theoretical and Applied Genetics*, **73**, 616–622.

Moran, G.F. and Hopper, S.D. (1987) Conservation of the genetic resources of rare and widespread eucalypts in remnant vegetation, in *Nature Conservation: the Role of Remants of Native Vegetation* (eds D.A. Saunders, G.A. Arnold, A.A. Burbidge and A.J.M. Hopkins), Surrey Beatty/CSIRO/CALM, WA, Australia, pp. 151–162.

Morgante, M. and Olivieri, A.M. (1993) PCR-amplified microsatellites as markers in plant genetics. *Plant Journal*, **3**, 175–182.

Mori, S.A., Boom, B.M. and Prance, G.T. (1981) Distribution patterns and conservation of eastern Brazilian coastal forest tree species. *Brittonia*, **33**, 233–245.

Morishima, H. (1989) Intra-populational genetic diversity in landrace of rice, in *Breeding Research: the Key to the Survival of the Earth. Proceedings of the 6th International Congress of SABRAO* (eds S. Iyama and F. Takeda), pp. 159–166.

Moseman, J.G., Nevo, E. and El-Morshidy, M.A. (1990) Reactions of *Hordeum spontaneum* to infection with two cultures of *Puccinia hordei* from Israel and the United States. *Euphytica*, **49**, 169–175.

Mosseler, A. (1992) Life history and genetic diversity in red pine: implications for gene conservation in forestry. *Forestry Chronicle*, **68**, 701–708.

Müller-Starck, G., Baradet, Ph. and Bergmann, F. (1992) Genetic variation within European tree species. *New Forests*, **6**, 23–48.

Mundt, C.C. (1994) Techniques to manage pathogen coevolution with host plants to prolong resistance, in *Rice Pest Science and Management* (eds P.S. Teng, K.L. Heong and K. Moody), IRRI, Philippines, pp. 193–205.

Munthali, M., Ford-Lloyd B.V. and Newbury H.J. (1992) The random amplification of polymorphic DNA for fingerprinting plants. *PCR Methods and Applications*, **1**, 274–276.

Muona, O. (1989) Population genetics in forest tree improvement, in *Plant Population Genetics, Breeding, and Genetic Resources* (eds A.H.D. Brown, M.T. Clegg, A.L. Kahler and B.S. Weir), Sinauer Associates, Sunderland, MA, pp. 282–298.

Murawski, D.A. (1995) Reproductive biology and genetics of tropical trees, in *Forest Canopies* (eds M. Lowman and N. Nadkarni), Academic Press, London.

Murawski, D.A. and Hamrick, J.L. (1991) The effects of the density of flowering individuals on the mating system of nine tropical tree species. *Heredity*, **67**, 167–174.

Murawski, D.A., Gunatilleke, I.A.U.N. and Bawa, K.S. (1994) The effects of selective logging on inbreeding in *Shorea megistophylla* (Dipterocarpaceae) from Sri Lanka. *Conservation Biology*, **8**, 997–1002.

Murawski, D.A., Hamrick, J.L., Hubbell, S.P. and Foster, R.B. (1990) Mating systems of two Bombacaceous trees of a neotropical moist forest. *Oecologia*, **82**, 501–506.

Murdoch, W.W., Evans, F.C. and Peterson, C.K. (1972) Diversity and pattern in plants and insects. *Ecology*, **53**, 819–824.

Murray, M.G., Cuellar, R.E. and Thompson, W.F. (1978) DNA sequence organisation in the pea genome. *Biochemistry*, **17**, 5781–5790.

Myers, N. (1988a) Tropical forest and their species: going, going ... ? in *Biodiversity* (ed. E.O. Wilson), National Academy Press, Washington, DC, pp. 28–35.

Myers, N. (1988b) Threatened biotas: 'Hot spots' in tropical forests. *Environmentalist*, **8**, 187–208.

Myers, N. (1990) *Deforestation Rates in Tropical Forests and their Climatic Implications*, Friends of the Earth, London, 116 pp.

Naess, A. (1984) Intuition, intrinsic value, and deep ecology. *Ecologist*, **14**, 201–203.

Nakamura, Y., Leppert, M., O'Connell, P. *et al.* (1987) Variable number of tandem repeat (VNTR) markers for human genome mapping. *Science*, **235**, 1616–1622.

Namkoong G. (1991a) Dynamics of *in situ* conservation: can fragmentation be useful? *Israel Journal of Botany*, **40**, 518.

Namkoong, G. (1991b) Biodiversity – issues in genetics, forestry and ethics. *Forestry Chronicle*, **68**, 438–443.

Namkoong, G. (1994a) Causes and effects of genetic erosion, in *Proceedings International Symposium on Genetic Conservation and Production of Tropical Forest Tree Seed* (eds R.M. Drysdale, S.E.T. John and A.C. Yappa), ASEAN–Canada Forest Tree Seed project, Muak-Lek, Thailand, pp. 139–143.

Namkoong, G. (1994b) An evolutionary concept of breeding. Marcus Wallenberg Prize Lecture, Stockholm, Sweden, 22 September 1994, 10 pp.

Namkoong, G. and Kang, H. (1990) Quantitative genetics of forest trees. *Plant Breeding Reviews*, **8**, 139–188.

Neale, D.B. (1985) Genetic implications of shelterwood regeneration of Douglas-fir in southwest Oregon. *Forest Science*, **31**, 995–1005.

Neale, D.B. and Adams, W.T. (1985) Allozyme and mating-system variation in balsam fir (*Abies balsamea*) across a continuous elevational transect. *Canadian Journal of Botany*, **63**, 2448–2453.

Nei, M. (1972) Genetic distance between populations. *American Naturalist*, **106**, 283–292.

Nei, M. (1973) Analysis of gene diversity in subdivided populations. *Proceedings of the National Academy of Science USA*, **70**, 3321–3323.

Nei, M. (1987) *Molecular Evolutionary Genetics*, Colombia University Press, New York.

Nei, M. and Li, W.H. (1979) Mathematical model for studying genetic variation in terms of restriction endonucleases. *Proceedings of the National Academy of Science USA*, **76**, 5269–5273.

Nei, M. and Murata, M. (1966) Effective population size when fertility is inherited. *Genetical Research*, **8**, 257–260.

Neuhaus, D., Kuhl, H., Kohl, J.G. *et al.* (1993) Investigation on the genetic diversity of *Phragmites* stands using genomic fingerprinting. *Aquatic Botany*, **45**, 357–364.

Nevo, E. and Beiles, A. (1989) Genetic diversity of wild emmer wheat in Isreael and Turkey: structure, evolution and application in breeding. *Theoretical and Applied Genetics*, **77**, 421–455.

Nevo, E. and Payne, P.I. (1987) Wheat storage proteins: diversity of HMW glutenin subunits in wild emmer from Israel. *Theoretical and Applied Genetics*, **7**, 827–836.

Nevo, E., Moseman, J.G., Beiles, A. and Zohary, D. (1984) Correlation of ecological factors and allozymic variation with resistance to *Erysiphe graminis hordei* in *Hordeum spontaneum* in Israel: patterns and application. *Plant Systematics and Evolution*, **145**, 79–96.

Nevo, E., Moseman, J.G., Beiles, A. and Zohary, D. (1985) Patterns of resistance of Israeli wild emmer wheat to pathogens. I. Predictive methods by ecology and allozyme genotypes for powdery mildew and leaf rust. *Genetica*, **67**, 209–222.

Nevo, E., Noy-Meir, I., Beiles, A. *et al.* (1991) Natural selection of allozyme polymorphisms: micro-geographical spatial and temporal ecological differentiations in wild emmer wheat. *Israel Journal of Botany*, **40**, 419–449.

Newbury, H.J. and Ford-Lloyd, B.V. (1993) The use of RAPD for assessing variation in plants. *Plant Growth Regulation*, **12**, 43–51.

Niñez, V. (1986) El huerto casero: un salavidas? *Ceres*, **112**, 31–36.

North, S.G., Bullock, D.J, and Dulloo, M.E. (1994) Changes in the vegetation and reptile populations on Round Island, Mauritius, following eradication of rabbits. *Biological Conservation*, **67**, 21–28.

Norton, B.G. (1994) On what we should save: the role of culture in determining conservation targets, in *Systematics and Conservation Evaluation* (eds P.L. Forey, C.J. Humphries and R.I. Vane-Wright), Clarendon Press, Oxford, pp. 23–29.

Norton, D. (1996) Ecological basis for restoration in mainland New Zealand, in *Proceedings of a Workshop on Scientific Issues in Ecological Restoration*, Landcare Science Series Publications, Landcare Research, Christchurch.

Norton, T.W. (ed.) (1994) Conserving biological diversity in temperate forest ecosystems – towards sustainable management. *Proceedings of the International Forest Biodiversity Conference, Canberra, Australia, 4–9 December, 1994*. Centre for Resource and Environmental Studies, Australian National University, Canberra, 223 pp.

Noss, R.F. (1991) Sustainability and wilderness. *Conservation Biology*, **5**, 120–122.

Noy-Meir, I. (1973) Divisive polythetic classification of vegetation data by optimized division on ordination components. *Journal of Ecology*, **61**, 753–760.

Noy-Meir, I. (1990) The effect of grazing on the abundance of wild wheat, barley and oat in Israel. *Biological Conservation*, **51**, 299–310.

Noy-Meir, I., Agami, M. and Anikster, Y. (1991a) Changes in the population density of wild emmer wheat (*Triticum turgidum* var. *dicoccoides*) in a Mediterranean grassland. *Israel Journal of Botany*, **40**, 385–396.

Noy-Meir, I., Agami, M., Cohen, E. and Anikster, Y. (1991b) Floristic and ecological differentiation of habitats within a wild wheat population at Ammiad. *Israel Journal of Botany*, **40**, 363–384.

NRC (National Research Council, USA) (1991) *Managing Global Genetic Resources: Forest Trees*, National Academy Press, Washington, DC.

NRC (National Research Council) (1993) *Managing Global Genetic Resources: Agricultural Crop Issues and Policies*, National Academy Press, Washington, DC.

Nunney, L. and Campbell, K.A. (1993) Assessing minimum viable population size: demography meets population genetics. *Trends in Ecology and Evolution*, **8**, 234–239.

O'Donnell, C.F.J. and Rasch, G. (1991) *Conservation of Kaka in New Zealand: a review of status, threats, priorities for research and implications for management*, Science and Research Internal Report, Government Printer, Wellington.

Oehrens, E. (1977) Biological control of the blackberry through the introduction of rust, *Phragmidium violaceum* in Chile. *FAO Plant Protection Bulletin*, **25**, 26–27.

Oka, H.I. (1988) *Origin of Cultivated Rice*, Developments in Crop Science 14, Japan Scientific Societies Press, Tokyo.

Oka, H.I. (1991) A survey of within-population genetic diversity in land races and wild races of tropical Asia. *Rice Genetics Newsletter*, **8**, 79–81.

Oldfield, M.L. and Alcorn, J.B. (1987) Conservation of traditional agroecosystems. *Bioscience*, **37**, 199–208.

Oldfield, S. (1988) *Buffer Zone Management in Tropical Moist Forests: Case Study and Guidelines*, International Union for Conservation of Nature and Natural Resources, Gland.

O'Malley, D.M. and Bawa, K.S. (1987) Mating system of a tropical rain forerst tree species. *American Journal of Botany*, **74**, 1143–1149.

Orstrander, E.A., Jong, P.M., Rine, J. and Duyk, G. (1992) Construction of small insert genomic DNA libraries highly enriched for microsatellite repeat sequences. *Proceedings of the National Academy of Science US*, **89**, 3419–3423.

Ortega, R. (1989) *Las Papas Amargas en Cusco-Peru*, I Mesa Redonda: Peru, Bolivia, La Paz, 7–8 Mayo, ORSTOM.

Owen, D.F. (1931) Mutualism between grasses and grazers: an evolutionary hypothesis. *Oikos*, **36**, 376–378.

Owen, D.F. (1980) How plants may benefit from animals that eat them. *Oikos*, **35**, 230–235.

Owen, D.F. (1981) Mutualism between grasses and grazers: an evolutionary hypothesis. *Oikos*, **36**, 376–378.

Özis, Ü. (1982) Development plan of the Western Tigris Basin in Turkey. *Water Resources Development*, **1**, 343–352.

Özis, Ü. (1983) The development plan for the Lower Euphrates Basin in Turkey. *Natural Resources and Development*, **16**, 73–82.

Painting, K.A., Perry, M.C., Denning, R.A. and Ayad, W.G. (1993) *Guidebook for Genetic Resources Documentation*, International Board for Plant Genetic Resources, Rome.

Palmberg-Lerche, C. (1994a) International programmes for the conservation of forest genetic resources, in *Proceedings of International Symposium on Genetic Conservation and Production of Tropical Forest Tree Seed* (eds R.M. Drysdale, S.E.T. John and A.C. Yappa), ASEAN–Canada Forest Tree Seed project, Muak-Lek, Thailand, pp. 78–101.

Palmberg-Lerche, C. (1994b) FAO programmes and activities in support of the conservation and monitoring of forest genetic resources and biological diversity in forest ecosystems. Paper to IUFRO Symposium on Measuring and Monitoring Biological Diversity in Tropical and Temperate Forests, Chiangmai, Thailand, 18 August– 2 September, 1994, 15 pp + appendices.

Park, Y.S. and Fowler, D.P. (1982) Effects of inbreeding and genetic variances in a natural population of tamarack (*Larix laricina* (du Roi) K Koch) in eastern Canada. *Silvae Genetica*, **31**, 21–26.

Parker, M.A. (1990) The pleiotropy theory for polymorphism of disease resistance genes in plants. *Evolution*, **44**, 1872–1875.

Parker, M.A. (1992) Disease and plant population genetic structure, in *Plant Resistance to Herbivores and Pathogens: Ecology, Evolution and Genetics* (eds R.S. Fritz and E.L. Simms), University of Chicago Press, Chicago, pp. 345–362.

Paul, N.D. (1990) Modification of the effects of plant pathogens by other components of natural ecosystems, in *Pests, Pathogens and Plant Communities* (eds J.J. Burdon and S.R. Leather), Blackwell, Oxford, pp. 31–45.

Pearce, D.W. and Morgan, D. (1994) *The Economic Value of Biodiversity*, Earthscan, London.

Pearce, D.W. and Turner, R.K. (1990) *Economics of Natural Resources and the Environment*, Harvester Wheatsheaf, New York.

Peet, P.K. (1974) The measurement of species diversity. *Annual Review of Ecology and Systematics*, **5**, 285–307.

Pellew, R.A. (1991) Data management for conservation, in *The Scientific Management of Temperate Plant Communities for Conservation. 31st Symposium of the British Ecological Society, Southampton, 1989* (eds I.F. Spellerberg, F.B. Goldsmith, M.G. Morris), Blackwell, Oxford, pp. 505–522.

Perlin, J. (1989) *A Forest Journey*, Norton, 445 pp.

Pernès, J. (1984) *Gestion des ressources génétiques des plantes, Tomes I et II*. Agence de Coopération Culturelle et Technique, Paris.

Perrino, P. and Hammer, K. (1982) *Triticum monococcum* and *T. dicoccum* Schubler (syn. of *T. dicoccon* Shrank) are still cultivated in Italy. *Genetica Agraria*, **36**, 343–352.

Perry, M.C. and Bettencourt, E. (1995) Sources of information on existing germplasm collections, in *Collecting Plant Genetic Diversity: Technical Guidelines* (eds L. Guarino, V. Ramanatha Rao and R. Reid), CAB International, Wallingford, pp. 121–129.

Perry, M.C., Painting, K.A. and Ayad, W.G. (1993) *Genebank Management System Software User's Guide*, IBPGR, Rome.

Person, C. (1959) Gene-for-gene relationships in host : parasite systems. *Canadian Journal of Botany*, **37**, 1101–1130.

PGRC/E (1992) *Coffee Genetic Resource Programme (Project Proposal)*, Plant Genetic Resources Centre/Ethiopia, Addis Ababa, Ethiopia.

Pham, J.L. (1991) Genetic diversity and intervarietal relationships in rice (*Oryza sativa* L.) in Africa, in *Rice Genetics II. Proceedings of the Second International Rice Genetics Symposium, IRRI, Los Baños, Philippines, 1990*, IRRI, pp. 55–65.

Pham, J.L. and Bougerol, B. (1993) Abnormal segregations in crosses between the two cultivated rice species. *Heredity*, **70**, 466–471.

Pham, J.L. and de Kochko, A. (1983) *Prospections de variétés traditionnelles de riz dans l'Ouest et le Sud-ouest de la Côte d'Ivoire du 5 au 9 septembre 1983*, Rapport multigr. ORSTOM, Abidjan.

Pham, J.L., Ghesquière, A. and Second, G. (1994) On-farm conservation of rice genetic resources based on the management of populations made with landraces, in *Discussion Workshop on On-Farm Conservation of Crop Genetic Resources, IRRI, Los Baños, Philippines, 24–26 March 1994*.

Philp, E.G. (1982) *Atlas of the Kent Flora*, Kent Field Club and Kent County Council.

Pickett, S.T.A. and Thompson, J.N. (1978) Patch dynamics and nature reserves. *Biological Conservation*, **13**, 27–37.

Pickett, S.T.A. and White, P.S. (eds) (1985) *The Ecology of Natural Disturbance and Patch Dynamics*, Academic Press, New York.

Pinard, M.A., Putz, F.E., Tay, J. and Sullivan, T.E. (1995) Creating timber harvesting guidelines for a reduced-impact logging project in Malaysia. *Journal of Forestry*, **93**, 41–45.

Plitmann, U. (1967) Biosystematical study in the annual species of *Vicia* of the Middle East. Unpublished thesis, The Hebrew University of Jerusalem.

Plucknett, D.L., Smith, N.H.J., Williams J.T. and Anishetty, N.M. (1987) *Gene Banks and the World's Food*, Princeton University Press, Princeton.

Poole, R.W. (1974) *Quantitative Ecology*, McGraw-Hill, New York.

Poore, D. (ed.) (1989) *No Timber Without Trees*, IIED/Earthscan, 252 pp.

Posey, D.A. (1985) Indigenous management of tropical forest ecosystems: the case of the Kayapó Indians of the Brazilian Amazon. *Agroforestry Systems*, **3**, 139–158.

Powell, A.H. and Powell, G.U.N. (1987) Populations dynamics of male englossine bees in Amazonian forest fragments. *Biotropica*, **19**, 176–179.

Prance, G.T. (1977) Floristic inventory of the tropics: where do we stand? *Annals of the Missouri Botanical Gardens*, **64**, 659–684.

Prance, G.T. (1989) Economic prospects from tropical rainforest ethnobotany, in *Fragile Lands of Latin America: Strategies for Sustainable Development* (ed. J.O. Browder), Westview Press, Boulder, CO, pp. 61–74.

Prance, G.T. (1994) Amazonian tree diversity and the potential for supply of non-timber forest products, in *Tropical Trees: the Potential for Domestication and the Rebuilding of Forest Resources* (eds B.R.B. Leakey and A.C. Newton), HMSO, London, pp. 7–15.

Prance, G.T. and Brown, K. (1987) Soil and vegetation, in *Biogeography and Quaternary History in Tropical America* (eds T.C. Whitmore and G.T. Prance), Oxford Monographs on Biogeography No. 3, Oxford Science Publications, Oxford, pp. 19–45.

Prance, G.T. and Campbell, D.G. (1988) The present state of tropical floristics. *Taxon*, **37**, 519–548.

Prance, G.T., Balee, W., Boom, B.M. and Carneiro, R.L. (1995) Quantitative ethnobotany and the case for conservation in Amazonia, in *Ethnobotany: Evolution of a Discipline* (eds R. Evans-Schultes and S. von Reis), Dioscorides Press, Portland, OR, pp. 157–174.

Prendergast, H.D.V (1995) Published sources of information on wild plant species, in *Collecting Plant Genetic Diveristy: Technical Guidelines* (eds L. Guarino, V. Ramanatha Rao and R. Reid), CAB International, Wallingford, pp. 153–180.

Prendergast, H.D.V., Smith, R.D., Linington, S. and Newman, M.F. (1991) Seeds for use and conservation: the role of the seed bank of the Royal Botanic Gardens, Kew. *Development in Practice*, **1**(2), 120–127.

Primack, R.B. (1993) *Essentials of Conservation Biology*, Sinauer Associates, Sunderland, MA.

Primack, R.B. (1995) *A Primer of Conservation Biology*, Sinauer Associates Inc., Sunderland, MA.

Prober, S.M. and Brown, A.H.D. (1994) Conservation of the grassy whitebox woodlands: population genetics and fragmentation of *Eucalyptus albens*. *Conservation Biology*, **8**, 1003–1013.

Pulgar Vidal, J. (1987) *Geografia del Peru. Las Ocho Regiones Naturales*, Novena Edicion PEISA, Promoción Editorial INCA S.A. Lima, Peru.

Qualset, C.O. and McGuire, P.E. (1991) Detecting desirable genetic variation in wild populations. *Israel Journal of Botany*, **40**, 513.

Quinn, J.F. and Harrison, S.P. (1987) Effects of habitat fragmentation on species richness: evidence from biogeographic patterns. *Oecologia*, **75**, 132–140.

Quinn, R.M., Lawton, J.H., Eversham, B.C. and Wood, S.N. (1994) The biogeography of scarce vascular plants in Britain with respect to habitat preference, dispersal ability and reproductive biology. *Biological Conservation*, **70**, 149–157.

Quiros, C.F, Brush, S.B., Douches, D.S. *et al.* (1990) Biochemical and folk assessment of variability of Andean cultivated potatoes. *Economic Botany*, **44**, 254–266.

Quiros, C.F., Ortega, R., van Raamsdonk, L. *et al.* (1992) Increase of potato genetic resources in their center of diversity: the role of natural outcrossing and selection by the Andean farmer. *Genetic Resources and Crop Evolution*, **39**, 107–113.

Rafalski, J.A. and Tingey, S.V. (1993) Genetic diagnostics in plant breeding: RAPDs, microsatellites and machines. *Trends in Genetics*, **9**, 275–279.

Raintree, J.B. and Taylor, D.A. (eds) (1992) *Research on Farmers' Objectives for Tree Breeding*, Winrock International, Washington, DC, 132 pp.

Ramkishima, W., Lagu, M.D., Gupta, V.S. and Ranjekar, P.K. (1994) DNA fingerprinting in rice using oligonucleotide probes specific for simple sequence repetitive DNA sequences. *Theoretical and Applied Genetics*, **88**, 402–406.

Rankin de Mérona, J.M., Hutchins, R.W. and Lovejoy, T.E. (1990) Tree mortality and recruitment over a five-year period in undisturbed upland rainforest of the Central Amazon, in *Four Neotropical Rainforests* (ed. A. Gentry), Yale University Press, New Haven, pp. 573–584.

Rassmann, K., Schlotterer, C. and Tautz, D. (1991) Isolation of simple sequence loci for use in polymerase chain reaction-based DNA fingerprinting. *Electrophoresis*, **12**, 113–118.

Raven, P.H. (1988) Our diminishing tropical forest, in *Biodiversity* (ed. E.O. Wilson), National Academy Press, Washington, DC, pp. 119–122.

Rebelo, A.G. (1994) Iterative selection procedures: centres of endemism and optimal placement of reserves. *Strelitzia*, **1**, 231–257.

Reid, W.V. (1994) Setting objectives for conservation evaluation, in *Systematics and Conservation Evaluation* (eds P.L. Forey, C.J. Humphries and R.I. Vane-Wright), Clarendon Press, Oxford, pp. 1–13.

Repetto, R. and Gillis, M. (eds) (1988) *Public Policies and the Misuse of Forest Resources*, World Resources Institute/Cambridge University Press, Cambridge, 432 pp.

Rerkasem, B. and Rerkasem, K. (1984) The agroecological niche and farmer selection of rice varieties in the Chiang Mai Valley, Thailand, in *An Introduction to Human Ecology Research on Agricultural Systems in Southeast Asia* (eds A.T. Rambo and P.E. Sajise), University of The Philippines, Laguna, Philippines, pp. 303–311.

Resurrecion, A.P., Villareal, C.P., Parco, A. *et al.* (1994) Classification of cultivated race into indica and japonica types by the isozyme, RFLP and two milled-rice methods. *Theoretical and Applied Genetics*, **89**, 14–18.

Rey, J.R. (1981) Ecological biogeography of arthropods on Spartina Islands in Northwest Florida. *Ecological Monographs*, **51**(2), 237–265.

Richards, P. (1986) *Indigenous Agricultural Revolution*, Hutchison, London.

Riggs, L.A. (1990) Conserving genetic resources on-site in forest ecosystems. *Forest Ecology and Management*, **35**, 45–68.

Riggs, L.R. (1982) *Douglas-fir Genetic Resources*, California Gene Resource Program, Sacramento.

Ritland, K. (1986) Joint maximum likelihood estimation of genetic and mating structure using open-pollinated progenies. *Biometrics*, **42**, 25–43.

Ritland, K. and Jain, S.K. (1981) A model for the estimation of outcrossing rate and gene frequencies using *n* independent loci. *Heredity*, **47**, 35–52.

Rivin, C. (1986) Analyzing genome variation in plants. *Methods in Enzymology*, **118**, 75–86.

Robinson, G.R. and Quinn, J.F. (1992) Habitat fragmentation, species diversity, extinction and design of nature reserves, in *Applied Population Biology* (eds S.K. Jain and L.W.J. Botsford), Kluwer Academic, Dordrecht, pp. 223–248.

Rogstad, S.H., Patton, J.C. and Schaal, B.A. (1988) M13 repeat probe detects DNA minisatellite-like sequences in gymnosperms and angiosperms. *PNAS US*, **85**, 9176–9178.

Rohlf F.J. (1992) *NTSYS-pc: Numerical Taxonomy and Multivariate Analysis System*, Exeter Software, New York.

Romm, J. (1991) Exploring institutional options for global forest management, in *Proceedings of Technical Workshop to Explore Options for Global Forestry*

Management (eds D. Howlett and C. Sargent), IIED/ITTO/ONEB, pp. 186–192.

Rowe, R. and Cronk, Q. (1995) Applying molecular techniques to plant conservation: screening genes for survival. *Plant Talk*, 1, 18–19.

Rowe, R., Sharma, N.P. and Browder, J. (1992) Deforestation: problems, causes and concerns, in *Managing the World's Forests: Looking for the Balance between conservation and development*, Kendall/Hunt & World Bank, pp. 33–45.

Russell, J.R., Hosein, F., Johnson, E. *et al.* (1993) Genetic differentiation of cocoa (*Theobroma cacao* L.) populations revealed by RAPD analysis. *Molecular Ecology*, 2, 89–97.

Ryskov, A.P., Jincharadze, A.G., Prosnyak, M.I. *et al.* (1988) M13 phage DNA as a universal marker for DNA fingerprinting of animals, plants and microorganisms. *FEBS Lett.*, 233, 388–392.

Salafsk, N., Dugelby, B.L. and Terborgh, J.W. (1993) Can extraction reserves save the rainforest? An ecological and socioeconomic comparison of non-timber forest product extraction systems in Petén, Guatemala and West Kalimantan, Indonesia. *Conservation Biology*, 7, 39–52.

Salazar, R. (1992) Community plant genetic resource management: experiences in Southeast Asia, in *Growing Diversity: Genetic Resources and Local Food Security* (eds D. Cooper, R. Vellvé and H. Hobbelink), Intermediate Technology Publications, London, pp. 17–29.

Sambrook, J., Fritsch, E.F. and Maniatis, T. (1989) *Molecular Cloning. A laboratory manual*, Cold Spring Harbor Laboratory Press.

Samways, M.J. (1994) *Insect Conservation Biology*, Chapman & Hall, London.

Sanchez, R. (1993) *Los Estudios sobre cambios tecnológicos en la Agricultura Campesina*, recopilación y análisis de bibliografía temática No. 3, CCTA, Lima Peru.

Sano, Y., Chu, Y.E. and Oka, H.I. (1979) Genetic studies of speciation in cultivated rice. 1. Genic analysis for F1 sterility between *O. sativa* L. and *O. glaberrima* Steud. *Japanese Journal of Genetics*, 54, 121–132.

Sano, Y., Chu, Y.E. and Oka, H.I. (1980) Genetic studies of speciation in cultivated rice. 2. Character variations in backcross derivatives between *O. sativa* and *O. glaberrima*: M-V linkage and key characters. *Japanese Journal of Genetics*, 55, 19–39.

Sano, Y., Sano, R. and Morishima, H. (1984) Neighbour effects between two occurring rice species, *Oryza sativa* and *O. glaberrima*. *Journal of Applied Ecology*, 21, 245–254.

Saunders, D.A., Hobbs, R.J. and Margules, C.R. (1991) Biological consequences of ecosystem fragmentation: a review. *Biological Conservation*, 5, 18–32.

Savolainen, O. and Kärkkäinen, K. (1992) Effect of forest management on gene pools. *New Forests*, 6, 329–346.

Sayer, J.A. and Whitmore, T.C. (1991) Tropical moist forests: destruction and species extinction. *Biological Conservation*, 55, 199–213.

Schoen, D. and Brown, A.H.D. (1991) Intraspecific variation in population gene diversity and effective population size correlates with mating system in plants. *Proceedings of National Academy of Science*, 88, 4494–4497.

Schoener, A. and Schoener, T.W. (1981) The dynamics of the species–area relation in marine fowling systems. Biological correlations of changes in the species–area slope. *American Naturalist*, 118, 339–360.

Schumann, D.A. and W.L. Partridge. (1986) *The Human Ecology of Tropical Land Settlement in Latin America*, Westview Press, Boulder, CO.

Schwartzman, S. (1989) Extractive reserves: the rubber tappers' strategy for sustainable use of the Amazonian rainforest, in *Fragile Lands of Latin America: Strategies for Sustainable Development* (ed. J.O. Browder), Westview Press, Boulder, CO, pp. 150–165.

Scott, R. (1988) DNA restriction and analysis by Southern hybridisation, in *Plant*

Genetic Transformation and Gene Expression; a Laboratory Manual (eds J. Draper, R. Scott, P Armitage and R. Walden), Blackwell Scientific Publications, Oxford and Edinburgh.

Scott, M.P. and Williams, S.M. (1994) Measuring reproductive success in insects, in *Molecular Ecology and Evolution: Approaches and Applications* (eds B. Schierwater, B. Streit, G.P. Wagner and R. DeSalle), Birkhäuser Verlag, Basel.

Second, G. (1982) Origin of the genetic diversity of cultivated rice (*Oryza* spp.): study of the polymorphism scored at 40 isozyme loci. *Japanese Journal of Genetics*, **57**, 25–57.

Second, G. (1985) Evolutionary relationships in the Sativa group of *Oryza* based on isozyme data. *Génétique, Séléction, Evolution*, **17**, 89–114.

Second, G. and Ghesquière, A. (1995) Cartographie des introgressions réciproques entre les deux sous-espèces *indica* et *japonica* de riz cultivé (*Oryza sativa* L.), in *Techniques et utilisations des marqueurs moléculaires, Montpellier (France), 29–31 March, 1994*, Institut National de la Recherche Agronomique (INRA), Paris, pp. 83–93.

Segal, A., Dorr, K.H., Fischbeck, G. *et al.* (1987) Genotypic composition and mildew resistance in a natural population of wild barley, *Hordeum spontaneum*. *Plant Breeding*, **99**, 118–127.

Segal, A., Manisterski, J., Fischbeck, G. and Wahl, I. (1980) How plant populations defend themselves in natural ecosystems, in *Plant Disease. V. How Plants Defend Themselves* (eds J.G. Horsfall and E.B. Cowling), Academic Press, London and New York, pp. 75–102.

Senior, M.L. and Heun, M. (1993) Mapping maize microsatellites and polymerase chain reaction confirmation of the targeted repeats using a CT primer. *Genome*, **36**, 883–889.

Shafer, C.L. (1990) *Nature Reserves: Island Theory and Conservation Practice*, Smithsonian Institute, Washington.

Shands, H.L. (1991) Complementarity of *in situ* and *ex situ* germplasm conservation from the standpoint of the future user. *Israel Journal of Botany*, **40**, 521–528.

Shands, H.L. (1994) Some potential impacts of the United Nations Environment Program's Convention on Biological Diversity on the international system of exchanges of food crop germplasm, in *Conservation of Plant Genetic Resources and the UN Convention on Biological Diversity* (eds D. Witmeyer and M.S. Strauss), AAAS, Washington, DC, pp. 27–38.

Sharma, N.P., Rowe, R., Openshaw, K. and Jacobson, M. (1992) World forests in perspective, in *Managing the World's Forests: Looking for the Balance between Conservation and Development* (ed. N.P. Sharma), Kendall/Hunt & World Bank, pp. 17–31.

Shepherd, G., Shanks, E. and Hobley, M. (1991) National experiences in managing tropical and sub-tropical dry forests, in *Proceedings of Technical Workshop to Explore Options for Global Forestry Management* (eds D. Howlett and C. Sargent), IIED/ITTO/ONEB, 70–112.

Silva-Dias, J.C., Monteiro, A.A. and Kresovich, S. (1994) Genetic variation of Portuguese Tronchuda cabbage and Galega kale landraces using isozyme analysis. *Euphytica*, **75**(3), 221–230.

Silvertown, J., Wells, D.A., Gillman, M. *et al.* (1994) Short-term and long-term after-effects of fertilizer application on the flowering population of green-winged orchid *Orchis morio*. *Biological Conservation*, **69**, 191–197.

Simberloff, D.S. and Abele, L.G. (1976) Island biogeography theory and conservation practice. *Science*, **191**, 285–286.

Simberloff, D.S. and Abele, L.G. (1982) Refuge design and island biogeography theory: effects of fragmentation. *American Naturalist*, **120**, 41–50.

Simmonds, N.W. (1962) Variability in crop plants, its use and conservation. *Biological Reviews*, **37**, 442–465.

Simons, A.J. and Dunsdon, A.J. (1992) *Evaluation of the potential for genetic improvement of* Gliricidia sepium. *Final Report, ODA Forestry Research Project R4525*, Oxford Forestry Institute, 176 pp.

Smedegaard-Peterson, V. and Stolen, O. (1981) Effect of energy requiring defence reactions on yield and grain quality in a powdery mildew resistant barley cultivar. *Phytopathology*, **71**, 396–399.

Smith, F.D.M., May, R.M., Pellew, R. *et al.* (1993) How much do we know about the current extinction rate? *Tree*, **8**(10), 375–378.

Smith, H. (1990) Signal perception, differential expression within multigene families and the molecular basis of phenotypic plasticity. *Plant Cell Environment*, **13**, 585–594.

Sokal, R.R. and Rohlf, F.J. (1981) *Biometry*, Freeman, San Francisco.

Soleri, D. and Cleveland, D.A. (1993) Hopi crop diversity and change. *Journal of Ethnobiology*, **13**, 203–231.

SoS/E (1993) *Programme Evaluation Report*, Seeds of Survival, Addis Ababa, Ethiopia.

Soulé, M.E. (1980) Thresholds for survival: maintaining fitness and evolutionary potential, in *Conservation Biology: an Evolutionary–Ecological Perspective* (eds M.E. Soulé and B.A. Wilcox), Sinauer Associates, Sunderland, MA, pp. 151–69.

Soulé, M.E. (ed.) (1986) *Conservation Biology: Science of Scarcity and Diversity*, Sinauer Associates, Sunderland, MA, 584 pp.

Soulé, M.E. (1987) *Viable Populations for Conservation*, Cambridge University Press, Cambridge.

Soulé, M.S. (1991) Conservation: tactics for a constant crisis. *Science*, **253**, 744–750.

Soulé, M.E. and Simberloff, D. (1986) What do genetics and ecology tell us about the design of nature reserves? *Conservation Biology*, **35**, 19–40.

Southern, E.M. (1975) Detection of specific sequences among DNA fragments separated by gel electrophoresis. *Journal of Molecular Biology*, **98**, 503–517.

Southwood, T.R.E. (1973) The insect/plant relationship – an evolutionary perspective. *Symposium of the Royal Entomological Society, London*, **6**, 3–30.

Southwood, T.R.E., Brown, V.K. and Reader, P.M. (1979) The relationship of plant and insect diversities in succession. *Biological Journal of the Linnean Society*, **12**, 327–348.

Southwood, T.R.E., Moran, V.C. and Kennedy, C.E.J. (1982) The richness, abundance and biomass of the arthropod communities on trees. *Journal of Animal Ecology*, **51**, 635–649.

Spellerberg, I.F. (1991a) Biogeographical basis for conservation, in *The Scientific Management of Temperate Communities for Conservation* (eds I.F. Spellerberg, F.B. Goldsmith and M.G. Morris), British Ecological Society by Blackwell Scientific Publications, Oxford, pp. 293–322.

Spellerberg, I.F. (1991b) *Monitoring Ecological Change*, Cambridge University Press, Cambridge.

Spellerberg, I.F. and Hardes, S.R. (1992) *Biological Conservation*, Cambridge University Press, Cambridge.

Spellerberg, I.F. Goldsmith, F.B. and Morris, M.G. (1991) *The Scientific Management of Temperate Communities for Conservation*, Blackwell, Oxford.

Sperling, L. and Loevinsohn, M.E. (1993) The dynamics of adoption: distribution and mortality of bean varieties among small farmers in Rwanda. *Agricultural Systems*, **41**, 441–453.

Sperling, L., Loevinsohn, M.E. and Ntabomvura (1993) Rethinking the farmer's role in plant breeding: local bean experts and on-station selection in Rwanda.

Experimental Agriculture, **29**, 509–519.

Srivastava, J.P. and Damania, A.B. (1989) Use of collections in cereal improvement in semi-arid areas, in *The Use of Plant Genetic Resources* (eds A.H.D. Brown, O.H. Frankel, D.R. Marshall and J.T. Williams) Cambridge University Press, Cambridge, pp. 88–104.

Stace, C.A. (1989) *Plant Taxonomy and Biosystematics*, Edward Arnold, London.

Stephens, P.R. (1976) Farmland, in *New Zealand Atlas* (ed. I. Wards), Government Printery, Wellington, pp. 126–137.

Stork, N.E. (1988) Insect diversity: facts, fiction and speculation. *Biological Journal of the Linnean Society*, **35**, 321–337.

Stork, N.E. (1991) The comparison of the arthropod fauna of Bornean lowland rain forest trees. *Journal of Tropical Ecology*, **7**, 161–180.

Strid, A. (1970) Studies in the Aegean flora, XVI. Biosystematics of the *Nigella arvensis* complex with special reference to the problem of non-adaptive radiation. *Opera Botanica Societatus Botanical Lund*, **28**, 1–169.

Strong, D., Lawton, J.H. and Southwood, T.R.E. (1984) *Insects on Plants: Community Patterns and Mechanisms*, Blackwell, Oxford.

Sumar, L. (1980) *Programa Kiwicha (*Amaranthus caudatus*) Centro de Investigacion en Cultivos Andinos*, UNSAAC, Cusco, Peru.

Summerhayes, V.S. (1968) *Wild Orchids of Britain*, 2nd edn, Collins, London.

Suneson, C.A. (1956) An evolutionary plant breeding method. *Agronomy Journal*, **48**, 188–191.

Sutherland, W.J. (1995) Introduction and principles of ecological management, in *Managing Habitats for Conservation* (eds W.J. Sutherland and D.A. Hill), Cambridge University Press, Cambridge, pp. 1–21.

Sylvan, R. (1985a) A critique of deep ecology: part 1. *Radical Philosphy*, **40**, 2–12.

Sylvan, R. (1985b) A critique of deep ecology: part 2. *Radical Philosphy*, **41**, 10–22.

Synge, H. (1992) Higher plant diversity, in *Global Biodiversity: Status of the Earth's Living Resources* (ed. B. Groombridge), Chapman & Hall, London, pp. 64–87.

Tan, A. (1992) Plant diversity and plant genetic resources in Turkey. *Anadolu, Journal of the AARI*, **2**, 50–64.

Tanksley, S.D., Young, N.D., Paterson, A.H. and Bonierbale, M.W. (1989) RFLP mapping in plant breeding: new tools for an old science. *BioTechnology*, **7**, 257–264.

Tanner, J.E., Hughes, T.P. and Connell, J.H. (1994) Species coexistence, keystone species, and succession – a sensitivity analysis. *Ecology*, **75**, 2204–2219.

Tapia, M. (1980) *Los Recursos Genéticos de los Andes Altos*, Proyecto de Investigacion de los Sistemas Agrícolas Andinos, Universidad del Cusco, Peru.

Tautz, D. (1993) *Notes on the Definition and Nomenclature of Tandemly Repetitive DNA Sequences, in DNA Fingerprinting: State of the Science* (eds S.D.J. Pena, R. Chakraborty, J.T. Epplen and A.J. Jeffreys), Birkhäuser Verlag, Basel, pp. 21–28.

Tautz, D. and Renz, M. (1984) Simple sequences are ubiquitous repetitive components of eukaryotic genomes. *Nucleic Acids Research*, **12**, 4127–4138.

Terborgh, J. (1986) Keystone plant resources in the tropical forest, in *Conservation Biology: the Science of Scarcity and Diversity* (ed. M.E. Soulé), Sinauer Associates, Sunderland, MA, pp. 330–334.

Terborgh, J. (1992) Maintenance of diversity in tropical forests. *Biotropica*, **24**, 283–292.

Tesemma, T. (1987) Improvement of indigenous wheat land races in Ethiopia, in *Proceedings of the International Symposium on Conservation and Utilization of Ethiopian Germplasm* (eds J.M.M. Engles, J.G. Hawkes and M. Worede), Cambridge University Press, Cambridge, pp. 232–238.

Tesemma, T. and Belay, G. (1991) Aspects of Ethiopian tetraploid wheats with emphasis on durum wheat genetics and breeding research, in *Wheat Research in Ethiopia: a Historical Perspective* (eds H. Gebre-Mariam, D.G. Tanner and M. Huiiuka), IAR/CIMMYT, Addis Ababa, pp. 47–72.

Teverson, D.M., Taylor, J.D. and Lenné, J.M. (1994) Functional disease resistance in *Phaseolus vulgaris* bean mixtures in Tanzania. *Aspects Applied Biology*, **39**, 163–172.

Thirgood, J.V. (1981) *Man and the Mediterranean Forest*, Academic Press, London, 194 pp.

Thompson, J.N. and Burdon, J.J. (1992) Gene-for-gene coevolution between plants and parasites. *Nature*, **360**, 121–125.

Tikhonov, V.L. and Gerasimova, S.M. (1990) Role of natural reserves in protection and restoration of protected plant species diversity, in *Natural Reserves of the USSR, their Present and Future. Part I*, Novgorod, pp. 191–193.

Times Books (1988) *Atlas of the World*, 7th edn, Times Books, London.

Tiwari, J.K. (1993) Vishnois: the saviours of desert wildlife. *Hornbill*, **3**, 2–6.

Torres, J. (1992) *Los Agroecosistemas Andinos del Peru. La oferta ambiental de los Andes y algunas sugerencias para optimizar su utilización, 2–4*, Centro Internacional de la Papa (CIP), Lima, Peru.

Townsend, C.C. (1967) Contributions to the Flora of Iraq: V., Notes on the Leguminosales. *Kew Bulletin*, **21**, 435–458.

UNCED (1992) *Convention on Biological Diversity*, United Nations Conference on Environment and Development, Geneva.

Urban, D.L., O'Neill, V. and Shugart, H.H. (1987) Landscape ecology. *BioScience*, **37**, 119–127.

Valladolid, R. (1988) *Agricultura Andina*, Universidad Nacional de San Cristobal de Huamanga, Ayacucho, Peru.

Van der Meijden, E. (1990) Herbivory as a trigger for growth. *Functional Ecology*, **4**, 597–598.

Van der Plank, J.E. (1963) *Plant Diseases: Epidemics and Control*, Academic Press, New York.

Vane-Wright, R.I., Humphries, C.J. and Williams, D.H. (1991) What to protect? Systematics and the agony of choice. *Biological Conservation*, **55**, 235–254.

Vaughan, D.A. and Chang, T.T. (1992) *In situ* conservation of rice genetic resources. *Economic Botany*, **46**, 368–383.

Vavilov, N.I. (1926) Studies on the origin of cultivated plants. *Bulletin of Applied Botany*, **26**, 1–128.

Vavilov, N.I. (1951) *The Origin, Variation, Immunity and Breeding of Cultivated Plants*, Chronica Botanica, Waltham, MA.

Vavilov, N.I. (1962) *Five Continents* (*Pyat' Kontinentov*, translated 1987), Moscow.

Vavilov, N.I. (1965) *Origin and Geography of Cultivated Plants*, (selected works in five volumes), Vol. 5, Nauka Publishers, Moscow, pp. 9–104.

Vergnaud, G. (1989) Polymers of random short oligonucleotides detect polymorphic loci in the human genome. *Nucleic Acids Research*, **17**, 7623–7630.

Verkaar, H.J. (1988) Are defoliators beneficial for their host plants in terrestrial ecosystems? A review. *Acta Botanica Neerlandica.*, **37**(2), 137–152.

Vilchez, V. and Ravensbeck, L. (1992) *In situ* conservation of important pine provenances in Nicaragua: example Yucul, in *Proceedings IUFRO S2.02–08 Conference, Breeding Tropical Trees, Cali, Colombia, 9–18 October 1991*, pp. 53–56.

Virk, P.S., Ford-Lloyd, B.V., Jackson, M.T. and Newbury, H.J. (1995a) Use of RAPD for the study of diversity within plant germplasm collections. *Heredity*, **74**, 170–179.

Virk, P.S., Newbury, H.J., Jackson, M.T., and Ford-Lloyd, B.V. (1995b) The identification of duplicate accessions within a rice germplasm collection using RAPD analysis. *Theoretical and Applied Genetics*, **90**, 1049–1055.

Vos, P., Hogers, R., Bleeker, M. *et al.* (1995) AFLP: a new technique for DNA fingerprinting. *Nucleic Acid Research*, **23**, 4407–4414.

Wahl, I. and Segal, A. (1986) Evolution of host–parasite balance in natural indigenous populations of wild barley and wild oats in Israel, in *The Origin and Domestication of Cultivated Plants* (ed. C. Barigozzi), Elsevier Science Publishers BV, pp. 129–142.

Wallace, A.R. (1910) *The World of Life. A Manifestation of Creative Power, Directive Mind and Ultimate Purpose*, Chapman & Hall, London.

Wang, Z.Y. and Tanksley, S.D. (1989) Restriction fragment length polymorphism in *Oryza sativa* L. *Genome*, **32**, 1113–1118.

Wang, Z.Y., Second, G. and Tanksley, S.D. (1992) Polymorphism and phylogenetic relationship among species in the genus *Oryza* as determined by analysis of nuclear RFLPs. *Theoretical and Applied Genetics*, **83**, 565–581.

Warren, M.S. and Key, R.S. (1991) Woodlands: past, present and potential for insects, in *The Conservation of Insects and their Habitats* (eds H.M. Collins and J.A. Thomas), Academic Press, London, pp. 155–211.

Watson, I.A. (1970) The utilization of wild species in the breeding of cultivated crop resistant to plant pathogens, in *Genetic Resources in Plants – Their Exploration and Conservation* (eds D.H. Frankel and E. Bennett), Blackwell, Oxford.

Watterson, G.A. (1962) Some theoretical aspects of diffusion theory in population genetics. *Annals of Mathamatics and Statistics*, **33**, 939–957.

Webb, N.R. (1989) Studies on the invertebrate fauna of fragmented heathland in Dorset, UK, and the implications for conservation. *Biological Conservation*, **47**, 153–165.

Weising, K., Nybom, H., Wolff, K. and Meyer, W. (1995) *DNA Fingerprinting in Plants and Fungi*, CRC Press, London.

Weising, K., Weigand, F., Driesel, A.J. *et al.* (1989) Polymorphic simple GATA/GACA repeats in plant genomes. *Nucleic Acids Research*, **17**, 10128.

Weissenbach, J., Gyapay, G., Dib, C. *et al.* (1992) A second-generation linkage map of the human genome. *Nature*, **359**, 794–801.

Wellhausen, E., Roberts, J., Roberts, L.M. and Hernández, X.E. (1952) *Races of Maize in Mexico*, The Bussey Institution, Havard University, Cambridge, MA.

Welsh, J. and McClelland, M. (1990) Fingerprinting gemomes using PCR with arbitrary primers. *Nucleic Acids Research*, **18**, 7213–7218.

Western, D. and Wright, R.M. (1994) The background to community-based conservation, in *Natural Connections: Perspectives in Community-Based Conservation* (eds D. Western, R.M. Wright and S.C. Strum), Island Press, Washington, DC, pp. 1–12.

Westneat, D.F. and Webster, M.S. (1994) Molecular analysis of kinship in birds: interesting questions and useful techniques, in *Molecular Ecology and Evolution: Approaches and Applications* (eds B. Schierwater, B. Streit, G.P. Wagner and R. De Salle), Birkhäuser Verlag, Basel.

Westoby, J.C. (1989) *Introduction to World Forestry*, Basil Blackwell, Oxford, 228 pp.

Whyte, R. and Julen, G. (1963) Proceedings of a technical meeting on plant exploration and introduction. *Genetica Agraria*, **17**, 573.

Wilcox, M.D. (in press) *Tropical Forest Resources and Biological Diversity: effects of deforestation on biological diversity and forest genetic resources*, FAO Forestry Paper, FAO, Rome.

Wilkes, H.G. (1967) Teosinte : the closest relative of maize. Unpublished PhD dissertation, Bussey Institute, Harvard University.

Wilkes, H.G. (1991) *In situ* conservation of agricultural systems, in *Biodiversity, Culture, Conservation and Development* (eds M.L. Oldfield and J.B. Alcorn), Westview Press, Boulder, CO, pp. 86–101.

Williams, J.G.K., Kubelik, A.R., Livak, K.J. *et al.* (1990) DNA polymorphisms amplified by arbitrary primers are useful as genetic markers. *Nucleic Acids Research*, **18**, 6231–6235.

Williams, J.T. (1985) A world network of crop genetic resources activities, in *15 Years Collection and Utilisation of Plant Genetic Resources by the Institute of Crop Science and Plant Breeding*, FAL, Braunschweig, Germany, pp. 41–48.

Williams, J.T. (1991) The time has come to clarify and implement strategies for plant conservation. *Diversity*, **7**, 37–39.

Williams, M. (1990) *Wetlands: a Threatened Landscape*, Blackwell Scientific Publishers, Oxford.

Williams, P.H. (1989) Screening for resistance to disease, in *The Use of Plant Genetic Resources* (eds A.H.D. Brown, O.H. Frankel, D.R. Marshall and J.T. Williams), Cambridge University Press, Cambridge, pp. 335–352.

Williams, P.H. (1992) *WORLDMAP – Priority Areas for Biodiversity*, Version 3, Natural History Museum, London.

Williams, P.H. and Humphries, C.J. (1994) Biodiversity, taxonomic relatedness and endemism in conservation, in *Systematics and Conservation Evaluation* (eds P.L. Forey, C.J. Humphries and R.I. Vane-Wright), Systematic Association Special Volume 50, Oxford University Press, Oxford, pp. 207–229.

Williams, P.H., Vane-Wright, R.I. and Humphries, C.J. (1991) Measuring biodiversity for choosing conservation areas, in *Hymenoptera and Biodiversity* (eds J. LaSalle and I. Gauld), CAB International, Wallingford, pp. 309–328.

Williams, P.H., Vane-Wright, D.I. and Humphries, C.J. (1992) Measuring biodiversity: taxonomic relatedness for conservation priorities. *Australian Systematic Botany*, **4**, 665–679.

Williamson, M.H. and Lawton, J.H. (1991) Fractal geometry of ecological habitats, in *Habitat Structure: the Physical Arrangement of Objects in Space* (eds S.S. Bell, E.D. McCoy and H.R. Mushohsky), Chapman & Hall, London, pp. 69–86.

Wilson, E.O. (1984) *Biophilia*, Harvard University Press, Cambridge, MA.

Wilson, E.O. (1992) *The Diversity of Life*, Penguin, 406 pp.

Wise, K.A.J. (1977) A synonymic checklist of the Hexapoda of the New Zealand sub-region: the smaller orders. *Bulletin of the Auckland Institute and Museum*, **11**, Government Printery, Wellington.

Withers, L.A. (1993) Conservation methodologies with particular reference to *in vitro* conservation, in *Proceedings of the Asian Sweet Potato Germplasm Network Meeting*, Guangzhou, China, CIP, Manila, The Philippines, pp. 102–109.

Witmeyer, D. (1994) The Convention on Biological Diversity changes rules of the game for international plant genetic resources regime. *Diversity*, **10**(3), 28–31.

Witmeyer, D. and Strass, M.S. (1994) *Conservation of Plant Genetic Resources and the UN Convention on Biological Diversity*, AAAS, Washington, DC.

Wolfe, M.S. (1992) Barley diseases: maintaining the value of our varieties, in *Barley Genetics VI, Vol. 2* (ed. L. Munck), Munksgaard International Publishers, Copenhagen, Denmark, pp. 1055–1067.

Wolfe, M.S. (1993) Can the strategic use of disease resistant hosts protect their inherent durability? in *Durability of Disease Resistance* (eds Th. Jacob and J.E. Parlevliet), Kluwer Academic, Dordrecht, pp. 83–86.

Worede, M. (1991) Crop genetic resources conservation and utilization: an Ethiopian perspective, in *Science in Africa: Achievements and Prospects*, AAAS, Washington, DC, pp. 103–123.

Worede, M. (1992) Ethiopia: a gene bank working with farmers, in *Growing*

Diversity: Genetic Resources and Local Food Security (eds D. Cooper, R. Vellvé and H. Hobbelink), Intermediate Technology Publications, London, pp. 78–94.

Worede, M. (1993) The role of Ethiopian farmers on the conservation and utilization of crop genetic resources. *International Crop Science Society of America*, **1**, 395–399.

Worede, M. and Hailu, M. (1993) Linking genetic resources conservation to farmers in Ethiopia, in *Cultivating Knowledge: Genetic Diversity Farmer Experimentation and Crop Research* (ed. D. de Boef), Intermediate Technology Publications, London, pp. 78–84.

World Commission on Environment and Development (1987) *Our Common Future*, Oxford University Press, Oxford, 400 pp.

Wratten, S.D. and Hutcheson, J. (1995) Restoration of ecological communities: potential for farmland sites on the Canterbury Plain, New Zealand, in *Proceedings of a Workshop on Scientific Issues in Ecological Restoration*, Landcare Science Series Publications, Landcare Research, Christchurch.

Wratten, S.D. and van Emden, H.F. (1995) Habitat management for enhanced activity of natural enemies of insect pests, in *Ecology and Integrated Farming Systems* (eds D.M. Glen, M.P. Greaves and H.M. Anderson), Wiley, Chichester, pp. 117–145.

WRI, IUCN and UNEP (1992) *Global Biodiversity Strategy*, World Resources Institute, Washington DC.

Wright, S. (1931) Evolution in Mendelian populations. *Genetics*, **16**, 97–159.

Wright, S. (1938) Size of population and breeding structure in relation to evolution. *Science*, **87**, 430–431.

Wright, S. (1939) Statistical genetics in relation to evolution. *Actualité scientifiques et industrielles*, **802**, 1–64.

Wright, S. (1943) Isolation by distance. *Genetics*, **28**, 114–138.

Wright, S. (1946) Isolation by distance under diverse systems of mating. *Genetics*, **31**, 35–59.

Wu, K.-S. and Tanksley, S.D. (1993) Abundance, polymorphism and genetic mapping of microsatellites in rice. *Molecular and General Genetics*, **241**, 225–235.

Xu, L.-X. and Nguyen, H.T. (1994) Genetic variation detected with RAPD markers among upland and lowland rice cultivars (*Oryza sativa* L.). *Theoretical and Applied Genetics*, **87**, 668–672.

Yahner, R.H. (1988) Changes in wildlife communities near edges. *Conservation Biology*, **2**, 333–339.

Yamamoto, T., Nishikawa, A. and Oeda, K. (1994) DNA polymorphisms in *Oryza sativa* L. and *Lactuca sativa* L. amplified by arbitrary primed PCR. *Euphytica*, **78**, 143–148.

Yang, G.P., Saghai Maroof, M.A., Xu, C.G. *et al.* (1994) Comparative analysis of microsatellite DNA polymorphism in landraces and cultivars of rice. *Molecular and General Genetics*, **245**, 187–194.

Yang, R.-C. and Yeh, F.C. (1992) Genetic consequences of *in situ* and *ex situ* conservation of forest trees. *Forestry Chronicle*, **68**, 720–729.

Yeh, F.C. (1989) Isozyme analysis for revealing population structure for use in breeding strategies, in *Breeding Tropical Trees – Population Structure and Genetic Improvement Strategy iin Clonal and Seedling Forestry* (eds G.L. Gibson, A.R. Griffin and A.C. Matheson), Oxford Forestry Institute, Oxford, and Winrock International, Arlington, Virginia, pp. 119–131.

Yonezawa, K. (1985) A definition of the optimal allocation of effort in conservation of plant genetic resources, with application to sample size determination for field collection. *Euphytica*, **34**, 345–354.

Young, A.G. (1995) Forest fragmentation: effects on population genetic processes.

Paper to IUFRO XX World Congress, Tampere, Finland, August, 1995 (in press).

Young, A.G., Merriam, H.G. and Warwick, S.I. (1993) The effects of forest fragmentation on genetic variation in *Acer saccharum* Marsh. (sugar maple) populations. *Heredity*, **71**, 277–289.

Zarucchi, J.L., Winfield, P.J., Polhill, R.M. *et al.* (1993) The ILDIS Project on the world's legume species diversity, in *Designs for a Global Plant Species Information System* (eds F.A. Bisby, G.F. Russell and R.J. Pankhurst), Oxford University Press, Oxford, pp. 131–144.

Zencirci, N. (1993) Relative genetic contributions of ancestral lines to winter and spring wheat gene pools of Turkey. *Doga – Turkish Journal Agriculture and Forestry*, **17**, 673–681.

Zhang, Q., Gao, Y.J., Saghai-Maroof, M.A. *et al.* (1993) Molecular divergence and hybrid performance in rice. *Molecular Breeding*, **1**, 133–142.

Zhang, Q., Gao, Y.J., Yang. S.H. *et al.* (1994) A diallel analysis of heterosis in élite hybrid rice based on RFLPs and microsatellites. *Theoretical and Applied Genetics*, **89**, 185–192.

Zhang, Q., Saghai-Maroof, M.A., Lu, T.Y. and Shen, B.Z. (1992) Genetic diversity and differentiation of *indica* and *japonica* rice detected by RFLP analysis. *Theoretical and Applied Genetics*, **83**, 495–499.

Zhao, X. and Kochert, G. (1992) Characterisation and genetic mapping of a short, highly repeated, interspersed DNA sequence from rice (*Oryza sativa* L.) *Molecular and General Genetics*, **231**, 353–359.

Zhao, X. and Kochert, G. (1993) Phylogenetic distrbution and genetic mapping of a (GCG) microsatellite from rice (*Oryza sativa* L.) *Plant Molecular Biology*, **21**, 607–614.

Zheng, K., Qian, H., Shen, B. *et al.* (1994) RFLP-base phylogenetic analysis of wide compatibility varieties in *Oryza sativa* L. *Theoretical and Applied Genetics*, **88**, 65–69.

Zhukovsky, P.M. (1950) *Cultivated Plants and their Wild Relatives*, (translated by P.S. Hudson), CAB, Farnham Royal, UK.

Zietkiewicz, E., Rafalski, A. and Labuda, D. (1994) Genome fingerprinting by simple sequence repeats (SSR)-anchored PCR amplification. *Genomics*, **20**, 176–183.

Zimmerer, K.S. and Douches, D.S. (1991) Geographical approaches to native crop research and conservation: the partitioning of allelic diversity in Andean potatoes. *Economic Botany*, **45**, 176–189.

Zobel, B.J. and Talbert, J.T. (1984) *Applied Forest Tree Improvement*, Wiley, New York, 505 pp.

Zohary, D. (1969) The progenitor of wheat and barley in relation to domestication and agricultural dispersal in the Old World, in *The Domestication and Exploration of Plants and Animals* (eds P.J. Ucko and G.W. Dimbleby), Duckworth, London, pp. 47–66.

Zohany, D. (1970) Centers of diversity and centers of origin, in *Genetic Resources in Plants – Their Exploration and Conservation* (eds O.H. Frankel and E. Bennett), Blackwell, Oxford.

Index

Page numbers in **bold** refer to figures and page numbers in *italic* refer to tables.